DayWater: an Adaptive Decision Support System for Urban Stormwater Management

DayWater: an Adaptive Decision Support System for Urban Stormwater Management

Edited by

Daniel Thévenot
Université Paris-Est, Cereve,
6-8 avenue Blaise Pascal, Cité Descartes, Champs sur Marne,
F-77455 Marne-La-Vallée cedex 2, France.
Phone: +331 64 15 36 25, Fax: +331 64 15 37 64.
e-mail: thevenot@cereve.enpc.fr

IWA Publishing

Published by IWA Publishing, Alliance House, 12 Caxton Street, London SW1H 0QS, UK

Telephone: +44 (0) 20 7654 5500; Fax: +44 (0) 20 7654 5555; Email: publications@iwap.co.uk
Web: www.iwapublishing.com

First published 2008

British Library Cataloguing in Publication Data
A CIP catalogue record for this book is available from the British Library

Library of Congress Cataloging- in-Publication Data
A catalog record for this book is available from the Library of Congress

ISBN: 1843391600

ISBN13: 9781843391609

Acknowledgement

The results presented in this publication have been obtained within the framework of the EC funded research project DayWater "Adaptive Decision Support System for Stormwater Pollution Control", contract no EVK1-CT-2002-00111, co-ordinated by Cereve at ENPC (F) and including Tauw BV (Tauw) (NL), Department of Water Environment Transport at Chalmers University of Technology (Chalmers) (SE), Environment and Resources DTU at Technical University of Denmark (DTU) (DK), Urban Pollution Research Centre at Middlesex University (MU) (UK), Department of Water Resources Hydraulic and Maritime Works at National Technical University of Athens (NTUA) (GR), DHI Hydroinform, a.s. (DHI HIF) (CZ), Ingenieurgesellschaft Prof. Dr. Sieker GmbH (IPS) (D), Water Pollution Unit at Laboratoire Central des Ponts et Chaussées (LCPC) (F) and Division of Sanitary Engineering at Luleå University of Technology (LTU) (SE).

This project was organised within the "Energy, Environment and Sustainable Development" Programme in the 5th Framework Programme for "Science Research and Technological Development" of the European Commission and was part of the CityNet Cluster, the network of European research projects on integrated urban water management.

Contents

Foreword

Andrea Tilche[1] and Zissimos Vergos[2]

Over the last 30 years significant evolutions have been seen around the world with regard to stormwater management. Older conveyance and storage based approaches are being enriched with new planning concepts and technological tools incorporating pollution control at the source, stormwater treatment, recycling, the exploitation of recharge potential, just to name a few. Devastating experiences over the last years – which are expected to increase in frequency due to climate change driven modification of rainfall patterns -clearly indicate the need for enormous investment to manage peak flows and safeguard environmental quality. Seen against the scarcity of resources and the situations faced in diverse socio-economic contexts, there is an undeniable need for optimizing investment choices through more informed decision-making in harmony with public awareness and perceptions. Similar to other aspects of urban water planning, stormwater management solutions would be judged against their contribution in reinstating natural water cycle conditions and tackling, previously underestimated, pollution fluxes.

And the challenges do no stop here. Current projections on the world's urban population growth, frequently anarchic in both industrialised and developing countries, give a clear indication of a looming crisis unless a knowledge-based paradigm shift is embraced, with stormwater management and pollution control among its highest priorities. This means that in practice, we shall need to examine each square meter of new and existing development against such new conceptual framework to contain ecological footprint and adverse impacts on our habitats and public health.

Hopefully, more than ever in the past, integrated concepts based on sustainability considerations and risk-based approaches are finding their ways around policy, planning and decision making. This is a direct result of our increasing expert knowledge and public awareness on global and local change drivers on water ecosystems, the complex hydro-morphology of urban catchments and the need to better manage our 'water and chemical literacy', and adjusting our everyday habits and lifestyles. Recent conceptual developments and tools on life cycle thinking approaches confirm the validity of such an evolution and are expected to further enhance such conceptual paradigm shift.

European research programmes alongside national initiatives remain in the forefront of such challenging transformation. Ahead of us lies the task of how to better manage the generated knowledge. This can only be done under a wider partnership bringing together research, industry, end-users communities and the wider public, with clear focus on the welfare of European, but also global citizens.

Back in 2002, the peer-review based selection and funding of the DayWater project, through the 5th Community Research Framework Programme (FP5, 1998-2002), demonstrated the importance attached to maintaining a critical mass of knowledge at European level to provide reliable responses to current and future challenges in the areas of

[1] Dr. Andrea Tilche, Head of Unit, European Commission, Directorate General for Research, Unit I.3 'Environnemental Technologies and Pollution Prevention'

[2] Eng. Zissimos Vergos, European Commission, Delegation of the to the Republic of Kenya, Head of Infrastructure Section and Ex-Project Officer for DayWater

stormwater management with focus on source control, pollution and flood prevention, and better mobilization of environmental technologies.

With 10 scientific partners from 8 European countries working on a day-to-day basis with 14 European end-users from 7 European countries, DayWater initiated its work by openly addressing burning questions linked to rigidities and contextual particularities of our current planning systems, founding its approach on a continuous feedback between scientific and practice partners. The resulting work gives new insights to issues ranging from the characterisation of decision making systems, stormwater source control methods, provision of structural and non-structural best management practices and models of analysis of pollution loads and environmental risk assessment.

DayWater has not acted in isolation but as part of a wider cluster of another five related projects which collaborated under the CityNet umbrella, also an FP5 funded project which worked on co-ordinating the integrated dissemination of knowledge in all aspects of urban water management (water supply, sewerage and drainage) including both system related but also key socioeconomic aspects.

This foreword on behalf of the European Commission and its Directorate General for Research is an expression of appreciation to the DayWater partners for their efforts. At the same time, it is an invitation to the interested reader, the researcher and the experienced planner to browse through Hydropolis, the Adaptive Decision Support System encapsulating the work achieved through DayWater. We are confident that it shall constitute an illuminating and state-of-the-art knowledge experience which can find its way to every day practical life of each of us.

Authors

Aftias Emmanuel, Assistant Professor, National Technical University Athens (NTUA), Faculty of Civil Engineering, Department of Water Ressources, 5 Iroon Polytechniou Str, Zografou (Greece), phone +30 210 77 22 835, fax +30 210 810 19 15, emaftias@central.ntua.gr ....chapters 1, 10, 26

Ahlmann Stefan, Research Engineer, DHI Water & Environment, Box 3287, Växjö SE-350 53 (Sweden), phone +46 470 75 27 60, fax +46, stephan.ahlman@dhi.se ....chapters 14, 18

Aldheimer Gudrun, Stockholm Water Co, Torsgatan 26, Stockholm SE-106 36 (Sweden), phone +46 8 522 122 28, gudrun.aldheimer@stockholmvatten.se ....chapter 20

Alfakih Elham, Research Engineer, INSA Lyon, Unité de Recherche du Génie Civil, Bâtiment J.C.A. Coulomb, 34 avenue des Arts, Villeurbanne CEDEX F-69621 (France)..chapter 23

Andrieu Hervé, Research director, Laboratoire Central des Ponts et Chaussées, Division Eau et Environnement, BP 4129, Route de Bouaye, Bougenais F-44341 (France), phone +33 2 40 84 58 77, fax +33 2 40 84 59 98, herve.andrieu@lcpc.fr ....chapter 21

Ashley Richard, Professor, University of Sheffield, Pennine Water Group, Department of Civil and Structural Engineering, Sir Frederick Mappin Building, Mappin Street, Sheffield S1 3JD (United Kingdom), phone +44 (0)114 222 5766, fax +44 (0)114 222 5700, R.Ashley@sheffield.ac.uk ....chapter 25

Barroca Bruno, Assistant Professor, Université Paris-Est, Laboratory of Urban Engineering, Environment and Building, Bâtiment Lavoisier, 5 boulevard Descartes, Champs sur Marne, Marne la Vallée Cedex 02 F-77454 (France), phone +33 1 60 95 73 40, barroca@univ-mlv.fr ....chapter 13

Baun Anders, Associate Professor, Director of innnovation, Coordinator of NanoDTU, Technical University of Denmark, Environment & Resources DTU, Bygningstorvet, Building 115, Lyngby DK-2800 Kgs. (Denmark), phone +45 4525 1567, fax +45 4593 2850, anb@er.dtu.dk ....chapters 11, 19, 20

Beyeler Claire, Chargée de mission, Syndicat Marne Vive, Hotel de Ville, Place Charles de Gaulle, Saint Maur F-94100 (France), phone +33 1 45 11 65 65 or 65 72, fax +33 1 45 11 65 70, marnvive@club-internet.fr ....chapter 16

Bonierbale Thomas, Maître de Contérences, Université Paris-Est, Laboratory of Urban Engineering, Environment and Building, Bâtiment Lavoisier, 5 boulevard Descartes, Champs sur Marne, Marne la Vallée Cedex 02 F-77454 (France), phone +33 1 60 95 73 45, fax +33 1 60 95 73 49, thomas.bonierbale@univ-mlv.fr ....chapter 13

Butler David, Professor, University of Exeter, Centre for Water Systems, School of Engineering, Computer Science & Mathematics, North Park Road, Exeter EX4 4QF (United Kingdom), phone +44 (0)1392 264064, fax +44 (0)1392 217965, D.Butler@exeter.ac.uk ....chapter 25

de Roo Colette, Research fellow in urban water management, Ecologic Institute for International and European Environmental Policy, Pfalzburger Str. 43/44, Berlin D-10717 (Germany), phone +49 30 86 880 123, fax +49 30 86 880 100, colette.deroo@ecologic.eu ....chapters 14, 15

Deroubaix José-Frédéric, Research Director, Université Paris-Est, Cereve, 6-8 Ave Blaise Pascal, Marne la Vallée Cedex 02 F-77455 (France), phone +33 1 64 15 36 31, fax +33 1 64 15 37 64, jfd@cereve.enpc.fr ....chapter 5

DEUTSCH Jean-Claude, Emeritus Research Director, Université Paris-Est, Cereve, 6-8 Ave Blaise Pascal, Marne la Vallée Cedex 02 F-77455 (France), phone +33 1 64 15 36 20, fax +33 1 64 15 37 64, jcd@cereve.enpc.fr .... chapters 1, 2, 4, 5, 26

ELLIS Bryan, Emeritus Professor, Middlesex University, Urban Pollution Research Centre, Queensway, Enfield EN3 4SA (UK), phone +44 208 411 5258, fax +44 208 411 54 40, b.ellis@mdx.ac.uk .... chapters 7, 8, 9, 10

ENGELHARD Carolina, Dr., University of Innsbruck, Institute of Infrastructure, Technikerstrasse 13, Innsbruck A-6020 (Austria), phone +43 512 507 6932, fax +43 512 507 2911, Carolina.Engelhard@uibk.ac.at .... chapter 24

ERIKSSON Eva, Assistant Professor, Technical University of Denmark, Environment & Resources DTU, Bygningstorvet, Building 115, Lyngby DK-2800 Kgs. (Denmark), phone +45 4525 1552, fax +45 4593 2850, eve@er.dtu.dk .... chapters 11, 19, 20

FÖRSTER Miriam, Former DayWater assistant project manager, Pôle de Compétitivité Ville et Mobilité Durables, c/o Polytechnicum, 6-8 Avenue Blaise Pascal, Cité Descartes, Champs-sur-Marne, Marne la Vallée F-77455 (France), phone +33 1 45 92 65 43, fax +33 1 45 92 66 99, forsterm@esiee.fr .... chapters 1, 26

GELDOF Govert, Director, Geldof c.s., PO Box 51, Bathmen NL-7437 ZH (The Netherlands), phone +31 570 54 22 28, fax +31 570 54 22 28, govert@geldofcs.nl .... chapters 1, 6, 26

GERMAN Jonas, Consultant in hydrology, Swedish Meteorological and Hydrological Institute (SMHI), Environment and Safety Services, Norrköping SE-601 76 (Sweden), phone +46 11 495 8596, fax +46 (0)11 495 8001, jonas.german@smhi.se .... chapter 18

HUBERT Gilles, Professor, Université de Cergy-Pontoise, Laboratoire Mobilités, Réseaux, Territoires et Environnement, UFR Lettres et Sciences Sociales, 33 boulevard du Port, Cergy Pontoise Cedex F-95011 (France), phone +33 1 34 25 64 04, Gilles.Hubert@lsh.u-cergy.fr .... chapter 13

HURLEY Louise, Research Associate, University of Sheffield, Civil and Structural Engineering, Pennine Water Group, Sir Frederick Mappin building, Mappin Street, Sheffield S1 3JD (United Kingdom), phone +44114 222 5766, fax +44114 222 5700, l.hurley@sheffield.ac.uk .... chapter 25

JACOBSSON David, Boras City, Va-proj Gatukontoret, Borås SE-50 (Sweden), phone +46 33 35 74 62, david.jacobsson@boras.se .... chapter 18

JIN Zhengyue, Research Assistant, Ingenieurgesellschaft Prof. Dr. Sieker mbH, Rennbahnallee 109a, Dahlwitz-Hoppegarten, D-15366 (Germany), phone +49 03 342 359 513, fax +49-03 342 359 529, zy.jin@sieker.de .... chapter 15

JONSSON Arne, Project Manager, Stockholm City, Environment and Health Administration, Environment Monitoring, PO Box 8136, Stockholm SE-104 20, (Sweden), phone +46 8 508 28 939, fax +46 8 508 28 808, arne.jonsson@miljo.stockholm.se .... chapter 19

KLUCK Jeroen, Specialist, TAUW, WRR, Handelskade 11, P.O. Box 133, Deventer NL-7400 AC (The Netherlands), phone +31 570699224, fax +31 570 699 666, jku@tauw.nl .... chapter 3

LARDIEG Emmanuel, Research Engineer, INSA Lyon, Unité de Recherche du Génie Civil, Bâtiment J.C.A. Coulomb, 34 avenue des Arts, Villeurbanne CEDEX F-69621 (France), emmanuel.lardieg@insa-lyon.fr .... chapter 23

LEDIN Anna, Professor, Technical University of Denmark, Environment & Resources DTU, Bygningstorvet, Building 115, Lyngby DK-2800 Kgs. (Denmark), phone +45 4525 1584, fax +45 4593 2850, anl@er.dtu.dk .... chapters 11, 19

LEGRET Michel, Director of Research, Laboratoire Central des Ponts et Chaussées, Division Eau et Environnement, Route de Bouaye, BP 4129, Bouguenais Cedex F-44341 (France), phone +33 2 40 84 58 66, fax +33 2 40 84 59 98, michel.legret@lcpc.fr .... chapters 1, 26

LEMS Pieter, Project Leader, Water Agency Groot Salland, P.O. Box 60, NL-8000 AB Zwolle (The Netherlands), phone +31 (0)38 455 7348, fax +31 (0) 38 453 0111, plems@wgs.nl .... chapter 6

MEMON Fayyaz Ali, Lecturer, University of Exeter, Centre for Water Systems, School of Engineering, Computer Science & Mathematics, North Park Road, Exeter EX4 4QF (United Kingdom), phone +44 (0)1392 264 048, fax +44 (0)1392 217 965, f.a.memon@exeter.ac.uk .... chapter 25

METELKA Tomas, Department head, DHI a.s., SW Department, Na Vrsich 1490/5, Praha 10 CZ-100 00 (Czech Republic), phone +420 267 227 119, fax +420 271 736 912, t.metelka@dhi.cz .... chapters 1, 2, 4, 26

MIKKELSEN Peter Steen, Professor, Technical University of Denmark, Environment & Resources DTU, Bygningstorvet, Building 115, Lyngby DK-2800 Kgs. (Denmark), phone +45 45 251 605, fax +45 4593 2850, psm@er.dtu.dk .... chapters 1, 11, 12, 19, 26

MIRAMOND Marcel, Professor, INSA-Lyon, Unité de Recherche du Génie Civil, Bâtiment J.C.A. Coulomb, 34 avenue des Arts, Villeurbanne CEDEX F-69621 (France), marcel.miramond@insa-lyon.fr ... chapter 23

MOUCHEL Jean-Marie, Professor, Université P. et M. Curie (Paris 6), UMR Sisyphe, 4 place Jussieu, Paris F-75005 (France), phone +33 1 44 27 51 26, fax +33 1 44 27 51 25, jean-marie.mouchel@ccr.jussieu.fr .... chapters 12, 13, 20

RAUCH Wolfgang, Professor, University of Innsbruck, Institute of Infrastructure, Technikerstrasse 13, Innsbruck A-6020 (Austria), phone +43 512 507 6932, fax +43 512 507 2911, wolfgang.rauch@uibk.ac.at .... chapter 24

REVITT D. Michael, Professor, Middlesex University, Urban Pollution Research Centre, Queensway, Enfield EN3 4SA (United Kingdom), phone +44 208 411 5308, fax +44 208 411 54 40, m.revitt@mdx.ac.uk .... chapters 1, 7, 8, 9, 20, 26

RODRIGUEZ Fabrice, Searcher, Laboratoire Central des Ponts et Chaussées, Division Eau et Environnement, BP 4129, Route de Bouaye, Bougenais 44341 (France), phone +332 40 84 58 78, fax +332 40 84 59 98, fabrice.rodriguez@lcpc.fr .... chapter 21

SCHOLES Lian, Searcher, Middlesex University, Urban Pollution Research Centre, Queensway, Enfield EN3 4SA (United Kingdom), phone +44 208 411 6361, fax +44 208 411 54 40, l.scholes@mdx.ac.uk .... chapters 7, 8, 9, 20

SEIDL Martin, Searcher, Université Paris-Est, Cereve, 6-8 Ave Blaise Pascal, Marne la Vallée Cedex 02 F-77455 (France), phone +33 1 64 15 36 20, fax +33 1 64 15 37 64, seidl@cereve.enpc.fr .... chapter 20

SIEKER Heiko, Director, Ingenieurgesellschaft Prof. Dr. Sieker mbH, Rennbahnallee 109a, Dahlwitz-Hoppegarten, D-15366 (Germany), phone +49 3342 35 95 15, fax +49 33 42 35 95 29, h.sieker@sieker.de .... chapters 1, 14, 15, 26

SVENSSON Gilbert, Professor, DHI, Water & Environment, Lilla Bommen 1, Gothenburg S-41104 (Sweden), phone +46 (0)31 80 87 94, fax +46 (0)31 152120, gilbert.svensson@dhi.se chapters 1, 14, 18, 26

THEVENOT Daniel R., Project coordinator & Emeritus Professor, Université Paris-Est, Cereve, 6-8 Ave Blaise Pascal, Marne la Vallée Cedex 02 F-77455 (France), phone +33 1 64 15 36 25, fax +33 1 64 15 37 64, thevenot@cereve.enpc.fr .... chapters 1, 26

VALKMAN Roel, Advisor, TAUW, WRR, Handelskade 11, P.O. Box 133, Deventer NL-7400 AC (The Netherlands), phone +31 57 06 99 167, fax +31 570 699 666, val@tauw.nl .... chapters 3, 6

VERHOEVEN Anton J., Senior policy adviser, Nijmegen Municipality, Water Department, Postbox 9105, Nijmegen NL-6500 HG (The Netherlands), phone +31 24 32 92 785, fax +31 24 32 99 019, t.verhoeven@nijmegen.nl .... chapter 17

VIKLANDER Maria, Professor, Luleå University of Technology (LUT), Dept. of Sanitary Engineering, Luleå S-971 87 (Sweden), phone + 46 070 330 1486, fax + 46 920 49493, maria.viklander@sb.luth.se .... chapters 1, 22, 26

WAHLBERG Cajsa, Environment chemist, Stockholm Water Co., Torsgatan 26, Stockholm S-106 36 (Sweden), phone +46 8 522 124 35, fax +46 8 522 133 02, cajsa.wahlberg@stockholmvatten.se .... chapter 19

WESTERLUND Camilia, Dr, Luleå University of Technology (LUT), Division of Architecture and Infrastructure, Department of Civil, Mining and Evironmental Engineering, Luleå SE-971 87 (Sweden), phone +46 (0)920 491494, fax + 46 920 49 28 18, Camilla.Westerlund@ltu.se........................chapter 22

ZUURMAN Antal H.J., Consultant in urban water management, Nijmegen Municipality, Department Stadtwerken, Postbox 9105, Nijmegen NL-6500 HG (The Netherlands), phone +31 24 32 92 743, fax +31 24 32 99 710, A.zuurman@nijmegen.nl .... chapter 17

Chapter 1

DayWater: the challenge to develop an Adaptive Decision Support System for urban stormwater source control

Daniel R. Thévenot, Miriam Förster, Jean-Claude Deutsch, Govert Geldof, Gilbert Svensson, Peter Steen Mikkelsen, D. Michael Revitt, Emmanuel Aftias, Tomas Metelka, Heiko Sieker, Michel Legret and Maria Viklander

INTRODUCTION

Runoff water concepts, such as stormwater source control, encounter difficulties with regard to adoption. The complex interaction with urban life and development implies complex decision-making processes (DMP) between numerous stakeholders (Deutsch and Deroubaix, 2003). In order to support such DMP, the collection of heterogeneous data and information became necessary.

An important aim of the DayWater project has been to build a prototype of an Adaptive Decision Support System (ADSS) related to urban stormwater pollution source control strategies (Scholes *et al.*, 2003; Förster *et al.*, 2004a and b; Sieker *et al.*, 2005). This decision support system (DSS) can be used for the selection of various source control techniques for local stormwater runoff control and treatment, such as detention and re-use, swales and soakaways, catch-basins, wetlands, ponds, porous paving etc. The adaptive aspect is designed to respond to both internal (variety of stakeholder viewpoint and knowledge) and external (spatial scale, climatic conditions) changes.

A total of 14 core end-users and 10 scientific partners were involved in the DayWater project preparation and the subsequent 3 year development of an ADSS. As a prime objective was to create a multi-disciplinary Decision Support System (DSS), the collaborating partners originated from a range of backgrounds and interests, including urban hydrology, engineering, risk assessment, social sciences and information technology. The diversity of climatic conditions and regulatory frameworks encountered in Europe were covered by the different countries of origin represented by the scientific partners.

As is often the case, the initial work plan proposed for DayWater did not correspond exactly with the actual work progress. This resulted mainly from the chosen diversity of involved disciplines and countries, as well as from the large number of involved institutions. As an example of a European research project dealing with a complex context, this chapter presents the ADSS development procedure within the DayWater project and describes the major results of this development at the end of the project, together with the encountered difficulties and progress achieved. Furthermore,

 ISBN: 9781843391609. Published by IWA Publishing, London, UK.

this chapter provides an introduction to this book by presenting the major contributions to the Final DayWater Conference held in Paris on 3-4 November 2005 and the structure and functions of the ADSS major components.

ADSS DEVELOPMENT PROCEDURE

DayWater project structure

As for any European research project, the DayWater description of work was divided into several work packages (WP) each including identified tasks and deliverables (Figure 1):

(1) Project coordination (WP1),
(2) Web-based tool production (WP2),
(3) Involvement of Urban Dynamics in the Urban Stormwater Management projects (WP3),
(4) Risk identification and assessment (WP4),
(5) Best Management Practice catalogue and Multi Criteria Comparator (WP5),
(6) Source and flux model (WP6) and
(7) Field testing on selected case studies (WP7).

The daisy type scheme used for presenting these work packages and their interaction, at the kick-off meeting, became the actual logo of the DayWater project (Figure 1).

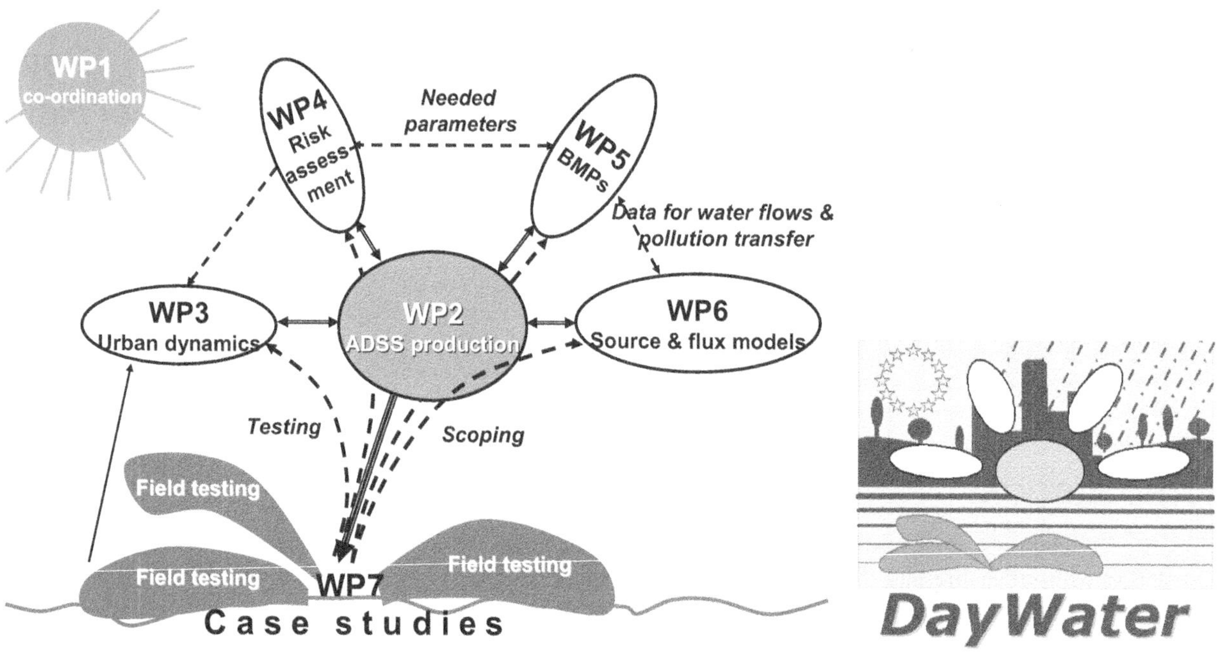

Figure 1: DayWater project structure in Work Packages (WP) and the evolution of this daisy into the project logo.

DayWater consortium

The consortium of ten scientific partners represents a broad knowledge domain involved in urban stormwater management (USWM) under various climatic conditions from the polar circle (Viklander *et al.,* 2003) to the Mediterranean Sea (Figure 2A). Nevertheless this scientific consortium, although dominated by engineers and urban hydrologists, contains a number of experts in socio-economics or information technology. Thus, the DayWater ADSS is not a traditional DSS, but is adaptive both to spatial scale and to the variety of stakeholder viewpoints and knowledge. It intends to summarise the professional experience of several scientists in urban stormwater management, including 3 private consultants (TAUW, DHI and IPS) and their respective contacts with end-users in this field.

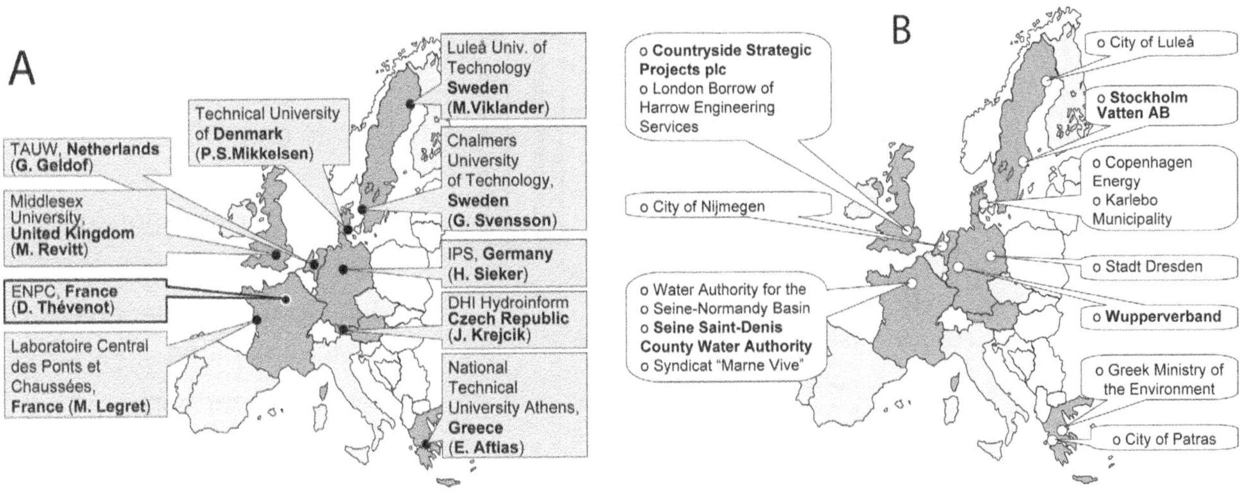

Figure 2: Location of DayWater major contributors: (A), (•) scientific partners and (B), (o) Core End-Users (CEU) where bold characters indicate the 4 sites for final ADSS testing.

In order to benefit from the common knowledge within the scientific consortium, most of the work packages (WPs) identified in the DayWater project (Figure 1) involved several or even all partners. Taking into account the important difference of national procedures, a common understanding of the goals and methods involved in each task, was needed to achieve representative results; this required close interactions within the research team. The role of the project coordinator was not limited to the harmonious organisation and development of parallel tasks, but was also needed when different opinions were expressed within the consortium and a consensus had to be found. Detailed interactions were enabled through a project web site, including an intranet part, and a file exchange platform (BSCW, OrbiTeam, Germany). The overall project progress was guided by annual Advisory Steering Board meetings and where specific bottlenecks in the project development appeared, these were discussed at smaller work meetings.

End-user involvement

As stated in the European Commission (EC) expert assessment report on the submitted DayWater project proposal, "the proposal is a research project with a strong (and healthy) focus on end-users and demonstration aspects". In order to collect their major concerns and needs in urban stormwater management, the scientific partners were in regular contact with 14 core end-users (CEU) (Figure 2B), during the DayWater project preparation as well as during its 3 year execution period. The CEU's contain representatives from a large variety of private companies, local, county or river basin institutions or associations. The number of end-users was further extended, at the end of the first project year, during 8 DayWater regional conferences (Figure 3) with one of the aims of these conferences being to increase the diversity of urban stormwater management projects considered in the ADSS development and to establish an extended end-user (EEU) group that could potentially participate in testing of the many components developed for inclusion in the ADSS. Among the more than 300 participants attending these 8 DayWater regional conferences, a large number represented municipalities or regional public organisations. Only a few private companies were involved, as they generally expressed a wish to see practical results before getting involved. For obvious geographic and linguistic reasons, local DayWater partners were in charge of interaction with their national CEU and EEU end-users. These interactions became particularly intense during the following 3 operations:

(1) the collection and analysis of urban stormwater management stakeholder needs,
(2) the testing of ADSS set of components,
(3) the final testing of the whole ADSS prototype.

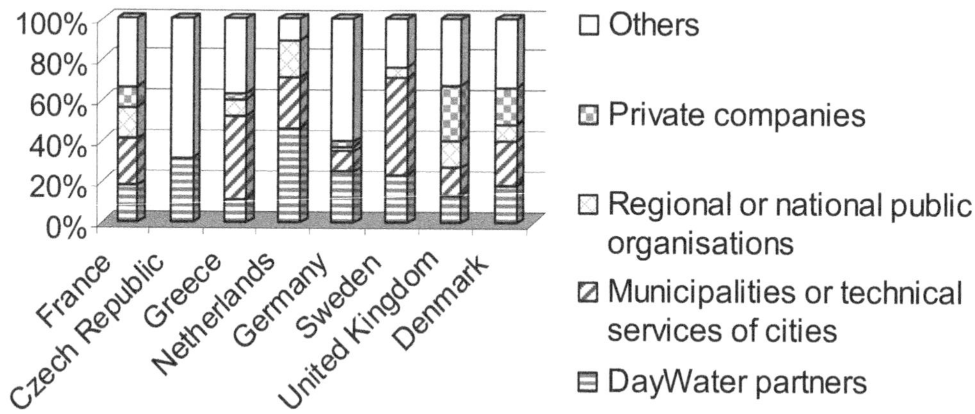

Figure 3: Affiliations of the 312 participants to the 8 regional DayWater conferences (October-November 2002)

Initial core end-user questionnaires

One of the most original challenges of the DayWater research project was the early involvement of future end-users of the ADSS prototype. At the commencement of the project, the scientific partners developed a questionnaire to collect details on the actual CEU needs, as well as the context of their integrated urban stormwater management (IUSWM) projects and related decisions (Table 1). These questionnaires were completed by the 14 CEUs, with the help of the scientific partners, and thoroughly analysed by all WP leaders. The resulting report defined the end-users needs for stormwater management which had to be met in the 'Terms of Reference' (ToR) of the ADSS (Figure 4) in order to clarify the objectives and characteristics of each component and the whole ADSS. Due to overlaps between the needs, they had to be classified and evaluated and this often led to a national focus, as partners concentrated on the needs described by their local end-users. Collecting questionnaires from a large group of end-users and subsequently analysing the results and agreeing on the structure and functionalities of the ADSS was a greater and more interactive challenge than anticipated, and this inevitably prolonged the process of framing the ADSS. The ADSS terms of reference were hence completed about a year later than planned (Figure 4).

Table 1: Structure of the questionnaire used for collecting information from the DayWater core end-users on their needs on integrated urban stormwater management (USWM) issues.

Questionnaire fields	**Number of questions**
End-user characteristics	15
Catchment's geographical characteristics and encountered urban water management problem	5
Specific stormwater problems	11
Stormwater plans and policies	4
Contact with other water management organisations	4
Available data and tools	10
On-going projects	6
Societal activity	6
European and national legislation	4
Economical aspects	3
Cultural and historical aspects	3
Other fields of urban policy	3

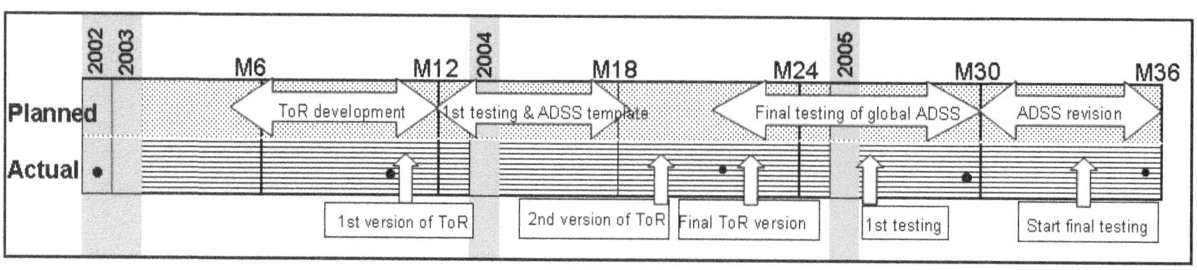

Figure 4: Comparison of initial work-plan (upper dotted part) and actual progress (lower striped part): (ToR) terms of reference, (•) annual project meeting.

Testing of ADSS components

The involvement of the end-users increased with the progress of the ADSS prototype development and the partner in charge of the field testing established a "Methodology for homogenous testing of ADSS components". With the help of this document, each partner was invited to conduct the testing of a set of components (Figure 4) with his local end-users (i.e. a selection of both the CEU and EEU) in order to collect feedback data on the following (Mousset *et al.*, 2008):

(1) global appreciation of the tested set of components, including both communication and scientific components,
(2) verification of the adaptability of the set of components to the context of the project,
(3) evaluation of the quality of the achieved results,
(4) evaluation of the usefulness of the achieved results,
(5) evaluation of the decision support function.

A major challenge encountered during these first testing operations was that many ADSS components were still in an early development stage at the beginning of the last project year. This challenge was partly overcome by presenting the ADSS component functionalities to the users as "mock-up" versions prepared in standardised presentation software (MS Power Point).

Testing of global ADSS

The final testing phase considered the whole ADSS prototype, which was planned to be tested on four case studies during the end of the second project year and the beginning of its last year (2005) (Figure 4). The case study sites were selected from the expressions of interest by CEUs willing to take part in this process. The selection of case study sites was also based on geographical, climatic and organisational diversity. Furthermore the following criteria were taken into account:

(1) the willingness and manpower of the involved end user,
(2) the availability of data necessary for testing the widest possible range of ADSS components,
(3) the diversity of the stages of development of the urban stormwater management projects and,
(4) the diversity of stakeholders involved in the same project.

The four selected end-users were:

- Countryside Properties, a private development company in United Kingdom,
- Seine Saint Denis County, a public water sanitation department working in the Seine Saint Denis district, in the east of Paris, in France,
- Stockholm Vatten, the public institution in charge of water management of the city of Stockholm, in Sweden,
- Wupperverband, an association in charge of managing the Wupper River basin in Germany.

The results from this full ADSS testing operation were expected to be used for revising the final version of the ADSS prototype. In order to allow such feed-back, the testing operations had to be finalised in the middle of the third project year (Figure 4). Unfortunately, as some of the ADSS components or external tools were not completely finalised at that stage, such recommendations were only available at the end of the DayWater project.

ADSS DEVELOPMENT

The DayWater project has followed both top-down and bottom-up approaches:

(1) top-down, as scientists provided up-to-date scientific results, related to the control and treatment of urban runoff;
(2) bottom-up, as the stakeholders in urban stormwater management projects, represented by DayWater end-users, were fully involved in the project, from the beginning of the ADSS design until the full testing of the prototype (Förster et al., 2004a and b).

This decision support system (DSS) intends to be adaptive to internal (variety of stakeholder viewpoint and know-how) and external (spatial scale, climatic conditions) variability.

Consideration of end-user needs

A unique aspect of the project was the end-user involvement during the project preparation and the development of the ADSS prototype. At the same time, this challenge became one of the most difficult tasks of the project. Consulting 14 core end-users (Table 1) throughout the project duration demanded a substantial effort for the team coordinating the field testing. Organisational procedures were often difficult, as, due to language barriers, each scientific partner was in charge of continuous contact with local end-users to ensure project progress. However, there were instances when end-users complained about not being sufficiently informed about the overall project advancement despite the preparation of a bi-annual DayWater newsletter. All CEU members were invited to the annual project meetings where all available

ADSS components were demonstrated but, in spite of this, the knowledge of some end-users about the DayWater project was limited to the component developed by their respective national scientific partner.

Common definition of ADSS structure and functionalities

The intrinsic difficulties of a multidisciplinary and multinational project are obvious. Even if scientific partners are working in the same field, there is a common language to be found for the project inherent terms. To overcome these problems and to establish a firm basis for moving forward, a series of work meetings were organised in the first and second project year, in which the ADSS functionalities were defined and agreed by all partners, in order to produce the terms of reference (Figure 4).

Development of ADSS components

The ambition and expectations for the ADSS prototype varied considerably during the first two project years. At the beginning of the project, it was difficult to define a realistic frame of all the interactions in the project due to its complexity. Each partner developed their components mostly independent of the other components and their mutual requirements. The terms of reference of the ADSS were expected to remediate this problem, requesting from every component developer a description of their requirements and their contribution to the other components and to the general ADSS structure. These processes were hampered due to the inter-relations only being discovered step by step, parallel to their development. This is in contradiction to the objective of the terms of reference, which should be the reference document for the prototype development. Furthermore, the initial work plan did not specifically mention the dates of delivery for each ADSS component, so the structure and functions were unclear when the project was planned. This omission made it difficult to make each partner understand the effect of delays in the elaboration of their component specification on subsequent implementation in the web based ADSS prototype. At this point, it is important to remember that research projects cannot be scheduled as production or demonstration processes, since the components are not available at the beginning and can only be described in general terms.

Interest in decision-making procedures in urban stormwater management

The analysis of the decision making procedure in urban stormwater management projects is a good example of this kind of intercultural and interdisciplinary misunderstanding (Deutsch *et al.,* 2004). Consulting engineers generally conceive the decision making procedure in a different way than researchers and south European countries do not consider decision making procedures as north Europeans do. Thus it was found necessary to integrate, in the ADSS system, several tools aiming at helping decision-makers to negotiate with all stakeholders involved in urban stormwater management and attempting to reach a consensus. In order to take into account all aspects of water and not only technical or financial ones (Chapter 6), two matrices for multi-criteria comparison were developed: a 'Multi-criteria Comparator' (MCC) of best management practices, based on pre-registered objective criteria and benchmarks (Chapter 9) and a 'Matrix of Alternatives' (MoA) including alternative measures and criteria which are selected, rated and weighted by each stakeholder involved in the urban stormwater management project, using his own data and opinion (Chapter 4).

Future actions

A remaining challenge of the DayWater project is how to maintain and further develop the ADSS prototype. Who will continue the ADSS development, enrich the databases and tools, and disseminate a viable product? All contacted end-users strongly expressed their concern on this point, stating that their involvement in the ADSS testing procedures requires a time perspective beyond the duration of the DayWater project, i.e. beyond the production of a prototype ADSS.

Furthermore the ADSS prototype has been developed in English, the common project language, but for an adoption in many European countries a translation of all documents, software and tools is necessary. Nevertheless, within the duration and the budget of the DayWater project, there was no possibility of translation into other European languages.

Taking into account the characteristics of European research and technology development (RTD) projects, there was no obvious method for supporting such demonstration and development activity. Thus, it was decided to keep the ADSS prototype in operation for some years after the end of the DayWater contract (November 2005), while suggesting that end-users seek national or European funding for demonstration, development or translation operations, of course with the assistance of the scientific partners.

ADAPTIVE DECISION SUPPORT SYSTEM MAJOR CHARACTERISTICS

ADSS functions

As demonstrated during DayWater field testing, the ADSS target user is an urban water manager with a good technical know-how, who is willing to promote stormwater source control strategies. A strong motivation is needed to convince the numerous stakeholders involved in any urban stormwater management project, of the effectiveness and sustainability of relevant 'Best Management Practices' (BMPs).

The ADSS is based on the typical structure of a decision support system and presents four main functions:

(1) the library function, organising information on stormwater source control matters (knowledge databases),

(2) the management function, supporting the definition of the urban stormwater problem,

(3) the analysis function, enabling a pre-selection of the data for the defined problem and a comparison of the different solutions,

(4) the communication function, supporting the negotiation between stakeholders and facilitating a consensus.

The ADSS is not an expert system but supports stakeholders in the decision making process by suggesting the most appropriate information or tools to use at the right time.

Figure 5: DayWater ADSS front page, part of graphical user interface called Hydropolis, available at URL: www.daywater.cz. For a free browsing, use 'guest' as login and password.

Structure of the ADSS

As described in the first part of this book, the ADSS is a combination of simulation models, assessment tools, databases and guidance documents (Chapter 2). It is designed to facilitate the information management and the preparation of decisions, both in free and guided mode, using the respective tiles on the graphical user interface (GUI) known as Hydropolis (Figure 5) (Chapter 3).

Figure 6: DayWater ADSS site map, identifying the 8 components described in Chapters 2 to 15.

A first set of components constitute knowledge bases to help the user to describe their urban stormwater management project through technical, economic, political and environmental databases. A second set of components, mostly external, but linked to the ADSS through XML file exchange, allow the user to compare the possible solutions for their urban stormwater management project and rapidly obtain a consensus with involved stakeholders. The ADSS site map (Figure 6) summarises the structure of the web based tool and its major components. A user manual can be found under the 'tutorial/start' bottom right tile, as well as a short presentation of each component.

Additionally several databases were specifically developed for the ADSS under the general term of library (Chapter 5). These libraries contain a classification of stakeholders possibly involved at different stages of an urban stormwater management project and of policy instruments available at regional, national or European scale. Numerous case studies on urban stormwater management projects have been entered by ADSS users willing to share their experience with others. Finally, a list of key terms have been developed for filtering the knowledge base and suggesting to each ADSS user the most appropriate set of data (BMPs, case studies, stakeholders, policy instrument information) and tools.

Tools associated to the ADSS

Following the common steps of decision making procedure in urban stormwater projects, i.e. identification and analysis of the problem, selection of possible techniques for the project construction and comparison and negotiation of alternative solutions, in its second part this book presents the tools associated to the ADSS.

Urban Dynamics

The success of an urban stormwater project depends not only on the technical, legal and economic feasibility. Equally important is that the stormwater project is compatible with the urban environment with its problems and opportunities, i.e. with its urban dynamic (Lems and Valkman, 2004). The Urban Dynamic part of Hydropolis comprises three components (Chapter 6). The aim of 'Water Aspects' is to broaden the user's view on stormwater projects and hopefully to provide new ideas. This helps not only in the communication with stakeholders but it can also give a new perspective on how to deal with stormwater. 'Coping with Urban Dynamics' illustrates how many problems encountered by water managers are related to problems on the edge of the urban water system and the urban context.

The 'Ambition Reflection' tool is dedicated to the urban stormwater manager in terms of thinking about attitude and choosing a more contextual approach, in order to implement successfully stormwater measures in the urban context.

Best Management Practices

A survey of the applicability and performance of different stormwater management practices within Europe has led to the development of a web based catalogue which includes national approaches to the use of 'Best Management Practices' (BMPs), applications of BMPs, design, operation and maintenance (O&M), costs, BMP performance, environmental advantages and sustainability (Chapter 8).

Pollutants

Chemical Hazard Identification & Assessment Tool (CHIAT) is a conceptual methodology for evaluation of chemical risks related to the handling of storm- and waste-water (Chapter 11) (Eriksson *et al.*, 2004). It can compare chemical risks when evaluating different handling strategies. It prioritises which compounds should be included in an investigation or a monitoring programme. Ranking and Identification of Chemical Hazard (RICH) is a tool, within the CHIAT methodology, that identifies and ranks organic substances which could pose a hazard based on their inherent characteristics (physico-chemical properties, environmental fate and chronic effects).

Risk and vulnerability

The traditional definition of risk in the engineering community is in practice difficult to use as well as to communicate to people. Therefore, the DayWater project has adopted a more symmetrical risk definition, where risk is a function of the hazard related to the risk source and of the vulnerability related to the risk object (Chapter 12). This interpretation of risk is advantageous because these concepts relate directly to possible actions to mitigate risks. Vulnerability identification and assessment focuses on two different application areas.. The ecological effects from discharges of chemical compounds in stormwater are discussed in Chapter 12. A flood vulnerability assessment tool (VAT) has been developed to allow users to reflect on what indicators of flooding in urban areas may be important in the local context (Chapter 13). Finally the FLEXT-tool can generate maps for on-site measures for rainwater in urban areas by using a knowledge-based decision support system integrated in a GIS (Chapter 15).

Multi-criteria comparator (MCC) and Matrix of Alternatives (MoA)

Starting from a definition of the problem, system (catchment, scale…) and context (economical, environmental and social) descriptions, the group of stakeholders, involved in an urban stormwater management project, is able to introduce relevant constraints (Chapter 4). A 'Multi-criteria Comparator' (MCC) allows an objective comparison of individual BMPs, using pre-defined sets of indicators and benchmarks which take into account the efficiency of each source control measure (Chapter 9) (Ellis *et al.*, 2004). Further to such classification, a 'Matrix of Alternatives' (MoA) tool facilitates the negotiation, within the group of involved stakeholders, through the comparison of selected alternatives, designed for the actual case, using selected criteria and weights for each of them (Chapter 4).

Tools

A prototype of a stormwater sources and flux model (SFM) for quantification of stormwater-carried material flows in urban areas has been further developed and linked to the ADSS (Chapter 14). The SEWSYS part of the software assesses flow quality and quantity on the basis of land use (GIS), using data collected throughout Europe, at different time scales, and taking into account the identified potential stormwater priority pollutants (PSPP) resulting from the CHIAT method. Using quality parameters derived from SEWSYS, the hydrological part of the model (STORM), generates water yields and pollutant fluxes and calculates results for different scenarios of stormwater source control. In addition to this SFM tool, simple design and costings assessment documents and tables have been developed for selected BMPs. This is especially useful for a preliminary stage of design of BMPs. Cost assessment documents and a methodology for life cycle cost evaluation and comparison of BMPs have also been included..

FINAL DAYWATER CONFERENCE AND WORKSHOP

In order to start the dissemination of the DayWater prototype and to facilitate the assistance of CEU and EEU in setting up demonstrations project, the consortium took an active part in the 10th International Conference in Urban Drainage (10 ICUD) in Copenhagen (21-26 August 2005) and organised an international conference dedicated to Decision Support for Integrated Urban Stormwater Management. This final conference, hosted by Ecole Nationale des Ponts et Chaussées at Marne-La-Vallée, in Paris vicinity, on 3 and 4 November 2005, was attended by 76 scientists and practitioners, half of them from outside the DayWater consortium. Three keynote speakers were invited to present their experiences in developing decision support systems in urban water management. Further invited speakers included both DayWater scientists and end-users, involved in the development and testing of ADSS components. Four scientific sessions were respectively dedicated to:

- Decision support for urban stormwater management,
- Risk analysis and impact assessment of urban stormwater,
- Selection criteria for urban stormwater source control,
- Specific issues in urban stormwater source control.

These 4 scientific sessions were followed by 4 demonstration workshops where participants could directly interact with the component developers, while using the on-line ADSS prototype on the internet:

- DayWater ADSS: a general visit via the Hydropolis interface,
- DayWater ADSS guided visit: from the 'Profile' definition to the building of the 'Matrix of Alternatives' (MoA) and the 'Multicriteria Comparator' (MCC),
- Chemical hazard identification and assessment tool and vulnerability mapping (FLEXT),
- Source and flux models for urban stormwater (STORM and SEWSYS).

Whereas CD-ROM proceedings of this final conference were sent to all participants, this book presents the major components of the ADSS prototype, as they appear on the front page of ADSS graphical user interface called Hydropolis (Figure 5) and allows the reader to be introduced to the major functions of the ADSS, while freely browsing its content available at URL: www.daywater.cz . The 5 parts of this book are respectively dedicated to:

(1) structure of the ADSS (Chapters 2 to 5),
(2) tools associated to the ADSS (Chapters 6 to 15),
(3) applications of the ADSS (Chapters 16 to 19),
(4) related research in stormwater management (Chapters 20 to 22),
(5) related decision support system (Chapters 23 to 25).

A final chapter presents the major DayWater conclusions and perspectives.

CONCLUSION

The diversity of the DayWater partners, i.e. 3 consulting companies and 7 academic institutions located from the polar circle to the Mediterranean sea, covers most geographical, climatic and administrative urban stormwater management conditions in Europe. Such diversity, created frequent discussions and delays, but also ensured that the varying situation across Europe was represented more truly. The involvement of end-users both prior to and during the development of the ADSS prototype, was challenging and time consuming, but was also a key issue for the success of the ADSS in achieving an adequate product for end-user needs and building the adaptive feature of the decision support system.

What are the lessons learned on how to develop an interactive prototype in a scientific environment? An agreement on the specification of functions and components is generally considered as necessary prior to the start of computer programming. When attempts were made to develop the computer design of the prototype before either functions or components of the system were fully agreed on, very time consuming discussions took place. For example, ADSS component developers had difficulties to specify all functions and possible links to other components, before the completion of their research work. The work schedule was very tight and a lot of time has been invested on internal discussions. However, such discussions and interactions between scientific partners are necessary in a research project, where the freedom for developing new ideas or tools is necessary. Furthermore, benefits to the project have been an improved broadness and usefulness for the final ADSS prototype.

In order to promote source control measures for integrated urban stormwater management, the major challenges of DayWater were to combine:

1) scientific knowledge on BMPs, their efficiency and cost, on risk and vulnerability, and on chemical hazard… with

2) operational tools for BMP dimensioning, for source and flux modelling, and for risk and vulnerability screening, and with

3) practical knowledge collected both by partners and end-users on different aspects of urban water, on negotiations between involved stakeholders and on decision-making procedures, into user-friendly software architecture.

Important innovation evolves from the integration of all these research areas into a learning process tailored to enable convergence of ideas and lines of thought. It is obvious that the role of the end-users was the core of the ADSS development. DayWater includes integration of alternative solutions into urban management and the ADSS prototype facilitates the involvement of stakeholders in the decision and realisation processes.

When the Water Framework Directive (WFD) is addressed while no stormwater source control is applied, the scaling-up of combined or separate sewer networks and of associated waste water treatment plants (WWTP) yields huge financial costs: the major economic relevance of this ADSS is clearly the control of such scaling-up costs. An important economic impact is also the effect of stormwater source control on limitation of flooding (addressed both in cold and Mediterranean climates) and pollution of receiving waters. Indeed, the existence of combined and separate sewer overflows (CSO and SSO) is a commonly important reason for non-abstraction of surface water and subsequent treatment to produce drinking water. The social relevance of this ADSS is also addressed through the increasing interest that citizens display both for surface water quality, for water architectural value within an urban district and more generally for an efficient environmental governance.

Finally, the DayWater results promote transparent negotiation and decision procedures on urban water related aspects. They are key issues for successful European and national policies related to water preservation in urban areas. Urban stormwater management case studies as well as numerous BMP examples, presented within the ADSS prototype, illustrate successful projects involving a large spectrum of stakeholders, from land or equipment owners, local government to sewer, road, open space and environmental officers and to urban planners. During the Final DayWater Conference, examples were given of successful urban stormwater management projects in the Netherlands involving elementary schools and children activities (Chapter 17).

This Chapter has been written following the completion of the DayWater project, i.e. at the end of the development of the ADSS components, their assessment by end-users and the development of the final ADSS prototype. The consortium of both scientists and end-users has been actively involved in the ADSS prototype promotion and dissemination. Besides several presentations made during the 10 ICUD conference in August 2005, and its specific sessions devoted to the CityNet research cluster, another important opportunity was the final DayWater conference and workshop, hosted at ENPC, on 3-4 November 2005. Although dissemination of the ADSS will require assistance by scientific partners and further development and translation tasks, its present prototype status has been found valuable by a large number of urban stormwater management technicians and decision makers, both within and outside the DayWater consortium. Publication, in this book, of the major results of the DayWater project, as well as of 3 other European related projects, (Chapters 23 to 25) will facilitate this dissemination process.

REFERENCES

Deutsch, J.-C. and Deroubaix, J.-F. (2003) Lutte contre la pollution des eaux de ruissellement urbaines: l'Europe à la rescousse, *Techni. Cité*, **58**, 17-19.

Deutsch, J.-C., Deroubaix, J.F., Chouli, E. and Carré, C. (2004) How and how far the computer can support the decision making process?, *Proc. 4th International Conference on Decision Making in Urban and Civil Engineering (DMUCE 4)* Porto, Portugal, 28-30 October 2004.

Ellis, J.B., Deutsch, J.C., Mouchel, J.M., Scholes, L. and Revitt, D.M. (2004) Multicriteria decision approaches to support sustainable drainage options for the treatment of highway and urban runoff, *Science of the Total Environment*, **334-335**, 251-260.

Eriksson, E., Baun, A., Mikkelsen, P.S. and Ledin, A. (2004) An approach for identification of problems related to chemical compounds and handling of rainwater locally, Novatech, *Proc. 5th International conference on sustainable techniques and strategies in urban water management*, 6-10 June 2004.

Förster, M., Thévenot, D., Geldof, G., Svensson, G., Mikkelsen, P.-S., Revitt, M., Aftias, E., Krejcik, J., Sieker, H., Legret, M. and Viklander, M. (2004a) Urban stormwater source control management in European countries: DayWater project, Novatech, *Proc. 5th International conference on sustainable techniques and strategies in urban water management*, 6-10 June 2004.

Förster, M., Thévenot, D., Geldof, G., Svensson, G., Mikkelsen, P.-S., Revitt, M., Aftias, E., Krejcik, J., Sieker, H., Legret, M. and Viklander, M. (2004b) Decision-making process in the context of urban stormwater source control management within European countries: DayWater project, *Proc. 4th International Conference on Decision Making in Urban and Civil Engineering (DMUCE 4)* Porto, Portugal, 28-30 October 2004.

Lems, P. and Valkman, R. (2004) Contextual water management, from necessity to opportunity, Novatech, *Proc. 5th International conference on sustainable techniques and strategies in urban water management*, 6-10 June 2004.

Mousset, C., Chouli, E., Carré, C., Förster, M., Deutsch, J.-C., Deroubaix, J.-F. and Thévenot, D. R. (2008) Contrôle à la source des eaux pluviales dans l'aménagement urbain : enjeux et perspectives de l'aide à la décision développée par le programme européen DayWater, *Revue des Techniques Urbaines*, in press.

Scholes, L.N.L., Revitt, D.M. and Ellis, J.B. (2003) A European project (DayWater) investigating the integration of stormwater source control into sustainable urban water management strategies, *Journal of Health, Social and Environmental Issues*, 2003, **4**(2), 37-40.

Sieker, H., Förster, M., Scheibel, M. and Zimmerman, U. (2005) DayWater: Ein Entscheidungshilfe-system für Regenwasserbewirtschaftungsmaßnahmen, *Wasserwirtschaft*, **03**, in press.

Viklander, M., Bäckström, M., Förster M. and Thévenot, D. (2003) Urban Stormwater Source Control Strategy within DayWater Project (FP 5 RTD): General Feature and Specific Issues in Cold Climate, *Proc. 1st International Conference on Urban Drainage and Highway Runoff in Cold Climate,* Riksgränsen, Sweden, 25-27 March 2003.

Part 1

Structure of the Adaptive Decision Support System

Chapter 2

Adaptive DSS – a new kind of DSS

Jean Claude Deutsch and Tomas Metelka

INTRODUCTION

Information plays a key societal-technical role in the overall decision-making process. Shared information constitutes a fundamental basis for co-ordination of work and common process language. Exchange of information increases the team knowledge, implies a generation of new information, prevents misunderstandings and brings trust and motivation. All these aspects provide the basis for success in the decision making process and constitute a platform for information technology support in the form of a Decision Support System (DSS).

The concept of DSS is one of those fuzzy concepts which have appeared when society has begun to believe that Science could solve societal problems. Two scientific branches have contributed to this concept: operational research and decision making theory. Other approaches have progressively joined these two first as complexity theory and business intelligence has been introduced in order to build a general theory supporting the implementation of DSS.

Power (1997) has defined DSS as "an interactive computer-based system intended to help managers make decisions; a DSS helps a manager to retrieve, summarise and analyse decision relevant data". Thus a DSS is intended to be an interactive, computer-based system utilising the data through specialized databases, communications technologies, simulation models, GIS systems, office documents, expert knowledge systems, etc. to identify problems and their potential causes and to stimulate the search for optimal resolution from the variety of potential decisions.

There is a large application of DSS within the Water Management field which has always been considered as a complex system. Water Management covers the complexity of interacting parameters such as: water quality, water quantity, natural environment, economical, social and cultural factors, as well as the institutional and political context. Beside the diversity of such influencing parameters, there is also a diversity of stakeholders intervening in the water management process. To face such complexity, the concept of Integrated Water Management (IWM) has emerged. This means that water management must be tackled within a global and integrated perspective.

A large number of Decision Support Tools have been developed to assist decision makers implementing IWM. These tools can be classified into two main types: DSS and Decision Support Methods (DSM). Two classes of DSS are generally considered: Environmentally Information Systems and Integrated Modelling Systems. DSM is structured mainly on a methodology which provides elements to support decision makers at whatever level of the processes involved. These methods include planning and development methods as well as Multi-Criteria Decision Methods techniques.

The Adaptive Decision Support System (ADSS) has been designed and implemented in the framework of 5th FP European research project DayWater in the field of Urban Storm Water Management (USWM) (Chapter 1). ADSS is understood as a computerized instrument, which will support decision making in USWM (Chapter 26). The ADSS adapts its structure, content and environment based on possible evolution in user's demands, interests and knowledge, and is developed as a web based client-server database application. The ADSS can be considered as a combination of

 DayWater: an Adaptive Decision Support System for Urban Stormwater Management. Edited by Daniel R. Thévenot. ISBN: 9781843391609. Published by IWA Publishing, London, UK.

DSS and DSM embedded into the field of USWM. It contains a combination of databases, modelling and GIS-based tools, as essential elements of decision-making process support methods. It additionally incorporates a variety of knowledge concepts on different topics and at different scale levels in order to provide support for decision makers facing various complex situations.

Adaptiveness must be understood from the user's point of view: an adaptive DSS enables the user to build an "information environment" based on his needs. Of course a DSS, whatever its level of adaptiveness, can only provide information that has been stored within it. When there are conflicts on the objectives and/or uncertainty on the way to solve an USWM problem, the possibility to provide help to the user relies on the capacity of the system to suggest new information not already provided by other policy instruments (e.g. technical and legal guidelines). To fulfil this objective, a link between various components and tools has been proposed by the DayWater scientific partners and tested by the end-users participating to the project (Chapter 16).

BASIC CONCEPTS OF A DECISION SUPPORT SYSTEM

Two basic types of DSS can be considered:

(1) Knowledge Management (KM). It is argued that KM can advance the understanding of decision support and the capabilities of decision support systems (Holsapple, 2001). This premise stems from several observations:

- KM is concerned with the representation and processing of knowledge by humans, machines, organizations, and societies,
- An aim of KM is to ensure that the right knowledge is available in the right format, for the right entities, at the right times and for the right costs,
- Decision making is a knowledge-intensive activity with knowledge as its raw materials, working process, by-products, and finished goods,
- Computer-based DSSs employ various KM techniques to represent and process knowledge of interest to decision makers, including descriptive knowledge (data, information), procedural knowledge (algorithms), and reasoning knowledge (rules),
- There are multiple techniques for usefully managing knowledge in the course of making a decision,
- Proficiency in KM is increasingly important to the competitiveness of decision makers as society rapidly moves to achieve global knowledge capabilities.

KM is a type of management which helps decision makers, mainly in private organisations, to control strategic change or following the activity of problem-solving.

(2) Business Intelligence (BI). The need for senior managers to follow the activity of an enterprise and to manage their relationship with clients has led to the concept of "computer based decision support systems". These systems have been implemented not only to give information, but also to give the tools of analysis useful for the decision making process.

The basic structure of these systems is constructed from similar components. Firstly, the structure contains a variety of databases concerned with applications: production, commercial. Data are treated with a tool named "Extract, Transform and Load" (ETL). The treated integrated data, classified as thematic and chronological data, are stored in a large database called the Data Warehouse. Multi-dimensional analysis can then be applied to this database in order to select relevant parameters corresponding to the request of the manager and to create strategic indicators for following the enterprise activity. The structure is completed by reporting tools.

From these two examples, it can be assumed that, in a DSS, there will be information and data, and something to process the information and data in order to yield by-products such as indicators for managing systems (for instance an enterprise) or assisting the decision making process.

A somewhat different approach has been made by workers dealing with integrated water management and difficulties arising from the complexity of the considered problem (Thomas *et al.*, 2001). The techniques and models need to be more complex in order to reflect adequately the required integrated water management, more refined to improve accuracy, while at the same time the model(s) should be as simple as possible, in order to control the development effort, computational time requirements and costs.

To address such problems, tools with different "shapes" have been implemented. Some of these tools are composed of groups of models such as hydrological models, water quality models, databases or geographic information systems GIS). Other tools are based on planning processes, methodologies or protocols in order to lead to a conclusion. A range of methods, called Multi-Criteria Decision Making (MCDM) methods are often part of these alternative tools: they are frequently used to help the decision maker in choosing alternatives, according to different parameters.

Trying to perceive the complexity associated to differing elements associated with a problem, including human, benefits, economics, environmental impacts, value judgments and technology, has led to another point of view: a solution which considers the critical questions from a holistic point of view and understands the contribution of the system parts in relationship to the whole, but set within the context of a range of view-points and value judgments. The approach using MCDM in conjunction with specific models comprises the definition and structure of the DSS that has been adopted for the DayWater project.

A further contribution to the development of the concept of an ADSS has been made by Chuang and Yadav (1998) wherein two main characteristics of an ADSS are explored: adaptive behaviour and knowledge and capacity building. The desired adaptive behaviour can be summarised as follows:

- Multiple modes: an ADSS should be able to support multiple presentation modes in order to accommodate the heterogeneity of users; given a particular user engaged in a particular problem situation, an individually tailored presentation mode should be provided;
- Multiple views: an ADSS should be capable of multiple views for the user to explore different underlying relationships amongst data;
- Multiple scenarios: an ADSS should be able to support the decision maker in performing "what if" analysis for various scenarios;
- Different problem situations: an ADSS should be able to generate various knowledge states for various problem situations;
- Self-moderation: after new knowledge is acquired and user interfaces for the problem situation are composed, an ADSS should be able to associate the new knowledge with the interfaces.

The following types of knowledge are required for supporting the identified behaviour:

- Meta-knowledge: knowledge about what we know or can know, or knowledge about how to use the knowledge that we have;
- User knowledge: knowledge about the user who is working on the problem domain within the ADSS; it is necessary for the system to present multiple modes because the system must adapt its presentation modes to the current status of the user facing a particular problem situation;
- Interface element knowledge: interface elements can be classified into four levels: the component level, the collection level, the screen layout level, and the application level;
- Problem domain knowledge: knowledge about the problem domain upon which the system is designed and the user is seeking support for decision-making;
- Presentation knowledge: knowledge that the system uses to organize interface layout for a particular user engaged in a particular problem solving session; it accommodates the dual nature of prescriptive and descriptive knowledge; given a problem situation, it exists as a set of the most effective user interfaces for the system to communicate with the user who has certain characteristics; this may refer to a prescriptive user interface in a given problem situation; however, the user may prefer another user interface to the prescriptive one; this specific user interface is descriptive by nature;
- Learning-methods knowledge: knowledge about the features and applicability of learning methods;
- Model knowledge: knowledge about various models, which are usually generic problem solving models or instrumental models.

This form of presentation is interesting because it gives a very formal approach to what must be an ADSS. The ADSS has to face the variability of the users and of the problems in a specific problem domain, but it assumes that, even in a multi-stakeholder environment, there is one final decision maker and that the main objective of the ADSS is to reach a decision.

The DayWater ADSS contains most of the components described above, but it assumes, than, even when there is only one single user, the kind of problems to be solved in urban stormwater source control are always multi-stakeholder problems. Thus the main objective is to give relevant elements to the user in order to support thinking about the best way to reach an agreement between the different stakeholders. It is why the ADSS must also give the possibility of a common representation of the problem for those interested in a solution.

APPLICATIONS OF DSS IN THE FIELD OF WATER MANAGEMENT

DSS EUREG – Czech case

One recent example of using DSS for decision support activities is related to the process of association of 10 candidate countries from the former eastern block to membership of the European Community (Figure 1). For these new countries the EU expansion required meeting the Copenhagen criteria – a set of standards to bring their government administrative systems into compliance with EU standards. The Czech Republic developed a DSS system to plan for the changes to be made for meeting the water-related environmental requirements of the EU Water Framework Directive (WFD) and specifically those of the Urban Waste Water Directive and the Directive for Quality of Fresh Water.

The DSS was developed at DHI Hydroinform in cooperation with the Carl Bro Company to assess distinct water policies and to select the most effective way of meeting WFD requirements on the national scale. The DSS includes data and modelling tools to provide an overview of river system water quality conditions, existing water supply and waste water treatment facilities as well as technical options for making improvements and tools for calculating associated costs. In addition, the DSS assessed the changes to water quality that would result from implementing various strategies and estimate corresponding investment for the implementation.

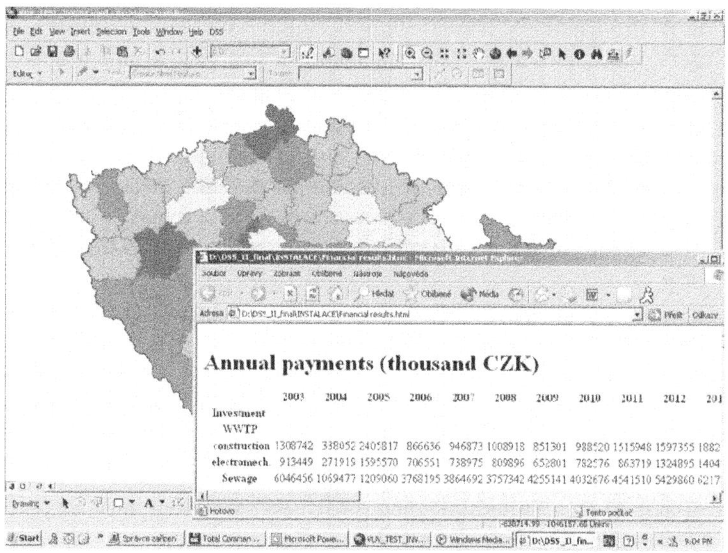

Figure 1: DSS EUREG environment.

The DSS was developed as an extension to ArcView (ARC8), although application also works on more comprehensive platforms such as ArcEdit or ArcInfo. The application has been developed in two parts:

(1) a graphical user interface which is closely connected to the ArcGIS, using its Application Programming Interface (API);

(2) the main functionality of the DSS (WWTP, water supply, non-point sources, ranking, economical, financial, models, communication with other models and the data sources) is placed into the individual Common Object Model (COM) objects; functions from the COM objects are executed by the Graphical User Interface (GUI).

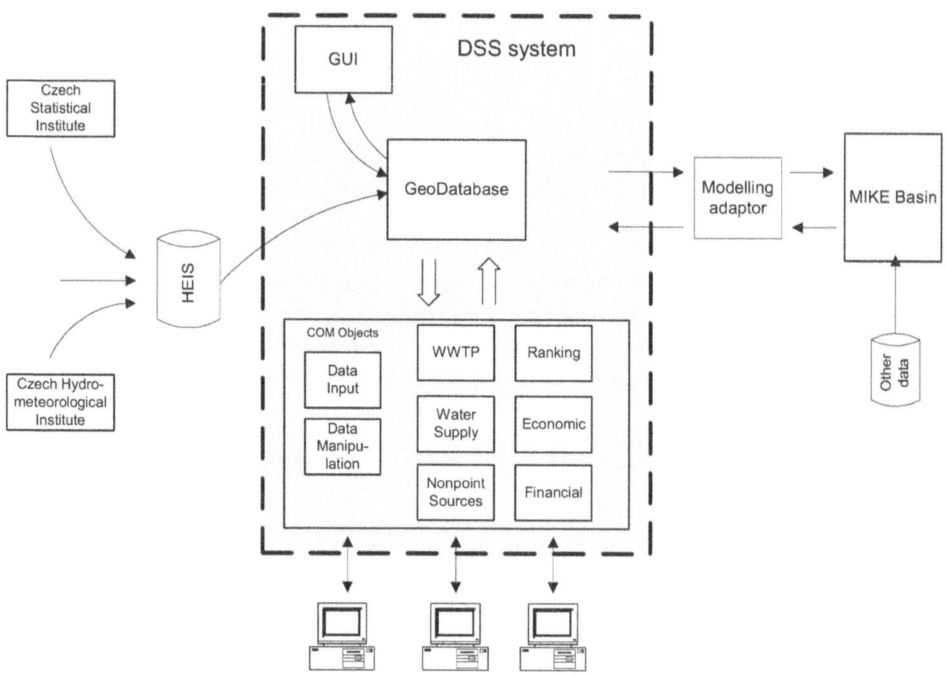

Figure 2: DSS EUREG architecture.

The ArcGIS geodatabase serves as the main data storage for all data (input data, parameters, results of the analysis etc) and graphical presentations are created using the ArcView facilities. The COM objects access the data directly. The DSS allows the general user to operate the functions without requiring extensive experience in databases, GIS and modelling tools whilst the more experienced user will be able to benefit from the built-in tools of ArcView. The use of COM objects allows the developer to use the functionalities of the various objects, even in other applications outside the framework of DSS.

It was intended to design a system which is expandable and to the largest possible extent, generic. The design allows for the inclusion of additional COM objects containing additional functionality. Use of more comprehensive platforms such as ArcGIS (ArcEdit, ArcInfo) allows for multi-user access to the system and for future web-based applications. The interfaces to data sources and the modelling tools were made generic in order to allow for potential use of the DSS within other data environments.

The DSS provided access to various databases and modelling tools through the GIS interface. The user could explore the range of options offered by different interpretations of the ultimate EU directives as well as alternative technical strategies and final time frames for meeting the requirements. The DSS consisted of different elements or "building blocks". The user got access to the functionality of the DSS via the GUI. The core functionality of the DSS was encapsulated in separate modules by use of the COM architecture (Figure 2).

WATSIDE: a methodology for WATer Supply Integrated Development

WATSIDE is a methodology that allows analysing and tackling the complexity of urban water supply systems, including stakeholders' intervention, in order to make to make more consensual and holistic decision (Thomas, 2001). This methodology considers the critical issues in an integrated way, so that the relationships can be established between upstream and downstream investments, remedial treatment and preventive quality control and the interest of separate parties. It provides a structure analysis of links between the components of the river basin upstream of the water intakes of a city supply system, the production and distribution system and the stakeholders involved. The overall objective is to give decision makers a better learning and understanding of the complex system that they are managing, in order to define a strategy for a sustainable development of the urban water supply.

The system analysis and the modelling phase of WATSIDE correspond to the classical first steps of a decision making processes: problem analysis, needs and objectives, system analysis, alternative definition. The position proposed is a minimalist but careful position; meaning that the first objective of the methodology remains to prepare a decision by providing a process of learning and understanding the system functioning. Depending on the accuracy of the system analysis and on the information available, it is possible that WATSIDE can be used to support the selection of alternatives.

The main core of this method is the development of a comprehensive water supply model, the basis of which is the system analysis of the definition and characterisation of each subsystem and elements of subsystem, according to the following attributes:

- Inputs and outputs (flows going into and leaving a subsystem);
- Parameters (constant or variable, defined by one or several criteria);
- Transfer function (how each element transform/transfers the inputs to outputs).

The transfer functions are made of three flows transiting through the different elements and relationships. These flows are the energy flow (money representing the costing and pricing of any action related to the water supply development), the matter flow (water characterised by its quality and volume) and the information flow (representing the transfer of data from one element to another). To make the links between upstream and downstream systems, a specific definition of the water quality and volume has been used with the water quality parameter comprising three criteria:

- Criteria Q1: the annual evolution of the mean value of the water quality (for the representative parameter);
- Criteria frequency F1: the number of hours per year when raw water, due to the water quality, necessitates treatment modification or the slow-down of the treatment plant production;
- Criteria frequency F2: the number of hours per year when raw water, due to the water quality, necessitates the shut down of the treatment plant.

The first criterion integrates the notion of capital investments to allow long term strategy whereas the last two integrate the operational costs. Moreover, similar to water pollution, flood or drought events affecting the functioning of the treatment plant are also considered:

- Criteria frequency F3: the number of hours per year when flood or drought necessitates treatment modification or the slow-down of the treatment plant production;
- Criteria frequency F4: the number of hours per year when flood or drought necessitates the shut down of the treatment plant.

ADAPTIVE DSS IN FRAME OF DAYWATER PROJECT

An Adaptive Decision Support System (ADSS) is generally understood as a DSS that is able to modify automatically or manually some aspects of its structure, functionality, or interface in order to meet different needs of its users. Within the DayWater project, the ADSS is described as a computerised instrument, which supports decision making in USWM in order to find the most appropriate measures. The ADSS should adapt its structure, content and environment, based on the variety and the changes in user demands, and should play an active role in supporting the decision-making process for stormwater source control.

General concept of ADSS

A decision process involves a wide range of stakeholders, interacting to produce a final decision. The stormwater problem needs first to be identified and analysed. Then the actual USWM project is constructed and possible techniques or measures are selected. Finally alternative solutions, i.e. designed measures, are compared in order to select those which are best supported by all stakeholders (Chapter 26). These stakeholders typically need to resolve conflicts and to find a mutually acceptable decision. However, in reality, each stakeholder has different scientific or technical backgrounds and languages as well as using different sources of information. All these differences complicate the process of reaching consensus and slow down the overall process while bringing fuzziness to decision making activities. At the same time stakeholders frequently hold the opinion that they know "best" about the local context and that they are the only ones appreciating the socio-economic situation as well as environmental and legislative restrictions. The ADSS should provide such stakeholders with general and theoretical scientific knowledge as well as practical information: this is the input of the various scientific partners of the DayWater project. The problem to be solved may concern hydrology and pollution topics, as well as decision-making processes or socio-economical and political contexts.

On one hand, the stakeholder can be presented with a hierarchy of the problems to be resolved, e.g.: impact of stormwater flooding in relation to urban vulnerability, chemical hazards in relation to the objectives given to the quality of the water environment or sanitary risks. The ADSS should, for example, give help to assess the flux and pollution production for a given rain event, and propose generic solutions.

On other hand, the stakeholders will often be familiar with their own institutional structure and operation, their relationships with other institutions, the financial capability for USWM, the local regulations and the way they are applied. Thus the ADSS should, for example, provide basic knowledge on decision theory, economics and political theory.

There is a dissymmetrical but complementary knowledge between these two stakeholder positions. The users will have knowledge and data relating to their local circumstances which can be used in the ADSS to derive relevant information of a more general (or specific) knowledge which they can use for further development of their project. In the case of the DayWater project, the ADSS should improve the general knowledge of the user concerning the implementation of source control pollution. However, the ADSS cannot solve the problem for the user, but provides only support in the decision making process.

The work has to be adaptive from the user's point of view, which means that the structure of the knowledge base within the ADSS enables the system to provide the user with the right level and type of information. The ADSS has to be user-friendly, ergonomic and have the ability to support all the information needed by the user.

In the consultation process, each stakeholder needs to have access to the same pool of information and should speak the same language. The ADSS should stimulate the decision-making based on the same reference platform and same data, although as stated previously, the ADSS should not deliver the decision but only support the search of a final decision.

The ADSS should also incorporate a variety of knowledge states in order to provide support for decision makers facing various problem situations. As shown in the Figure 3 based on Christensen (1985) (Chapter 3), the ADSS should be able to give answers to urban stormwater managers essentially when the state of the knowledge is stabilised and when the stakeholders are able to define clear objectives (quadrant I: 'Optimising, Programming, Modelling' of Figure 3). But the ADSS also aims at rationalising the decision-making process when the knowledge is partly missing (quadrant III: 'Learning, Innovation') or when there are conflicts on the definition of objectives (quadrant II: 'Negotiation'). Last but not least, the ADSS should be able to suggest new ways to tackle the issue of water management when there is a lack of stabilised knowledge and a conflict on the objectives (quadrant IV: 'Chaos').

Finally, the ADSS should comply with basic functional requirements of any DSS e.g. library functions, management functions, analysis functions and communication functions.

These basic assumptions were defined by the DayWater project team as theoretical fundamentals for defining the functionality of the ADSS. The main goals of the ADSS were then summarised as follows:

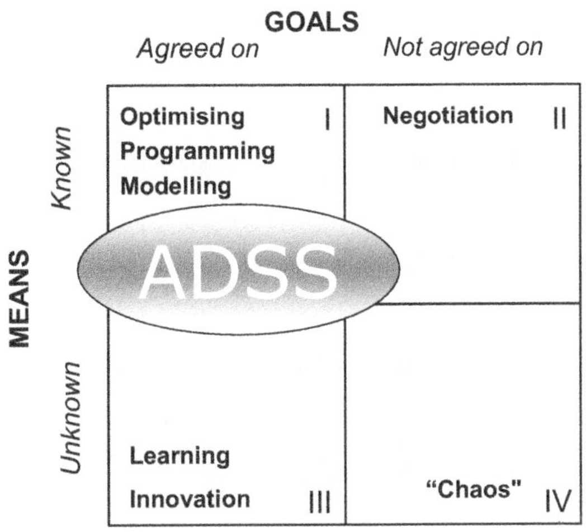

Figure 3: The DayWater ADSS concept, according to identified goals and means (adapted from Christensen, 1985).

- to focus on problems concerning hydrology and pollution as well as decision making process within socio-economical and political context,
- to offer and improve knowledge through a systematic representation of USWM,
- to provide the functionality for adapting the structure, content and environment of the ADSS to the user knowledge and interest.

Based on these general ADSS concepts, the basic ADSS characteristics were then analysed.

Basic ADSS features

Fundamental ADSS features are based on defined goals and focussing on general concept of decision support processes with the principal features defined by the project team as follows. The ADSS should:

- contain a set of knowledge databases and components, be highly modular in structure as well as co-ordinating information exchange procedures between components,
- tackle changes in stakeholders or projects and to manage interactions between stakeholders and projects,
- produce a representation of the problems and solutions shared by all the stakeholders involved in the project,
- support usual DSS functions.

As stated above, the ADSS needs to contain information on the explicit objectives of the user, but also on various problems that he has encountered, possible links between the stormwater problem and contextual issues, the stakeholder role in USWM, relationships among stakeholders and the results of modelling tools. It was agreed that the ADSS should be dedicated to potential users who have some technical knowledge and who already favour urban stormwater source control solutions. Thus the ADSS can be used in two ways defining the means of solving an USWM problem and tuning the different objectives of stakeholders.

ADSS Components and functions

The ADSS has been developed as a web-based, client-server database application. The system is developed as a set of database tables, software tools, specialised components and overall system logics (relations, functions, procedures etc).

The database core of the ADSS is composed of a number of tables and relations (Figure 4). The tables are logically grouped into areas ("logical databases") representing specialised information sources which are to be used during the decision support process: Best Management Processes (BMP) (Chapter 8) and their Multi Criteria Comparator (MCC), (Chapter 9), case studies (Chapter 5), water aspects (Chapter 6), policy instruments and stakeholders database (Chapter 5), cost and risk assessment (Chapters 10 and 12). A developed data model then defines the logical relationships among these areas as a basis for integrated use for project identification, analysis and construction.

A substantial part of the database is devoted to the collaborative decision making, especially during the last step of the decision making process, i.e. comparison between alternative designed solutions and negotiation between involved stakeholders. This is accomplished by means of digital web-based questionnaire which is predefined by the project manager for each project case. Then cooperating partners are invited to consult the questionnaire and to give their personal opinion on the issue described in the questions. Once submitted, the project manager can analyse the differences and potential conflicts and can arrange subsequent negotiation with his team of partners. The management of the complex functionality behind this is implemented in the form of a specialised ADSS component called "Water Expert".

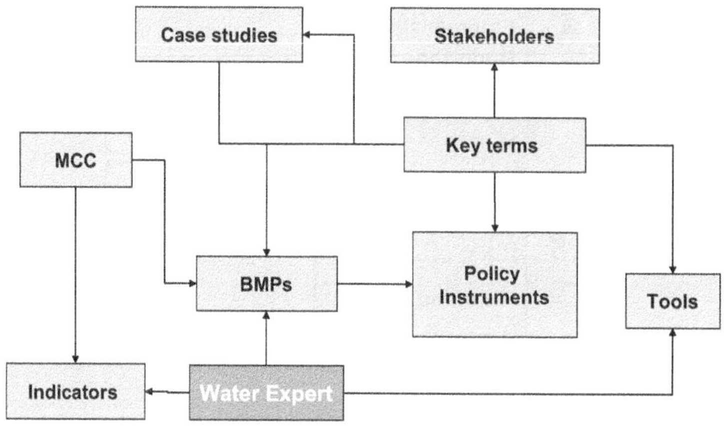

Figure 4: ADSS data structure: assistance to problem identification and project construction (see text for component description).

In addition to the core databases, there is a number of specialised (internal or external) components constituting the ADSS bringing assistance to the USWM project building and to the comparison of possible solutions. First of all, there are specialised tools developed to support the decision making process in the area of source and flux modelling (STORM & SEWSYS) (Chapter 14), chemical identification and risk assessment (CHIAT) (chapter 11), vulnerability assessment (FLEXT and FVAT) (Chapters 13 and 15), design and multi-criteria comparison of alternatives (Matrix of Alternatives or MoA) (Chapter 4). Some of these tools are external and need to be downloaded before using (e.g. STORM, SEWSYS, and CHIAT), others are integrated in the body of the ADSS (e.g. cost and dimensioning tools) (Chapter 10). These tools are (more or less) interfaced with the core database and able to exchange the data with other components. There are also other data available in the ADSS which are developed in a form of interlinked web pages, MS Power Point presentations, reports or publications, etc.

The management of the database tables and components is accomplished by a web-based Graphical User Interface (GUI) called HYDROPOLIS (Chapter 3) which represents a structured map with mouse-sensitive windows and main menu. By clicking on sensitive places the user enters distinct areas of interest which carry distinctive information and functionality. Based on given user rights, the user can utilise specific modes of ADSS usage or, in other words, use distinct navigational pathways through HYDROPOLIS.

Modes of ADSS usage

The ADSS GUI presents three modes of usage, i.e. free, semi guided and guided mode, enabling an easy swap between them.

Free ADSS mode

The use of the free ADSS mode is very similar to any other web-based application. In a free mode the user can visit different areas indicated in the HYDROPOLIS map. Free browsing through databases and components will support the user in finding support for the decision making process inside these areas through themes such as urban dynamics, BMPs, risk and vulnerability, and a number of other "logical" databases. The user can also use his private "Caddy" to store important data. Later he can return to track the history of his visit and can also search in the ADSS using predefined keywords and/or develop private archive of documents, photos, office files, etc. The main purpose of this mode is seen as constituting a form of learning library.

Smart ("semi") guided ADSS mode

This mode can be used by more experienced users while searching for information within the linked components of the ADSS. This inter-linkage can be understood also as a "SMART GUIDANCE" functionality showing the user the context of the information that he is searching and broadening his understanding regarding conceptual water management. The ADSS suggests a number of hyperlinks among distinct components and databases. The user can then modify the ADSS functionality according to his needs by using the KEY TERMS filtering tool which represents the response of the ADSS to the definition of the specific USER PROFILE terms, selected within the predefined user questionnaire. The filtering procedure is applied on all the ADSS databases in order to select only relevant information for the personal user needs.

Guided ADSS mode

The most advanced mode of ADSS usage is called the guided mode (Chapter 4). Using this mode the user can build up his own set of projects, based on the available questionnaire template, and to compare proposed solutions using a matrix of alternatives. This mode guides the user through a chain of activities related to the way of solving his problem. The procedure starts with an expert consultation on the given user stormwater problem based on questionnaire (or set of questionnaires) analysis: this is called dynamic thinking. While giving or selecting answers to the ADSS, key terms are defined and the user is suggested relevant measures to be analysed (in the form of alternative BMPs) as well as appropriate tools and criteria/indicators to consider. The user can then analyse in detail all proposed alternatives and fill-up a predefined matrix of alternative (MoA) with his own expert view on these alternatives. Once the matrix is filled-up, the user can then perform a multi-criteria analysis (MCA) in order to compare the proposed measures to be implemented. The user can work on an arbitrary number of such projects based on the same template questionnaire, in order to assess the differences between optimal measures using different views and emphases on the local stormwater problem. The guided mode is developed as a team support approach to a project. The template questionnaire can be shared by a number of stakeholders while building their own individual projects. The team leader or "manager" can then observe the results of this work and compare the differences that have emerged within the project team. The result of this work can be used as an input, stimulating further discussion in the stakeholder's team and eventually commonly searching for an optimal and mutually acceptable solution.

The work of the user is stored in the ADSS database and a tracking function is made available for helping the user in analysing his previous actions.

ADSS users and User rights

The ADSS is a multi-user/multi project system, even though it is not supporting transactioning and resolving conflict procedures. This is why the presence of distinct user access rights to the HYDROPOLIS GUI are controlled by login and password. Four basic types of ADSS users are available: Administrative, Manager, User and Guest.

The basic rights of a "Guest" are restricted to free browsing (free mode) in the ADSS. The "Guest" access is freely accessible to all visitors of the ADSS prototype at: www.daywater.cz .

The "User" inherits the rights of a "Guest" and in addition can work with a personal HYDROPOLIS and edit the data in databases.

The "Manager" also inherits the rights of a "User" and, in addition, can build up the template with questionnaires, assign this template to a number of team partners ("Users") and can also analyse the results of their work either by viewing their projects or by analysing the results of their multi-criteria analysis. By comparing the indicators, the "Manager" can identify different outcomes and discuss them with the project team.

The "Administrator" inherits the rights of a "Manager" and, in addition, can modify the data structure and procedures implemented. The "Administrator" is also giving the user rights to "Manager" which can be passed on to their "Users".

Compliance with basic ADSS concept

During the system requirement identification and functional analysis step of the ADSS development, a number of demands and wishes were expected by end-users and scientific partners as ADSS functionalities. It has to be clearly stated here, that, mostly due to time limitations, not all were finally implemented in the on-line prototype.

The project team has attempted to build a common understanding and agreement concerning the principles of the ADSS structure and functions throughout the duration of the DayWater project. Intensive discussions and numerous work meetings gradually developed a common consensus within the different scientific partners: sewer modelling team, IT specialists, chemists and biologists, hydrologists, database experts or experts on Web technology. The whole process of software development was then based on a "prototyping" approach with a number of detours, trials and errors, hampered by lack of time, resources and testing opportunities. In addition, some of the expectations proved to be unrealistic from an IT viewpoint or impossible to structure in algorithmic form. In addition, some ADSS components were developed in parallel with the ADSS core and their final form sometimes did not fully fit the expected ADSS behaviour. However, most of the functional expectations were met by the final form of ADSS prototype. The basic decision support functions of the ADSS, as defined during the analytical part of DayWater project, are supported in the following way:

Library function

The ADSS presents a comprehensive data structure for storing, managing and updating information and knowledge which is structured in a data form, according to distinct themes, and is accessible to all users of the ADSS, even in a free browsing mode. The use of Key Terms can personalise the library according to the needs of user, after he has defined his personal profile. Users can add their own information and knowledge into the ADSS over a period of time and build up a continuously updated and extended library of USWM. The personalisation of the ADSS library can be also accomplished via use of a "Caddy" or of personal archives.

Management function

The multi user/multi project approach can be referred to as one of key managerial functions of the DSS. This approach develops the background for the use of a guided ADSS mode, of parallel work on the defined project by involved stakeholders and of final analysis of team work results for any subsequent negotiation process.

The possibility of storing distinct information in the project "Document library" (e.g. documents, maps, photos, etc.) supports the project management. The tracking function brings the opportunity to view previous actions and their consequences for both types of user rights: "User" and "Manager".

Analysis function

The various multi-criteria analyses implemented in the ADSS, support the user in deciding between numerous potential mitigation measures.

The Multi-Criteria Comparator (MCC) approach uses sensitivity analysis for identifying and comparing all available BMPs (Chapter 9). The MCC uses a performance matrix which contains data relating to how a range of possible techniques perform in relation to a range of criteria and/or indicators (Chapter 7). In the MCC, the solutions are BMPs and the criteria are significant factors such as: Technical, Environmental, Operation/Maintenance, etc... The indicators provide more detailed discriminators for each of the criteria (e.g. flood storage, pollution control and system adaptability for the Technical criterion). Weightings which reflect the importance of each criterion/indicator to the decision-makers (i.e. the stakeholders in the development process) can then be freely applied to the data in this performance matrix, enabling the BMP options to be sorted into an order of preference, according to their ability to meet the requirements of the stakeholder group as a whole.

The Multi Criteria Analysis (MCA) is then an integral part of the guided ADSS mode, but can also be accessed and run via the free mode (Chapter 4). This MCA can be applied to the matrix of alternatives (MoA) generated by the ADSS, after consulting with the user by means of questions and answers. At this level, the user is required to enter both quantitative as well as qualitative values, as well as weights related to the selected set of criteria/indicators. The MCA then performs the scoring of proposed BMPs based on the given data.

Of course, the analytical functionality is also supported by a number of tools ranging from Excel "macros" to simulation models.

Communication function

The various ADSS users are working on the same Web environment, with the same data pool, and they are using the same "ADSS semantics". This avoids misunderstanding, information delay and brings the storage of data to one place, accessible to all relevant users at any time. The system of project management and the evaluation of its results motivate the project partners to discussions about controversies and their search for a commonly acceptable solution.

Role of ADSS adaptability

The adaptability of the ADSS represented a particular challenging task for the DayWater project team. Based on the above mentioned definition by Chuang and Yadav (1998), the ADSS is intended to modify automatically or manually some aspects of its structure, functionality, or interface to meet different needs of its users. By referring to this definition it is argued that the ADSS is adaptive at least in terms of functionalities.

Adaptability in personalising the HYDROPOLIS

The ADSS structure and functionality is adapted to user demands once the USER PROFILE is defined. The content of the personal HYDROPOLIS will then vary from user to user and can be changed over a period of time.

Adaptability in template questionnaire development

This functionality represents one of the most complex ADSS features. In principle, each manager has permission to build up a template questionnaire and to implement his particular knowledge and experience regarding the problematic area, before assigning it to the involved users for building up projects. The questionnaire can be focussed to whatever stormwater purposes are needed, whilst using the ADSS databases for making a viable proposition. While building the questionnaire template, the Manager can connect the particular answer of a specific question to an additional information e.g. proposal for action to be taken. Such proposal holds information regarding possible BMPs, criteria/indicators to be considered and tools to be used, while supporting means of addressing the stormwater management issue. Using this functionality, the manager implements his "Domain knowledge" to the template questionnaire. This template can be then used by distinct users, while working on their own projects. The functionality is based on the concept of "TREES and NODES", as basic components holding the overall logic (Figure 5). The Manager can develop user friendly questionnaire templates without any knowledge of software development.

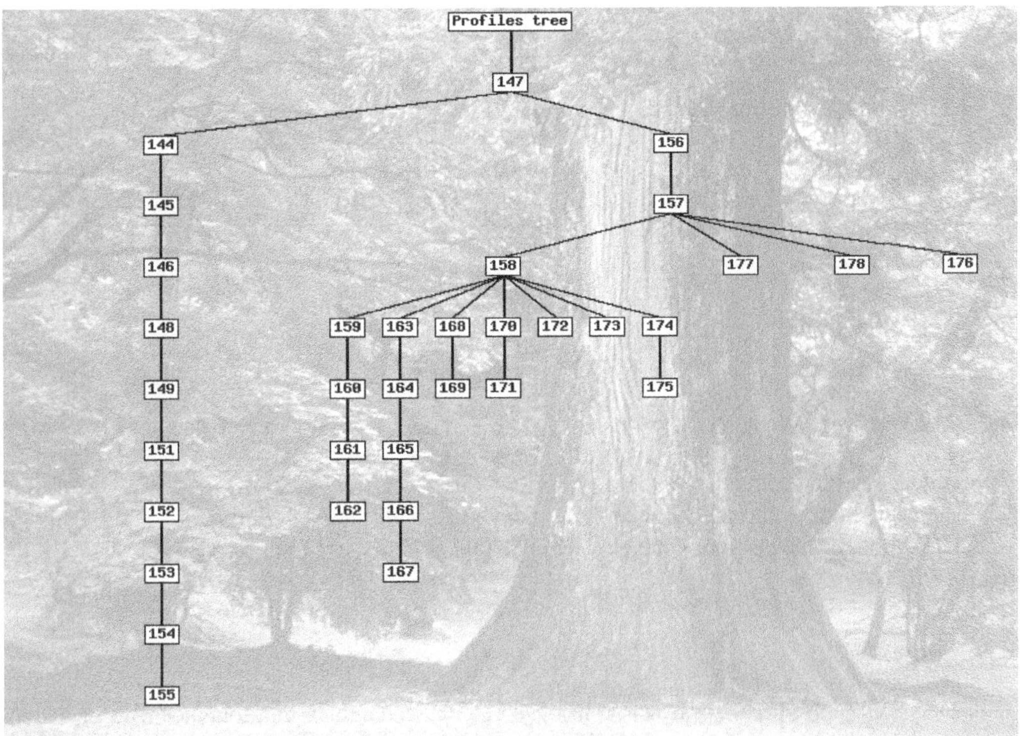

Figure 5: DayWater concept of trees and nodes.

The functionality implemented in the ADSS works as an independent engine for building up the templates within particular domain knowledge. Gradually, the ADSS can extend the number of questionnaire templates and broaden the consultation part within the guided mode and, in this way, the ADSS can adapt its in-built knowledge.

Adaptability in user navigation, during the consultation process, in guided mode

The use of the guided mode also brings a level of adaptability to users. Once the user creates his own project based on the selected template, he starts consulting the ADSS, through a question/answer routine. In terms of background trees and nodes, the user is following a specific path which allows him to build up his matrix of alternatives and to reach a decision concerning a proposed mitigating measure. However, the user can learn by free browsing within the ADSS and can return back to the guided mode to consult the ADSS on a higher level of knowledge. He can then adapt his specific path through the tree and nodes, by tracking back and modifying his actions.

Adaptability in ADSS data updates

The Edit, Insert, Update and Delete functionalities are made available to those users having special user rights. These users can adapt the content of the ADSS databases according to new achievements in methodology, to new available information or based on new project results. The adaptation of ADSS content can be done, via the ADSS, through the Internet connection.

Adaptability in building personal archives

The last functionality which refers to the ADSS adaptability is related to the building of personal user archives. The user has rights to adapt HYDROPOLIS by bringing his own information to the personal archive, project document library (for each project) and/or personal "Caddy" storing the most important pages that he has visited.

ADSS Software architecture

From the technological view point the ADSS is a relational database management system (RDMS) complemented by Web technology. The system is built on top of MySQL client-server database using Apache Web Server PHP and Java programming languages (Figure 6). The interactive GUI is optimised for Microsoft Internet Explorer, although it can be accessed (with some functional limitations) also by other Internet browsers.

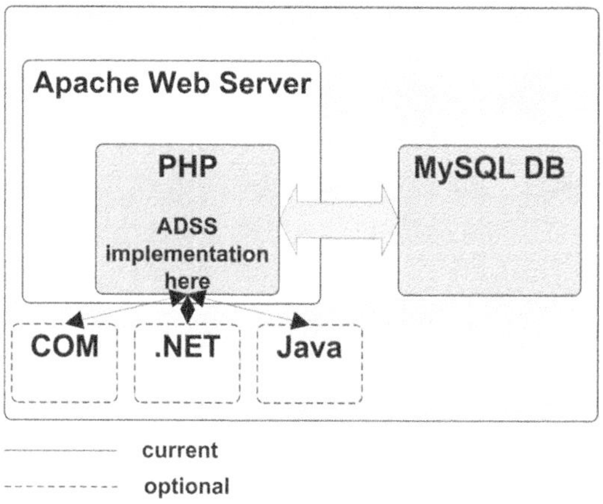

Figure 6: ADSS software structure.

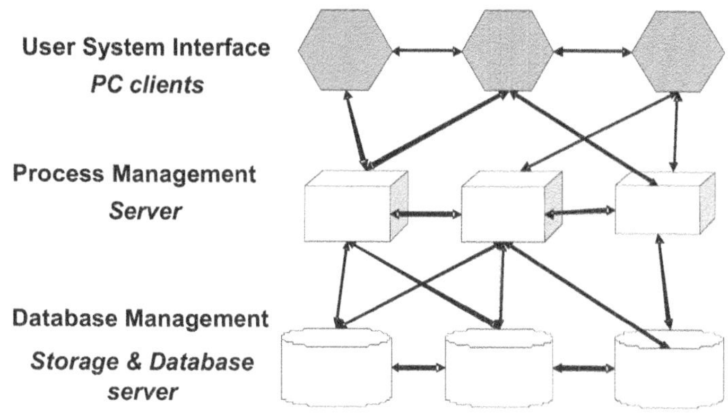

Figure 7: Three tier ADSS architecture.

From the architectural viewpoint the ADSS is developed as three tier system (Figure 7). On the lower level the "Database management" system and functionality is implemented, with the ADSS processes and functions being placed at the intermediate "Process Management" level. The GUI functionality or "User System Interface" is then located in the upper tier.

CONCLUSIONS

The DayWater Decision Support System research has used computer engineering, together with geochemical and eco-toxicological modelling, embedded in user behaviour and urban policy assessment, to develop a multi-stakeholder approach to USWM. The DSS represents a shortcut between research and operational fields. Two challenges have been faced considering that the ADSS cannot represent a breakthrough in directly solving urban stormwater problems: (1) the ADSS rather provides an enhancement of strategic discussion activity of those stakeholders charged with solving such problems, as well as addressing the interests of those non-responsible groups (such as the public and NGOs) wishing to participate in the decision-making process; (2) it is a way to centralise, to present and to select relevant information and to propose appropriate alternative solutions to a given problem that the end users have to resolve within a given set of constrained contexts.

The ADSS prototype has been developed and implemented. Numerous elements are still missing to provide a full operational tool, but the first tests with selected end users have shown that this tool is able to solve a variety of urban stormwater management problems (Chapters 16-19, Chapter 26).

References

Christensen, K.S. (1985) Coping with uncertainty in planning. *J American Planning Association* **51,** 63-73.

Chuang, T.-T. and Yadav S.B. (1998) The development of an adaptive decision support system. *Decision Support Systems* **24**, 73-87.

Holsapple, C.W. (2001) Knowledge management support of decision making. *Decision Support Systems* **31**, 1-3.

Power, J. (1997) What is DSS? *Jr Data-Intensive Decision Support*, 1 (3).

Thomas, J.-S. (2001) Watside: a decision methodology for Water Supply integrated development, *PhD Thesis*, RMIT University, Australia, pp.266 and annexes.

Chapter 3

Welcome to Hydropolis

Jeroen Kluck and Roel Valkman

INTRODUCTION

The main object of the DayWater project is to help urban water managers to manage stormwater in the urban environment (Chapter 1). The central issue in the decision support is the way a water manager has to deal with uncertainties in urban water management. One way of dealing with these uncertainties is the use of a Decision Support System (DSS) which mainly serves to reduce uncertainties by e.g. providing data on operation and cost of possible solutions. In the DayWater project, an Adaptive Decision Support System (ADSS) has been created, which not only reduces the uncertainties, but also supports the water manager to cope with them. This is important, because different types of problems require different ways of dealing with uncertainties and therefore require different levels of support in investigation and navigation. For the navigation level a map of a model city and its water cycle could be very useful: this is why the front page of the DayWater ADSS contains a mouse-sensitive picture of a city called Hydropolis (Figure 5). This paper first gives account of the scientific discussion between the partners of the DayWater project, on how to support the urban water manager. Then, it introduces the major functionalities of Hydropolis.

DEALING WITH UNCERTAINTIES IN DECISION SUPPORT

Two types of uncertainties

In the last few years the use of DSS has grown enormously, partly because of the growing capacity of personal computers, but also because the problems to tackle became more complicated. The aim of a DSS is 'to support decision making in complex problems by structuring the information and by analyses and evaluation of the effects of different alternatives'. The complexity of urban water associated problems is related to the uncertainty in the behaviour of the system that causes the problem. The uncertainty exists because of lack of knowledge of the functioning of the system e.g. effects of certain source control measures, or because of structural uncertainties in the system e.g. when the success of a measure is dependent on the cooperation of stakeholders. Decision making can be supported by reducing these uncertainties: this can be achieved by structuring the information needed to get a clearer understanding of the problem and by modelling the processes which cause the problems. For a deeper understanding of uncertainty issues, see a whitepaper report on Risk Perception, Risk Assessment and Risk Management in the DayWater Context (Hauger *et al.*, 2004).

The A of ADSS

One of the main characteristics of the decision support system, developed in DayWater, is the A of the ADSS, which stands for 'adaptive'. Adaptive has been defined as the quality of the ADSS to adjust the information and knowledge given to the local issues e.g. scale of problem, culture, country and time scale. As a result, an ADSS should incorporate a variety of knowledge on different topics and different levels of scale, in order to provide support for decision makers facing urban stormwater management (USWM) problems. It means that the structure of knowledge within the ADSS enables the system to provide the user with the right level of information. Next it should be possible to adjust and update the ADSS to local situations and advancing knowledge (Chapters 2 and 4).

It is however not possible to reduce the uncertainties sufficiently. People have to face the fact that the problems involve uncertainties and that they somehow have to cope with it. An adaptive DSS helps to offer the appropriate tools to cope with these different types of uncertainties. In the arrangement of knowledge within the ADSS the problems have been categorized in four quadrants.

Problems in four quadrants

The four quadrants model characterisation of problems, according to their respective objectives and means (Christensen, 1985) (Figure 1) helps to distinguish the kind of problems that the ADSS supports. The four quadrants also represent four ways of dealing with the uncertainty regarding the clarity of objectives and/or means.

(1) When both the objectives and means are clear (1st quadrant), a modelling process might lead to the optimal solution of the problem. The ADSS contains information that supports the user in this modelling process.
(2) When the objectives are clear but the means are not (3rd quadrant), e.g. uncertainty about the long term performance of porous pavement, results of monitoring programmes may help the user in his decision process. The ADSS also contains this type of innovative information.
(3) When stakeholders do not agree on known objectives (2nd quadrant) a negotiation process is needed.
(4) This is also the case when stakeholders do not agree on the objectives and the means are unknown (4th quadrant): the water manager cannot solve his "chaos" problem by technical information only.

In these two last quadrants the problems are complex. Solving these problems, as if they were first quadrant ones – as engineers tend to do - is called premature consensus. In these cases, the ADSS provides information about the *how*, offering information on the decision process.

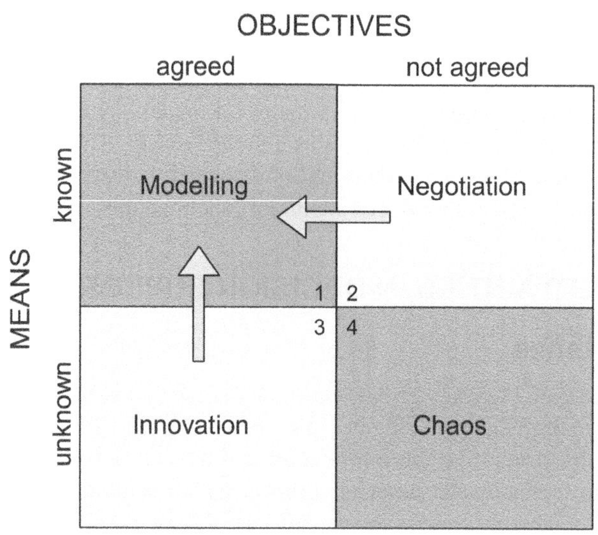

Figure 1: Characterisation of problems in four quadrants depending upon the related objectives and means (adapted from Christensen, 1985).

The way the ADSS will be used depends on the requirements of the user. Some users require support in the 1st quadrant with modelling, clear information of some best management practices (BMPs) (Chapter 8) or a comparison of alternatives techniques (Chapter 9). Others require support on how to implement a solution or how to negotiate with other stakeholders (Chapter 4). Most users will require both.

The ADSS mainly focuses on first quadrant problems. Components such as a BMP catalogue (Chapter 8), source and flux models (Chapter 14) and multi-criteria comparator (Chapter 9) have been developed for such assistance. The other quadrant problems are addressed in components such as ambition reflection and urban dynamics (Chapter 6) which helps the water manager to take into account all involved processes, factors and stakeholders.

Coping with uncertainties: two level approach

First of all, only the ADSS user may know the relevant details of his USWM problem: the (tacit) knowledge of the water manager should not be underestimated, even if it is not always comprehensive. A second main point is that the user's mind is far more sophisticated than any search engine that we can offer in the ADSS in order to provide the user with the right information at the right time. Therefore it appeared to scientific partners of the DayWater project that the ADSS should guide the user intuitively, by offering him a logical arrangement of information. The navigation through the information in the ADSS can be described by analogy of finding one's way in a city (Figure 2).

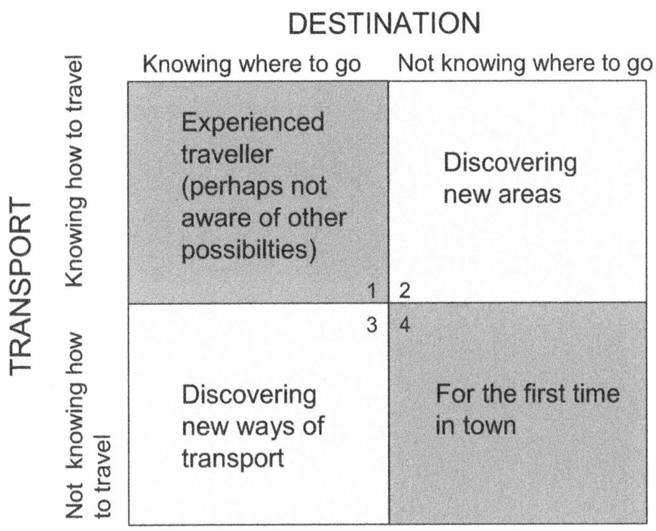

Figure 2: The four quadrants approach of travelling in a city.

When arriving in a city for the first time, one can find his way by looking at a map. A map reflects the spatial characteristics of a city. A map is not necessarily true or false, but it is a pragmatic tool that helps to orientate. By looking around and focussing on the map, one can discover where he is and where he might want to go. In combination with information about busses, metro, etc... one can make plans for touring around. The four quadrants of Figure 2 present different approaches, depending upon the respective knowledge of the destination and of the city transport system.

Arriving at a hotel, more detailed investigation can be achieved in a deterministic way. A clear goal can be formulated and specific actions taken e.g. asking for a room with a view on the medieval church. The learning process of getting to know the city can be characterised as a process of switching between two levels: pragmatic navigation and local deterministic investigation (Figure 3). Step by step one learns "to read the city landscape", to recognise different parts of the city, to know the most convenient way to go from one place to another one and to get detailed knowledge on the finest pubs in the city (Table 1).

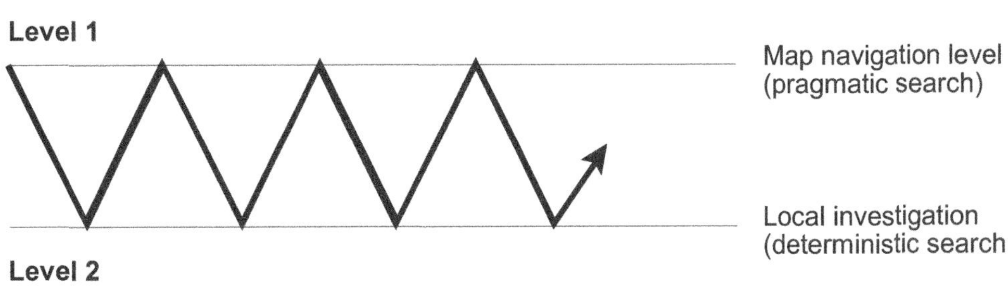

Figure 3: The two levels of navigation within a city: pragmatic and deterministic searches.

Table 1: Characterisation of the two levels of navigation and learning.

Level 1: the whole	Level 2: its parts
Value oriented	Goal oriented
Pragmatic	Deterministic
Learning by doing	Optimisation
Integration	Specialisation
Implicit and explicit knowledge	Explicit knowledge
Meta needs	Basic needs

This two level approach of a problem is very helpful when trying to intervene in stormwater processes within a complex urban situation. For example, in an urban renewal project with different actors, unclear goals and measures (thus bearing many uncertainties), one can facilitate a social learning process by switching between these two levels. Such dual approach was the idea behind the first draft of a city map, called Hydropolis, initially considered for the ADSS front page (Figure 4).

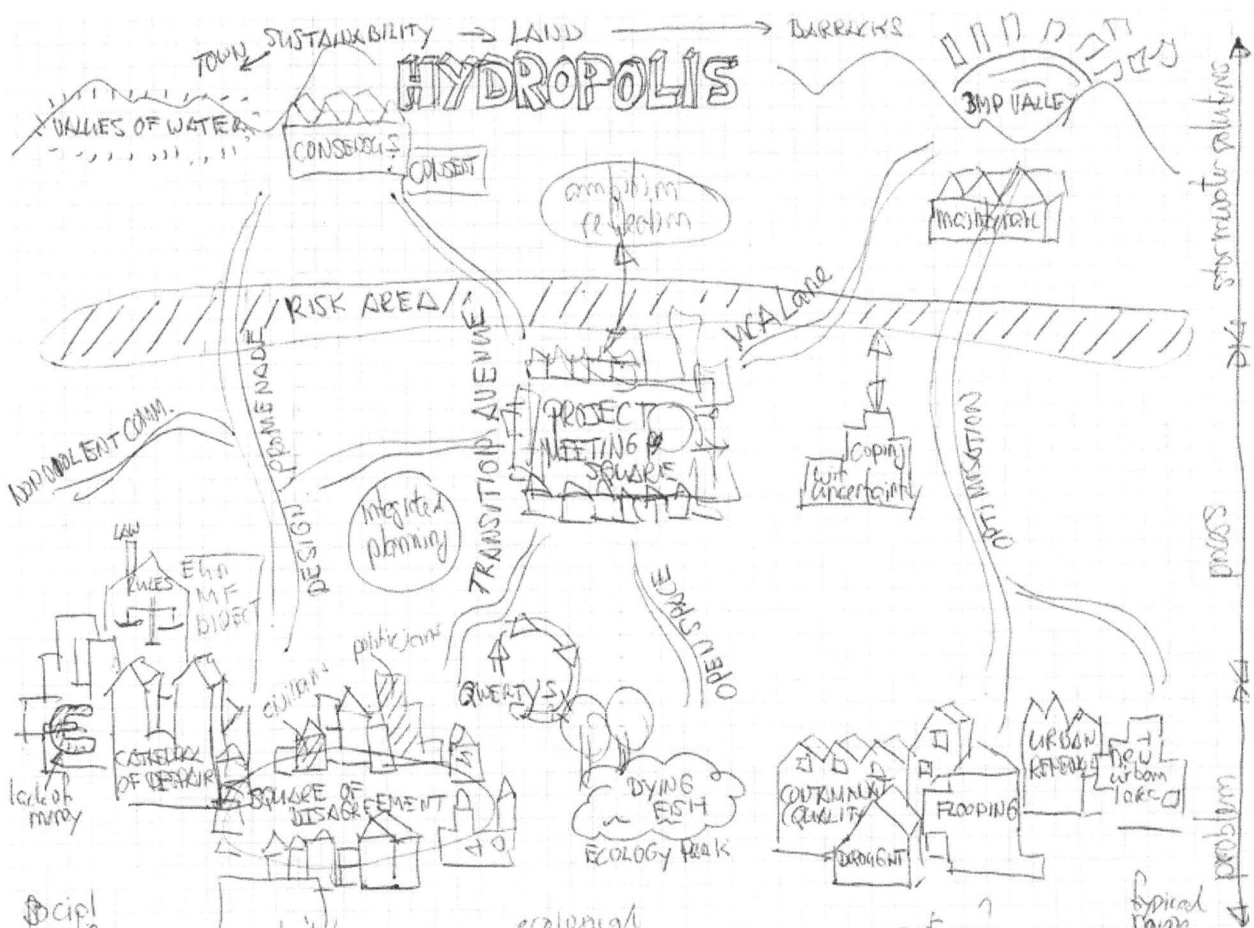

Figure 4: A first sketch of the DayWater ADSS front page, called Hydropolis (see text for explanations).

The sketch of this city map is organised according to two axes:

(1) along the X-axis are classified the various context issues of stormwater source control: social, context, ecological, system and physical;
(2) along the Y-axis are classified the various approaches to stormwater source control: problem, process, stormwater solutions.

Around the project "meeting square", roads lead to various city districts:

(1) open space leads to ecology park, including dying fish,
(2) multi criteria analysis (MCA) lane leads to BMP (Sun) valley,
(3) transitional avenues goes from the square of disagreement (bottom) to consensus and consent (top),
(4) design promenade goes from the district around the cathedral of despair, including lack of money, law, rules,

(5) optimisation goes from contamination (pollutants), drought, flooding, urban renewal, new urban lake to BMP valley.

It appears that starting from a system-related problem, the real urban stormwater problems are often context-related and that the ADSS provides some road-maps for taking the context into account and enables reaching consensus while passing over dangerous risk areas.

This sketch was used for organising the information presented in the ADSS. By clicking on active parts of the sketch the user is able to access to specific information e.g. a museum giving access to case studies and a technical department building with information on BMPs. The transformation of this original sketch into an attractive homepage of the ADSS required artistic capacity beyond the ability of the project partners. Therefore at the end of the DayWater project, this sketch was replaced by a nice photograph of a city (Figure 5) including mouse sensitive tiles. Unfortunately, with this photograph instead of the original sketch, the front-page lost its original purpose of guiding the user, as within an unknown city.

Figure 5: DayWater ADSS front page called Hydropolis.

The level of support was an important point of discussion within the DayWater team. It was very tempting to develop an expert system, where the user presents his problem and the system replies with a clear answer suggesting the "best" measure for USWM for his situation. This was found unfortunately impossible, since it is a too complex task. Furthermore it would present risks of loosing its ability to adapt to the problem and context, and to the user. Less ambitiously, the DayWater ADSS prototype provides relevant information in order to support the decision-making process. The idea of an expert system has been investigated and has finally led to a tool aiming at selecting relevant information and tools for the ADSS user: this is the "guided mode" of operation (Chapter 4). Taking more influence in the decision-making process seems not applicable for the wide variety of situations. Furthermore it might take away the responsibility of the user(s) in the final decision.

FUNCTIONALITIES OF HYDROPOLIS

Basic principles of Hydropolis

The final prototype of Hydropolis is a web based application (Chapter 2) containing databases and tools related to urban stormwater management, information on the user's profile (data that makes Hydropolis adaptable to individual user's needs and situations) and system settings (data necessary for the running of the web application). Information may be pre-selected, if the user wishes to create a user profile and answer a questionnaire. A web based application has been preferred, instead of traditional software installed on user's computer, because of its accessibility, low software requirements and the easy way to update the application for all users and projects. Hydropolis' ability to develop by entering new state of the art information and new case studies was a major concern (Chapter 5).

Tiles

The Hydropolis front page (Figure 5) is the main entry point to all its ADSS components which are accessible via the selectable tiles briefly described in Table 2, together with the reference for the chapter where they are presented in details. To further facilitate the use of Hydropolis, search and "caddy" functions have been designed together with a help function.

Table 2: Description of the major database and tools which may be accessible through Hydopolis mouse-sensitive tiles (Figure 5).

Tiles	Description
MCC: Multi Criteria Comparator	The MCC offers advice on suitability of BMPs and contains a list of indicators and of available techniques for stormwater source control (Chapter 9). The user can replace his own set of scores for indicators or use default values, in order to rank selected BMPs.
BMP: Best Management Practices	The BMP catalogue provides information on a wide range of both structural and non-structural BMPs (Chapter 8). Characteristics of each BMP is accompanied by information on sources & loads, design, performance, operation & maintenance and photographs.
Pollutants	This component helps to determine which pollutants are relevant and describes selected stormwater priority pollutants and their behaviour within major BMPs (Chapter 11).
Risk and vulnerability	Several tools allow chemical risk identification and assessment, ecosystem vulnerability and flooding assessment (Chapter 12, 13 and 15)
Urban Dynamics	This component helps to investigate the USWM problem and to reflect on the possibilities, using 3 modules: Aspects of Water, Coping with urban dynamics, and Ambition reflection (Chapter 6).
Tools	This tile provides access to internal and external tools available for BMP design and modelling: • BMP cost assessment (Chapter 10) • Chemical risk assessment methodology (CHIAT) (Chapter 11) • Hazard identification tool (RICH) (Chapter 11) • BMP dimensioning tool • Source and Flux Model (SEWSYS & STORM) (Chapter 14) • External Risk screening tool (FLEXT) (Chapter 15) • BMP Costing tools (Chapter 10) • Flood vulnerability assessment tool (Chapter 13)
Libraries	This tile gives direct access to several databases (Chapter 5): • Case studies: examples of USWM projects • Policy Instruments Information: information on laws and regulations • Tools: tools relevant to USWM in DayWater project context • Stakeholders: information on all stakeholders important in USWM • Key Terms: index to all databases which are used to make a preselection of databases and tools relevant to the users USWM problem and to the user interest and knowledge • Criteria and Indicators: allow comparison of techniques (BMPs) using Multi Criteria Comparator, or designed solutions (guided tour and Matrix of Alternatives)
Guided Tour – Matrix of Alternatives (MoA)	This system helps to build an USWM project (Chapter 4): based on answers to predefined questions, the system filters relevant solutions, tools and databases. The more questions are answered the more precise the proposal will be.
News	Presents what is new in Hydropolis ADSS and DayWater project
Site maps	All parts of Hydropolis can also be accessed from this site map
Tutorial	This tool presents a set of short documents and illustrations related to each ADSS database and tool

CONCLUSIONS

The central issue in the decision support for stormwater source control is the way water manager has to deal with uncertainties in urban water management. Different types of problems require different ways of dealing with uncertainties and therefore require different levels of support: investigation and navigation.

One way of dealing with these uncertainties is the use of a Decision Support System (DSS). A DSS mainly serves to reduce uncertainties (e.g. by providing data on functioning and cost of possible solutions). In the DayWater project we have created an Adaptive DSS which not only reduces the uncertainties, but also supports the water manager to cope with them.

REFERENCES

Christensen, K.S. (1985) Coping with uncertainty in planning. *Journal of the American Planning Association* **51,** 63-73.

Hauger, M.B., de Roo, C., Geldof, G.D., and Mikkelsen P.S. (2004) D4.1 Whitepaper on Risk Perception, Risk Assessment and Risk Management in the DayWater Context, pp.50.

Chapter 4

DayWater ADSS guided tour: a feature for the Decision Support System adaptability to its user

Jean-Claude Deutsch and Tomas Metelka

INTRODUCTION

The Adaptive Decision Support System (ADSS) was developed within the framework of the DayWater project as a multipurpose internet portal serving distinct user needs in terms of library function, management function, analysis function and communication function (Chapter 1). In addition, the ADSS was expected to provide functionality for coping with a variety of knowledge states, problems, scales and environments: thus, in addition to traditional decision support system (DSS) functionalities, the adaptability of ADSS to change its content, structure and environment had to be supported (Chapter 2).

The ADSS prototype delivered at the end of the project is providing a number of functionalities which support the above requirements. These functionalities vary from simple "Google" browsing, use of "pop-up messages", "search" and "caddy" options, through the use of "key terms" for data linking, "filtering" and "User profile" definition, to sophisticated methods for encapsulation and later use of particular domain knowledge and definition of so called "user fingerprint". Free, smart (semi-) guided and guided modes of ADSS usage were defined (Chapter 2) and implemented to handle these distinct system functionalities in logical groups.

This chapter is aiming at describing the system functionality developed, in the framework of the decision making process, for storing the expert knowledge and using it with the knowledge of the stakeholders involved in a specific urban stormwater management (USWM) project, in order to lead to a measure or solution accepted by all.

ADSS GUIDED MODE STRUCTURE

Out of three modes of usage implemented in the ADSS, the "guided mode" is the one directly focussed on deep interaction with the ADSS user. A specialised expert system engine has been developed to encapsulate expert knowledge for later consultation with users. This expert system is designed to be used not only in the area of USWM but in fact within any domain of knowledge.

The technique used for the expert system implementation is based on graph theory. Particular knowledge domain is represented by a system of knowledge trees and nodes. Each node is holding particular pieces of domain knowledge in relation to the topics encapsulated in this node. The user is moving throughout the tree and presents to the expert system his particular socio-technical problem.

Knowledge trees

The system for storing and retrieving the expert domain knowledge is represented in ADSS by means of knowledge trees (Figure 1). The expert knowledge is stored in a form of "Nodes" composing a particular "Tree" related to a specific domain expertise e.g. Best Management Practices (BMPs), social aspects of USWM, surface runoff modelling, etc.... Branches are connecting nodes in order to develop a knowledge tree based on logical relationship between the nodes

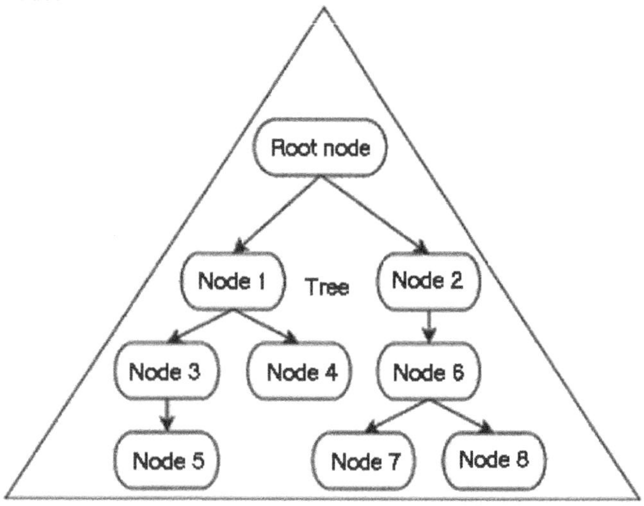

Figure 1: Knowledge tree.

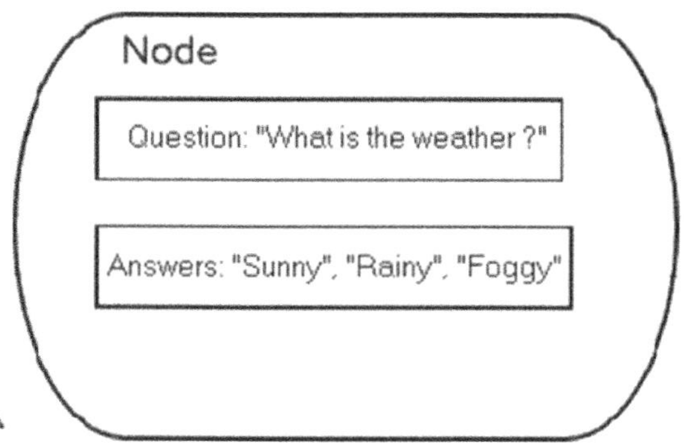

Figure 2: Node structure.

The main information, which the node holds, refers to a domain specific question (e.g. size of impermeable area, equivalent inhabitant number, etc...) and alternative answers to this question in the form of YES/NO, number, selection from list, etc. (Figure 2). Based on the answer given by the user, a particular branch is defined between two nodes. Each node then "knows" its parent and children. The knowledge tree starts up from a "root node" with no parent. There are no loops allowed in the structure so that each node has only one parent whilst taking care of many children.

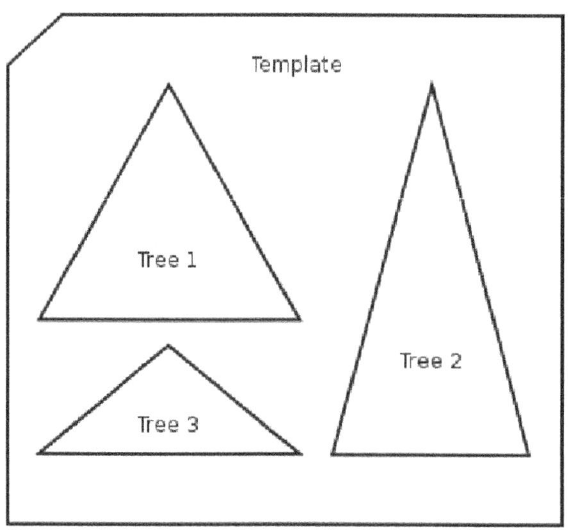

Figure 3: Tree template.

In order to cover more domain knowledge subjects which are to be used during the consultation process, an arbitrary number of knowledge trees can be developed. These trees can be then wrapped into special envelopes called "Templates" (Figure 3). This means that, for instance, a template for implementation of BMPs in the south of France can comprise:

(1) a BMP knowledge tree,
(2) a socio-economical tree and
(3) an environmental tree.

These trees are packed in the tree template “BMP implementation in the south of France” to be later used by a particular stakeholder during his consultation process. In addition, tree 1 can be used in parallel within other templates if relevant. Both trees and templates are built up by experts, over the internet, using a specific Graphical User Interface (GUI) without any need for programming (see next chapters).

The users are not working with templates but with their particular copies of templates (similarly to the use of a template in the MS Word text processor). The consultation is carried out in a form of questionnaire which can be filled out by the user. This situation simulates the expert discussion with his client: the client is asked by the expert about the project status, system, context as well as potential problems, before the expert gives his recommendations. The graphical representation of the questionnaire is structured in a form of web pages with questions, places for user answers and additional information (Figure 4) which may help the user in his own decision making (e.g. pictures, hyperlinks, text fields, animation, etc…).

Basic features and functions of knowledge trees

From the technological viewpoint, the system of knowledge trees is developed on top of a database engine (mySQL in this DayWater project). The presence of a database makes sure that data storage and update are correctly performed, and allows data personalization, by means of login and password-based user access to the data. In order to simplify and optimise the development of such knowledge trees (Figure 1), a specialized GUI has been developed in order to allow the remote access for the knowledge tree development and use. This GUI provides the user with a set of functionalities for tree and node development (e.g. insert, edit, update, delete), then with a set of functionalities for implementing the domain knowledge into the system (e.g. visibility rules, proposal development, references and books development) and finally with a set of functionalities for building the graphical features of the web page questionnaire. All these functionalities are made accessible, over the internet, with a standard internet browser (optimised for MS Explorer 6.0) under special user access rights (login and password). Neither programming technique nor particular compiler is necessary for the development of such knowledge trees, templates and questionnaire graphical interfaces.

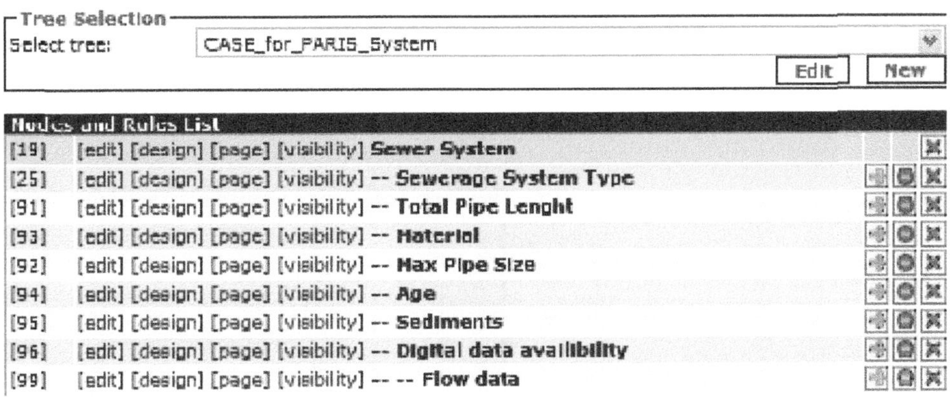

Figure 4: Tree and node design interface.

The basic tree and node design interface is presented in Figure 4. Using this interface, the access to node editing, design of the graphical node interface, design of questionnaire page, and visibility rules design are available. By clicking on a particular feature, the user validates respective functionalities and interfaces for building the knowledge tree. Node editing mode allows the development of the actual internal node information (e.g. question, set of answers, answer type, parent node, etc...) as well as connection to the set of “references” and “books”. This means that each node (representing a particular piece of knowledge from a particular knowledge domain) can be accompanied by additional relevant information in the form of publications, pictures, hyperlinks, tables, etc..., useful for the client user during his consultancy with the ADSS. This information can be handled either separately or it can be grouped into “books”, based on subject, territory, data type, etc... During consultancy with ADSS the user receives this information when answering this question at this selected node.

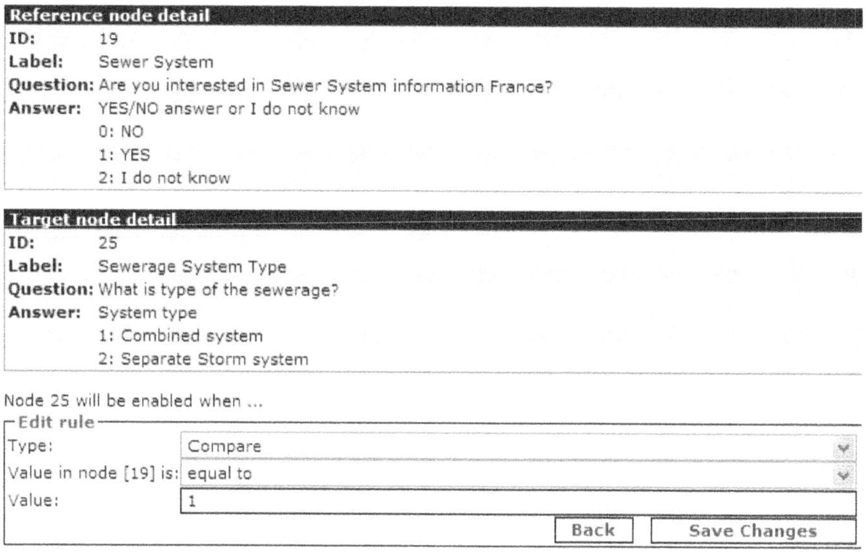

Figure 5: User interface for visibility rule definition.

Visibility rules (Figure 5) are implemented in the system to cope with the situation when the presence of some branch or node is not relevant to the actual consultation state. Visibility rules are based on the relationship between reference and target node. Based on the particular answer at the reference node, the visibility of target node receives three distinct modes based on a system of visibility rules: "enabled", "greyed" or "disabled". There are distinct mathematical operators used for defining the relationship used for such visibility rule selection e.g. equal, less then, greater then, interval, etc…

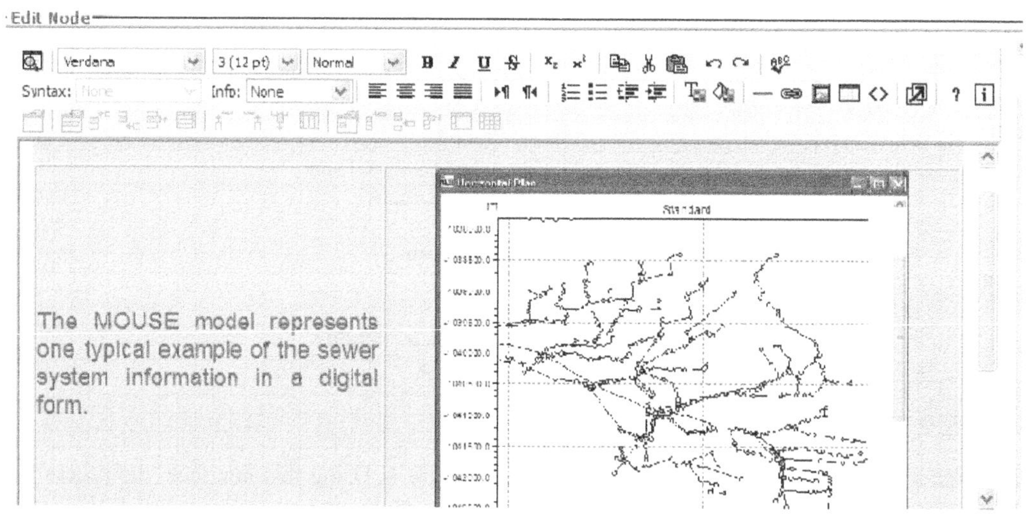

Figure 6: Node design graphical editor.

The design on node is handled in a specialised graphical editor, where graphical shape and features are given to the node (Figure 6). The user can edit colors, format text; insert tables, images, internet hyperlinks, while working on graphical node features. Predefined "Question", "Answer" and "Submit" words refer to the actual node content. They stand for the node's question, answer field and submit button respectively. By implementing the graphical node interface, the expert user can implement add-on value to particular domain topics carried by the node. The presence of text comments, picture example and other possible information helps the user during the consultation process to understand particular topics and their context.

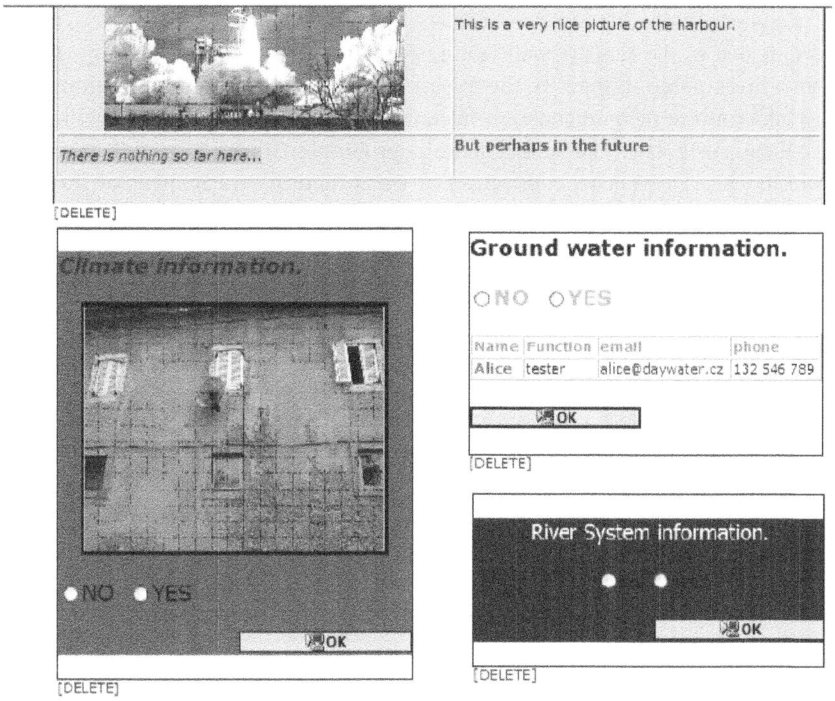

Figure 7: Page design graphical editor.

The localization of the actual position of the node on the questionnaire page is supported by page design interface (Figure 7). Each node is represented by its graphical features in a form of particular box on page. By clicking and dragging a particular node box can be moved to a desired position. This graphical editor is used for finalisation of the questionnaire skin, which will be finally accessed by the client user during his consultancy with the ADSS.

The basic guided mode feature is providing recommendations based on consultancy with user clients (by means of questionnaires). A special functionality is implemented in the ADSS, called "proposal development", in order to meet these requirements (Figure 8). Each node can be coupled with the set of "proposals" defining the expert recommendations, suggestions or comments. This is accomplished by reference to specific ADSS databases (BMPs, Criteria, Tools) which indicate the preferences of experts, in relation to the actual state of the node in the tree. Picture bellow shows the editor for proposal development and coupling with Criteria, tools and BMPs.

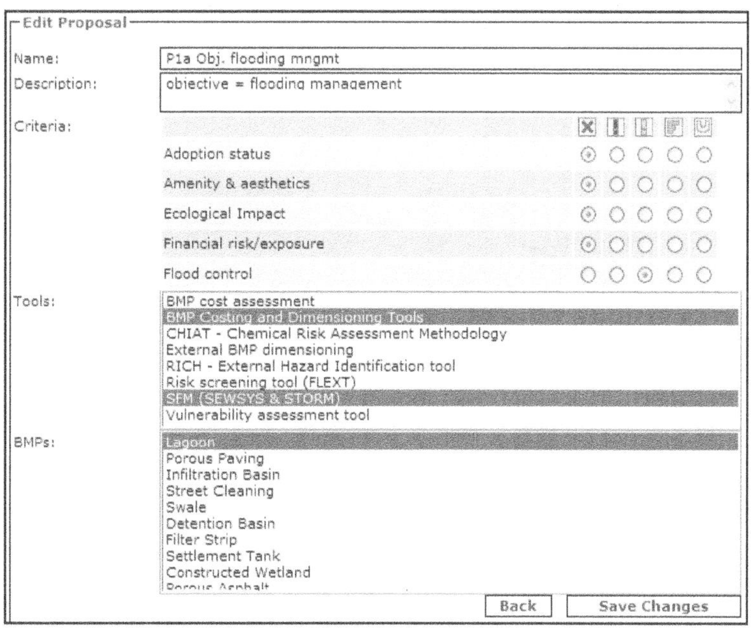

Figure 8: Proposal design editor.

A proposal body is composed of a list of Criteria (including indicators and benchmarks), defined within the scope of the DayWater project, as one of the fundamental features for many ADSS functionalities. By clicking on criteria in a selected column from Criteria table (Figure 8), the expert implements his recommendations related to the state of the node under focus. In addition the recommendation in terms of suitable software tools and BMPs is accomplished by selection from the Tool and BMP lists. The connection of a proposal to a particular node (or nodes) is managed by links called "proposal realisations". This makes connection of one proposal to a number of nodes possible. The proposal realisation editor allows the tree and node of interest to be connected to the proposal. The actual state of node during consultation plays an important role. Distinct proposals can be connected to one node, based on distinct user answers to the question in this node. Then the recommendation will differ, based on actual user situations as described in his user questionnaire.

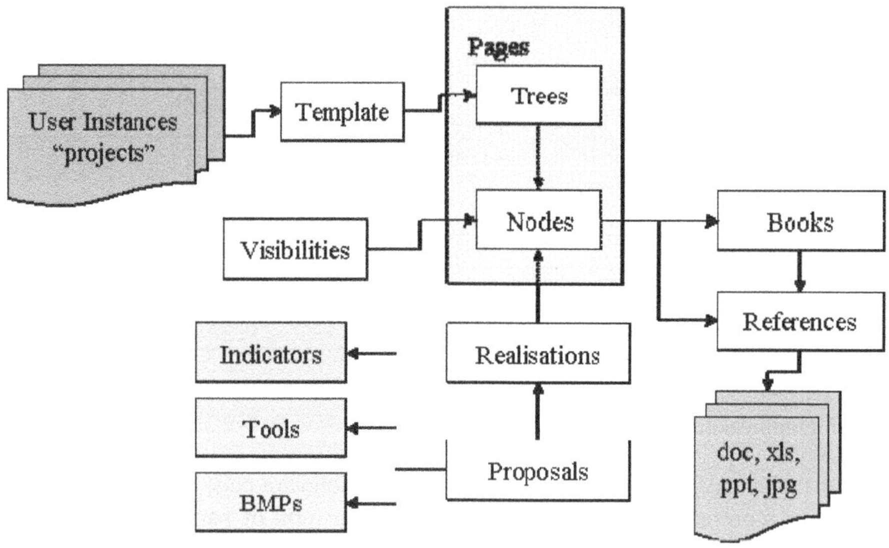

Figure 9: Basic ADSS structure developed for guided mode.

The conceptual data structure developed for the ADSS guided mode is presented in Figure 9. The users have access to this structure though the co called "instances" of template developed by the project expert. These "instances" are personalized, based on user access rights: they hold the user personal data for subsequent use.

USING KNOWLEDGE TREES FOR DECISION MAKING PROCESS

There are different approaches to the way the decision making process is conducted in urban issues: unilateral, consultation, concerted action, negotiation are recognised alternatives (Rowe and Frewer, 2000). The multi-stakeholder aspect of projects in USWM has led DayWater partners to select negotiation as the best way to achieve decision making process in this field. The negotiation concept is schematically shown on Figure 10 and described in next paragraph.

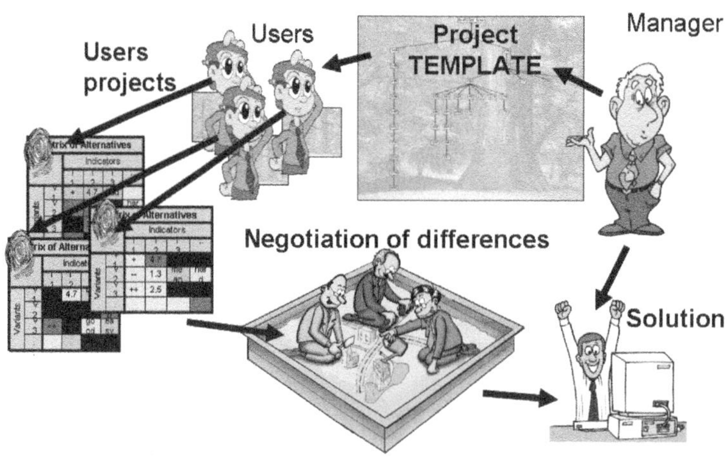

Figure 10: Concept of consensual decision making.

Negotiation concept in ADSS

The negotiation concept used in the ADSS aims to lead to a solution accepted by all stakeholders involved in the USWM project (Figure 10). Starting from a project template build by the project manager, different stakeholders with distinct interests, experience and responsibilities need to come to a solution which is, at the time of decision, the best consensus between all the interests, or which, at least, do not create frustrations among some stakeholders, thinking that their point of view has not been taken in account. To be able to reach such an agreement, at first they need to define controversies and disagreements among them. The conditions for analysing and negotiating controversies and for searching consensus among stakeholders are: (1) clearly defined subject of interest and its boundaries, (2) the use of similar semantics and (3) operating in the same working environment (Geldof, 2005). The ADSS is providing such functionalities for consensual decision making among a group of stakeholders, using the system of knowledge trees as a basic technological background for achieving this requirement. A special software engine "Water Expert" was developed to support multi-user negotiation by means of knowledge trees.

Water Expert engine

The Water Expert software engine (Figure 11) is represented by a set of functionalities for defining (1) the system of trees and nodes, (2) for implementing expert knowledge in the form of proposals and visibility rules (3) as well as for the graphical skin of node and page for the particular questionnaire template. In addition, the possible printout of the questionnaire related to consultation with selected users (stakeholders) is supported (Figure 12), as well as the final comparison or evaluation of different stakeholder views and disclosing controversies (Figure 13).

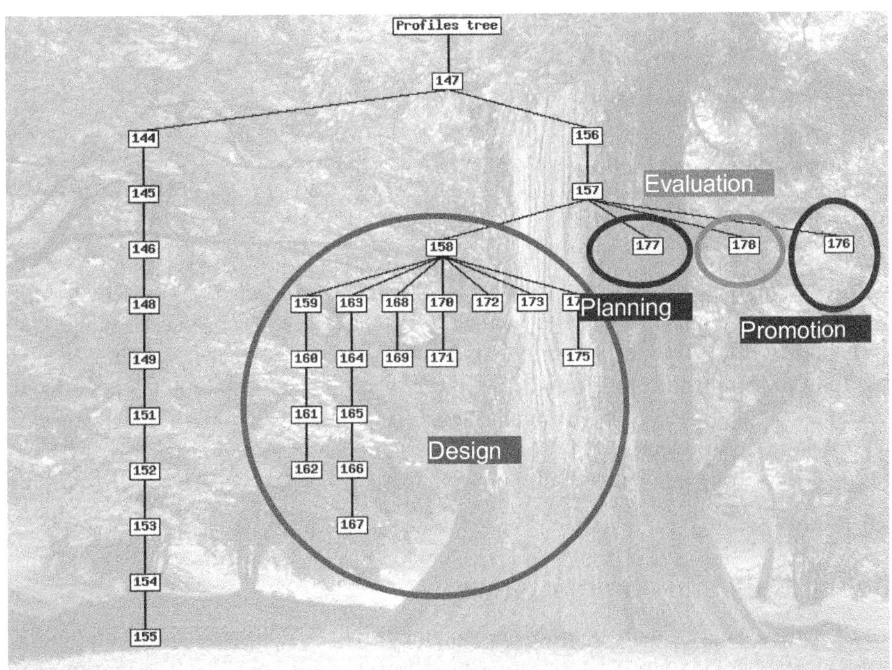

Figure 11: Water Expert view on trees.

The system is built in a way, that the expert knowledge tree can be developed by the expert user, here called project manager, without any programming needs and even regardless of his actual location i.e. just via internet. The project manager defines the template questionnaire describing a particular domain of interest or problem area (e.g. design, planning, promotion, evaluation, etc.), and then invites other partners in the team, i.e. regular users, to consult ADSS and answer to the questionnaire in giving their views on this particular topic. Once the user (stakeholder) starts consultation with the ADSS, he obtains his own "instance" of the template for personal use. By filling up the questionnaire, the stakeholder defines his personal user profile or "fingerprint". The term "fingerprint" (Figure 10) is used for the automatic construction of a "matrix of alternatives" (MoA) (Figure 12) including a proposed set of measures or BMPs (columns) and indicators (rows). The user inserts in this MoA his own priorities while scoring and weighting each indicator or benchmark in order to rank these measures. The project manager can compare the results of each user (stakeholder) and conduct the subsequent negotiation process (Figure 13). This process can be repeated after the negotiation of differences or/and after a better understanding of the USWM issue by stakeholders, when they learn from other ADSS components, functions, tools or documents.

Matrix of alternatives

The matrix of solution, measure or BMP alternatives (MoA) is built up according to proposals selected by the Water Expert engine, for a particular user, using his interaction with ADSS, while answering questions (Figure 12).

Matrix of Alternatives

	A	B	C	D	E	F	G	H	I	J	K	L
1				Street Cleaning		Soakaway		User 1		User 2		
2	Category	Indicators	Benchmarks	Value	Score	Value	Score	Value	Score	Value	Score	Weight
3			Pollutant concentration probability exceed		2		1		1		1	5
4			First-flush capture potential (10/15mm effe		3		3		1		1	5
5	Technical	Pollution co	%age pollution capture for given RI storms		2		5		1		1	1
6			Draw-down times		1		1		1		1	1
7			Downstream erosion		3		1		1		1	1
8			Groundwater recharge		5		4		1		1	1
9	Environmental	Impact on r	Downstream flow protection value		1		4		1		1	10
10			Contribution to urban sustainable develop		4		2		1		1	5
11			Role in Agenda 21		1		1		1		1	1
12			Role in Biological Action Plans (BAPs)		4		1		1		1	10
13			Additional benefits offered by different BM		1		2		1		1	5
14			Material use: aggregate/concrete/top-soil u		3		2		1		1	5
15	Social and Urban	Sustainable	Energy use: construction, operation and m		1		3		1		1	5
16			Design and capital costs		1		1		1		1	1
17			Operational & maintenance costs		3		4		1		1	10
18			Sediment disposal costs		2		1		1		1	20
19			Site decommissioning costs		1		4		1		1	1
20			Stormwater fees		2		1		1		1	0
21	Economic	Life Cycle C	O&M fees		2		3		1		1	1
22			Fulfilment of European regulations		2		1		1		1	5
23			Fulfilment of national regulations		1		1		1		1	1
24			Fulfilment of local regulations		1		1		1		1	1
25	Legal and Urban		Urban storn	Fulfilment of legislation relating to constru		1		1		1		1	5
26			Total score		2.23		2.07		1		1	100%

Figure 12: Matrix of alternatives.

The matrix rows contain indicators, each indicator consisting of several benchmarks. Its columns contain alternative BMPs, solutions or measures, which were included in the matching proposal. Two columns are generated for each alternative, the first one is termed "Value" and the second one "Score". The user fills up the value using his subjective score i.e. the way he considers the indicator from the point of view of his personal interest:, marks range from 1 to 5, for each indicator/benchmark. In addition, each benchmark has its own relative importance, expressed in the last column of the matrix, termed 'Weight'. The weight defines the percentage importance of each row, the sum of these weights reaching always 100. After scoring alternatives, the priority list is proposed to each user based on their answers in the questionnaire.

Analysing the multiple stakeholder views

The main purpose of consensual decision support implemented in the ADSS is seen in inviting a number of distinct ADSS users i.e. project stakeholders, to work on the same subject implemented in the questionnaire template (Figure 13). Each invited user works with his own copy of the questionnaire while consulting the ADSS. He receives his own MoA which he fills up, based on his own knowledge and experience, educational background, interest and affiliation. The purpose of the project manager is then understood as: (1) collecting these distinct views, (2) analysing the differences in the scoring and weighting of the indicators, (3) identifying if the "best" solution alternative is different for some stakeholders, and finally (4) inviting stakeholders to discuss the indicators where the difference in the marks are typically larger than 2. This discussion aims to specify if the differences are only misunderstandings or are a true conflict of interest. In this latter case a negotiation is engaged in order to find a compromise. This idea is supported in the Water Expert engine by a specialised functionality for analysing multiple stakeholder views. The manager can evaluate the differences between different views of team experts by analysing their questionairs. For instance, the different proposed BMPs can be scored to see prevailing opinion of project team, the differences in estimated impervious area can give the insight on different approaches for newly built urbanised district, etc. (Figure 13).

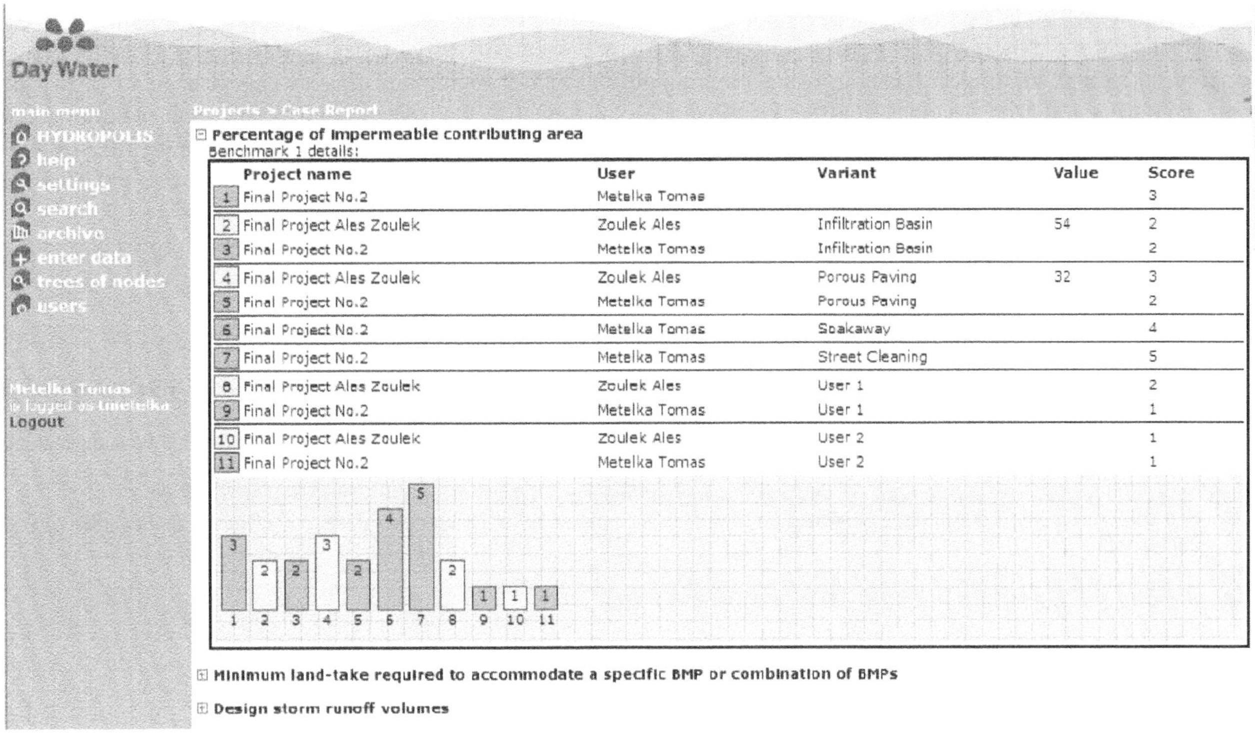

Figure 13: Analysis of multiple stakeholder's views on solution, measure or BMP alternative, called here variant.

Guided tour

At present, two general template questionnaires are designed for decision support purposes in the ADSS. They are respectively called "dynamic thinking planning" and "dynamic thinking design". They have been designed in order to support the decision making process, in the first case for a planning project, and in the second case for a design project (Deutsch *et al.*, 2004). Use of these templates is made possible in the special ADSS component called "Guided tour". By selecting one of these questionnaire templates, the user enters the area, where he can build up his own project by means of distinct instances indicated in the questionnaire. He can consult the ADSS on a particular subject by: (1) giving answers to these questions, (2) building up the MoA and (3) scoring each indicator or benchmark of each BMP or solution alternative and (4) finally ranking these alternatives. He can save this project at any time or make a copy for subsequent use after visiting other ADSS components and learning more details from there.

CONCLUSIONS

The guided tour ADSS component presented in this chapter is a tool which is at the same time a computer expert tool and a conceptual approach of the kind of negotiation occurring when multiple stakeholders are involved in the same project. It requires that every user involved in the decision making process considers the process as a kind of "serious" game. As with any game, it is necessary to know and to respect some rules. It means, in our case, that each participant actually wished to find a solution, and that he is ready to present his interests in the project, in a transparent way. No participant can have a hidden interest, which does not concern directly the project.

Other conditions are desirable for a successful use of this ADSS component: (1) building before the beginning of this procedure a common understanding, among all the stakeholders, of the USWM problem which needs to be solved, (2) participation of every stakeholder at all 3 steps of the decision making process using the guided tour, (3) this "game" should be led by someone outside the stakeholder team but accepted by them. If all these conditions are fulfilled, the guided tour will be successfully used: it has been tested, in 2006, with a group of technicians with different skills on the problem of re-opening a brook, the Bièvre River, transformed into a sewer since half a century in the Paris area (Conseil général du Val de Marne, 2005).

Finally this ADSS component presents an alternative approach to the Multi-Criteria Comparator (MCC) (Chapter 9): MCC is a multi-criteria analysis of all possible individual BMPs and all listed indicators or benchmarks, using either default scores suggested by the software, or personal scores for each BMP. This tool intends to make rapid comparison of individual BMPs, the user being mostly involved in the weighting of each indicator or benchmark. Alternatively, the guided tour MoA allows: (1) the possible selection of a combination of BMPs within a solution or measure, (2) its possible design at the project location, using for example the ADSS source and flux models (Chapter 14), (3) the selection of significant solutions by the involved stakeholders, (4) the selection of significant indicators and benchmarks, (5) their scoring using either intuitive or deterministic tools such as BMP costing (Chapter 10), (6) their weighting, yielding the ranking of these solutions, and, finally, (7) the comparison of all MoA "instances" produced by all involved stakeholders. Thus, it appears that MCC is an important tool for the awareness building of involved stakeholders, whereas the guided tour MoA is devoted to the last stage of decision making in an USWM project.

References

Conseil général du Val de Marne, Laboratoire départemental d'archéologie, Direction des Services de l'Environnement et de l'Assainissement, Direction des espaces verts et du Paysage (2005) Un nouveau regard sur la vallée de la Bièvre, September 2005, ISBN 2-913078-25-7, 28 p.

Deutsch J.-C., Deroubaix J.-F., Chouli E. and Carré C. (2004) How far and how the computer can support the decision making process, *Proceedings of DMUCE4,* Porto, Portugal, 11 p.

Geldof G. (2005) *Coping with complexity in integrated water management.* TAUW BV publisher, ISBN 90-76098-04-2.

Rowe G. and Frewer J.L. (2000) Public Participation Methods: A Framework for Evaluation, *Science Technology and Human values,* **25**(1), 3-29.

Chapter 5
Libraries

Jean-Claude Deutsch and José-Frédéric Deroubaix

INTRODUCTION

The libraries of the DayWater Adaptive Decision Support System (ADSS) consist in five components:

- Case Studies (CS) database,
- Policy Instruments Information (PII) database,
- Stake Holders (SH) database,
- Key Terms (KT) database,
- Criteria and Indicators (CI) list.

KT and CI are lists rather than typical databases and "act" as linkages between several components of the ADSS. As already mentioned in Chapters 2, 3 and 4, there are three different ways of navigating through the ADSS, from a "guided" to a "free browsing" mode, including a "smart guided" mode. These three options apply to the ADSS libraries as one can freely collect as much information as wished, without any help from the guiding system, or, alternatively, fill-up questionnaires in order to enable the system to suggest relevant information. This filtering is operated with the help of Key Terms which value (corresponding to the user answer) will determine relevant information in several databases. If the user is already involved in a decision making process (either a planning or an already designed project), he will be proposed other types of questionnaires and will get, resulting from this set of questions, several proposals, including relevant Best Management Practices for stormwater management (Chapter 8), useful tools (Chapters 9 to 15) and relevant indicators for comparing alternative measures (Chapter 7). The adaptive feature of the ADSS relies in the possibility for the user to add (i) his own indicators in the comparison of these alternatives and (ii) his own data into the CS, PII and SH databases.

Each component is presented below taking into account: (i) its position (location) within the ADSS system, (ii) the functionality that it assumes.

CASE STUDIES (CS) DATABASE

Purpose of the CS database

The purpose of this CS database is to present, in an analytical perspective, the "story" of previous or ongoing Urban Stormwater Management (USWM) projects, while considering their technical, financial and urban contexts.

User's involvement in the construction of the CS database

Each CS relies on the point of view of its editor/author, i.e. the main stakeholder involved in the project. Everyone benefiting user's rights can start the procedure of description of his CS, either using some pre-defined characteristics or answering open questions. He is responsible for what is selected/written in the case description and is the only user allowed to modify these data, shared with all ADSS registered users.

A CS can be further selected, either clicking on a red dot of a map of Europe indicating the site location (Figure 1), or selecting the case study name within a list.

Fields of the CS database

The **CS editor** section includes 13 different fields including the project name, the type of institution, the status of the institution (public or private), its competence in the USWM field (compulsory and/or added function), the type of knowledge that his action is based on, the number of people involved in USWM, the resources invested (annual budget for USWM, annual budget for BMP), the tools used for the project elaboration/implementation/monitoring, the type of experiments conducted.

The **territory** section includes the following fields: global sewerage characterisation, local sewerage characterization, type of sewerage (combined or separate), % of area covered by sewerage (urban catchment), sewerage manager name, type of climate, mean annual rainfall (mm/y), 10-year return period maximal rainfall (mm).

The **project** section consists in a detailed description of the project through three items: problem characterization, solution analysis, and financial aspects.

Last but not least, the **partner** section is a description of the main involved stakeholders, and the **document** section enables to store and link all available graphical or text documents related to the USWM project.

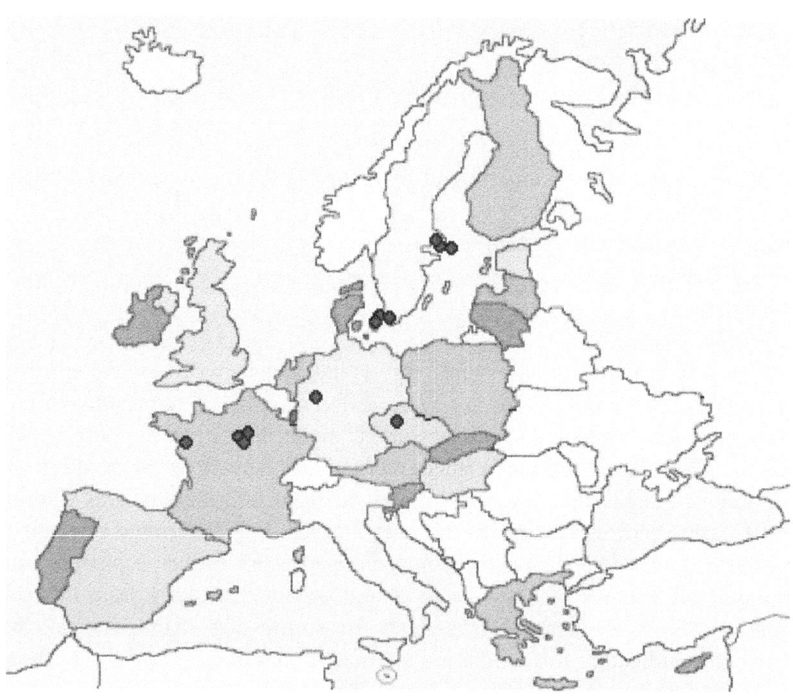

Figure 1 : Location of Case Studies described in the ADSS prototype.

The CS database is linked with the other ADSS databases and to the BMPs catalogue (Chapter 8) using the key terms (KT) (see below). The empirical data in the CS database can be considered both, as a good way to access more theoretical information contained in other databases, and as a good way to "exemplify" this theoretical information. The link with the BMPs catalogue is, for example, a good way to show the user the possibility to implement (or the necessity to adapt) a same BMP to various spatial and organisational contexts. A link is also established with the policy instrument(s) database as, in the Project section of the CS database, one of the fields of the "Solution analysis" is precisely the regulation and policy instrument context.

Criteria and indicators used in this CS database in order to report on the project are the same as the ones used for the Matrix of Alternatives (MoA) which is one of the tools proposed for comparing various alternatives (Chapter 4).

POLICY INSTRUMENTS INFORMATION (PII) DATABASE

This PII database is dedicated to fulfil the needs of four categories of users, according to the context of their decision making:

- Developers, public or private, managing projects in a new urban area: their need is to generally to minimise costs and to save time; the PII objective is to provide them with regulatory constraints and assessment of the procedures generating "delays";
- Developers managing projects in an already urbanised district (rehabilitation, renovation): they may find, in this PII database, information about feasibility constraints, means of consultation and negotiation and public incentives (technical and financial);
- Sewage managers whose main need is information on regulation related to sewage management: who is responsible for what? who pays? who can initiate a project? what are the risks associated to the project (sanitary, flooding)? what are the links with urbanism (planning integration)?
- Other public organisations, such as departments in charge of the road infrastructure, services responsible for parks and other open public spaces maintenance: they may find useful information about the way to successfully integrate sewage considerations into their urban projects e.g. what standards do they have to take into account? what benefits can they get integrating USWM into their projects?

The Policy instruments described in the PII database are listed in Table 1.

Table 1: Policy instruments information within the PII database.

Categories of policy information	Domain/purpose of policy information
Regulatory information	Sewer management for USW treatment Integration of sewer management and urbanism Development projects and urban drainage Risk management (flooding, sanitary risks, USW re-use, etc.) (Table 2) Relation between environment protection and sewer equipment Relation between environment protection and special projects (industry, public work, polluting activities) Summary of regulatory information
Authorities to be contacted with the necessary documents	Impact of the USWM project on the natural environment Risk management (sanitary and flooding control, etc...) Urban planning Summary of regulatory procedures to be followed with specific delays General comments on regulatory procedures
Financial incentives	Environmental protection Risk management (sanitary and flooding control, etc...) Stormwater management General comments on financial incentives
Technical incentives	Environmental protection Risk management (sanitary and flooding control, etc...) Stormwater management General comments on technical incentives
Other incentives	Environmental protection Risk management (sanitary and flooding control, etc...) Stormwater management General comments on other incentives
Information for all public	Sustainable development Environmental protection Water management BMPs
Information for some stakeholders (technical services, land owners, etc.)	Sustainable development Environmental protection Water management BMPs
General comments	

The information provided for each country is most of the time in English. Still, as scientific partners from all around Europe contributed to the filling of the present database, the information concerning some countries is in the corresponding national language (Table 2): this is undoubtedly one of the limits of an ADSS at the European scale, a limit which needs to be over passed through a translation effort (Chapter 26).

Table 2: Examples of regulatory information on risk management in Sweden (in Swedish), France and Greece.

Sweden *Local* ABVA for Stockholm (Local regulations for drinking water and sewerage in Stockholm) §34 i ABVA Bolaget är inte skyldigt att ta emot dag- och dränvatten frÍn fastighet, om avledning av sÍdant vatten med större fördel kan ske pÍ annat sätt, t ex genom lokalt omhändertagande av dagvatten (s.k. LOD-anläggning). Bolaget har rätt att föreskriva att dagvatten fördröjs inom fastighet, sÍ att av bolaget angivet maximiflöde i förbindelsepunkt inte överskrids. (Stockholm Water can request a property owner to decrease the maximum stormwater flow by controlling this flow) http://www.stockholmvatten.se/pdf_arkiv/kundkontakt/ABVA-brosch.pdf
France *National* USW discharge can be treated in a specific hazard prevention plan (PPR ruissellement péri-urbain) or inside the town and country plan. USW re-use in France is not much developed. Re-use is permitted for car wash and land watering. Domestic re-use must be approved by the direction of sanitary and social affairs (DDASS), and will be precisely defined in a decree in 2008. Doc 6 - PPR ruissellement peri urbain.doc Doc 7 - autorisation - declaration.doc www.eau-artois-picardie.fr/IMG/pdf/amenagement-urbain.pdf www.carteleau.org/faq/q94.htm www.graie.org/graie/graiedoc/jtam5synth.pdf
Greece *National* Civil code: The land owner can use the water on his land as he pleases, he shouldn't change the natural flow of water, he shouldn't deprive downstream communities from the existing water resources, he shouldn't flood the downstream communities. In case of septic tanks, they should have no leaks. In case of USW re-use, it depends from the local urban culture (there exist still communities that use USW for domestic use, many communities use it for irrigation) and the inhabitants are responsible of their action. There is no control of the quality. The same is true for private wells. No prescriptions for flood protection. Basements are forbidden to be used for housing. For important projects USWM systems and flood protection works can be requested by the relevant authority: municipality, prefecture or Ministry.

How to select the information within the PII database?

The PII database is classified in 3 categories:

- Mandatory policy instruments: This type of policy instruments includes all kind of provisions which main characteristic are to be compulsory. They can be standards with national or local regulations, as well as procedures of administrative control for industrial infrastructures, as stated in these regulations. .
- Incentives: This type of policy instruments includes all type of procedures which can be used by BMP promoters in order to favour participation to their project. These procedures whether legal, financial or technical are not compulsory: there describe different ways to improve the concern and the participation of stakeholders not readily interested in stormwater source control.
- Information: This type of policy instruments can be implemented in order to make a source control solution more attractive: it provides information about the various benefits (economical, organisational, social, political…) which can be expected.

Incentives and informative policy instruments can be further detailed by end-users because they are largely site-specific.

Policy instruments can also be indexed according to their geopolitical influence at:

- European,
- National or
- Regional - Local scales.

The PII database includes information on the European and national level. The regional-local level can also be completed by the involved end-user, according to his specific context. A country can have many regional-local Policy Instruments items. This way, every ADSS user can:

- Get some general information about Europe, his country and other countries,
- Build his own specific database following the database structure provided by the ADSS prototype,
- Other users can benefit from the experience of users who have previously filled the data related to their own situation and USWM project.

In order to bridge the gap between English and the national languages, the PII database includes a brief English summary of each item and a link with the existing document in the national language. The existing content can provide the user with relevant data from French, Greek and Dutch cases, including e.g. regulatory information on risk management (Table 2). There are links with the CS and the Stakeholders (SH) databases. Selection of specific PII may be performed using key terms: this means that the PII database is indexed with the same key terms than the CS and SH ones.

STAKE HOLDERS (SH) DATABASE

The SH database presents an ideal set of Stakeholders, based on their potential involvement in USWM projects (Figure 2).

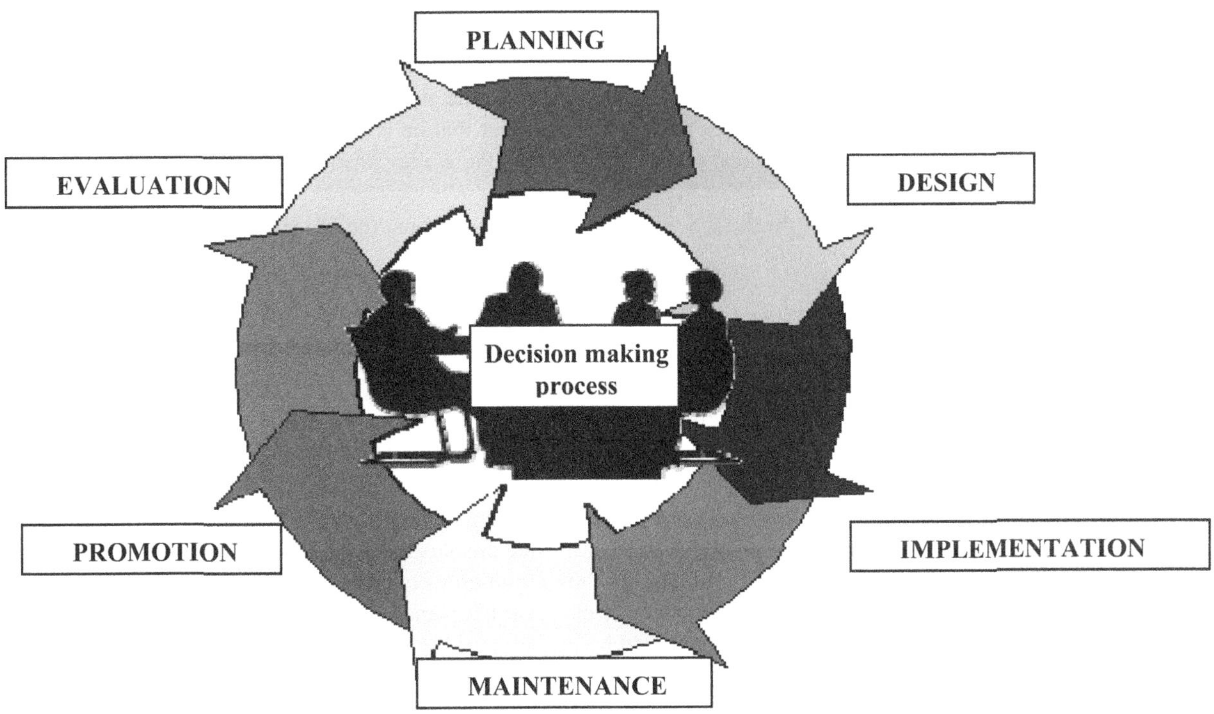

Figure 2 : Phases of the common USWM project cycle, as shown in the SH database.

As shown on the SH database presentation viewer, all the phases of an USWM project are presented as a circle (Figure 2). When clicking on one phase of the project, it is possible to visualise the categories of stakeholders associated to the selected phase.

Urban planer
Services de l'urbanisme (Fr)
Uzemní planovánii (Cz)

Figure 3 : SH database: example of Urban planer.

A second page gives the global list of all categories of stakeholders involved in USWM projects:

- Land owner,
- Equipment owner,
- Local government,
- Developer,
- Project owner - Contracting authority,
- Regulatory bodies,
- Consulting companies,
- Contractor,
- Territory association,
- Environmental association,
- Architect,
- Landscape architect,
- Sewer manager,
- Sewer office,
- Road office,
- Open spaces office,
- Environment office,
- Urban planer (Figure 3),
- Researcher.

The name of the stakeholders is presently given in three languages: English, French and Czech. When clicking on a specific stakeholder category, the user is provided with information on the stakeholder typical involvement in USWM projects, his specific competence, the scale and the phase of the project that he is expected to be involved in, the tools that he commonly uses. The system also displays the type of BMPs, policy instruments and criteria, which this stakeholder category is preferably using.

As mentioned above, there are links with the CS and the PII databases: thus, the SH database is indexed with Key Terms.

KEY TERMS (KT) LIST

21 KT have been presently implemented in the ADSS prototype (Table 3). They need to be determined in order to collect relevant information for the project that the user is working on, within the different ADSS databases and libraries. Each KT can take different values, either a numerical value or a word. A value can be given to each KT, either by selecting it directly in the list of Key Terms and their possible values or by answering the "specific Information" questionnaire displayed in the Guided Tour proposed when accessing to the ADSS (Chapter 4). When using this questionnaire, each question is related to a specific KT. The KT value enables the ADSS to sort out, within its libraries, any information relevant to the specific USWM problem. The more numerous questions are answered, the more specific information the user will get. After such procedure, the user can visit his personal Hydropolis (Chapter 3) in which he will get data from the components which have been filtered by the selected KT values.

KT 16 to 21 are technical ones: the threshold values used are the same than those used for the building-up of the Multi Criteria Comparison matrix (Chapter 9). No answer, one or several answers can be given for each KT question. The SH, the PII and the CS databases have been indexed with these KT. The BMPs catalogue (Chapter 8) is indexed through the CS database.

Table 3: Major characteristics in the 21 key terms presently implemented in the ADSS prototype.

KT number	Major KT characteristics	Major KT values
1	Phase of the project	Planning, design, implementation, maintenance, evaluation, promotion
2	Domain of knowledge that the user is working in	Sewer, road, urbanism, landscape, building, public organization, environmental protection
4	Pollutant Type	Total suspended solids (TSS), biological oxygen demand (BOD), nutrients, metals, organic micropollutants, pathogen, herbicides...
5	Scale of the project	From small (<1 ha.) to very large (more than 2000 ha)
6	Urban Area	Either not-constructed or urbanised
7	Land use	Public building, public housing, private dwelling, city centre, industrial, commercial/retails, open space, rural
8	European Climate	Oceanic, Mediterranean, polar influence, mountainous, continental
9	Risk	Groundwater pollutions, surface water pollutions, oil contamination, flooding, sanitary or legal security
10	Type of funding that the user can obtain for an USWM project	From public general budget to private funding
11	Tools available for the project	Planning control, regulation implementation, modelling, data acquisition, public participation, cartography
13	Policy Instruments types	Same terms as the types described in the PII database
14	Policy Instruments topics	Stormwater management, urban development, risk management, sewer management and environmental protection
15	Public Involvement	High, low or none
16	Design flow volume	
17	Land area available for BMPS	
18	Water table level	
19	Soil hydraulic conductivity	
20	Soil type	
21	Infiltration rate	

CRITERIA AND INDICATORS (CI) LIST

This list is divided in criteria, primary indicator, secondary indicator and benchmarks (units or threshold values) (Chapter 7). Criteria are divided in primary and secondary indicators. Benchmarks are given in order to build a quantitative assessment of each indicator. Criteria and corresponding primary indicators are listed in Table 4.

Table 4: Criteria and primary indicators (see Chapter 7 for examples of benchmarks).

Criteria	Indicators
Site Characteristics	Site Area Soil Characteristics
Technical Criteria	Flood control Pollution control System flexibility & potential for retrofitting Impact on drainage system,
Environmental Criteria	Impact on receiving water volume Impact on receiving water quality Ecological Impact
Operation and Maintenance	System reliability and durability Maintenance & servicing requirements
Social and Urban Community Benefits	Amenity & aesthetics Public/community information & awareness Stakeholder acceptability Public health & sanitary risks Sustainable development
Economical Criteria	Life cycle costs Financial risk/exposure Long term affordability Land cost
Legal and Urban Planning	Urban stormwater management regulations Local building and development issues Adoption status

The criteria, indicators and benchmarks are used to fill the lines of the two matrices provided by the ADSS: Multi Criteria Comparison (MCC) (Chapter 9) and Matrix of Alternatives (MoA) (Chapter 4). They are given also as proposals, resulting from answers to the Dynamic thinking design and planning questionnaires in order to be used in the MoA.

CONCLUSION

As already presented in Chapter 4 (ADSS guided tour), there are different methods for navigating through the ADSS. The system has been designed in order, either to browse freely through the libraries or, alternatively, to be guided through them, when completing questionnaires identifying the user's context and project. Whatever the chosen navigation mode, libraries are structured in a way to promote integrated water management, even when the user has a very specific problem for stormwater management, such as only comply with regulation on pollution control, or solve a specific stormwater runoff problem.

The information provided in the various ADSS libraries and databases is multidisciplinary and both theoretical and empirical. For instance the PII database incorporates not only data on the standards that the sewer manager has to conform, but also the useful instruments for integrating his project into a more global and urban strategy. The linked examples (documents) provide very practical information on the resources and constraints embedded in each BMP or measure that can be used in order to promote stormwater source control solutions.

The CS database is especially useful for comparing a given USWM project to similar ones. As the user can describe his USWM project, a collective social learning process is expected: thus a new ADSS user can benefit from the successful experiences and failures experimented by previous users. The CS database has also been designed in order to favour cooperation with others stakeholders involved in the same "urban project".

Consultation of the CS database should logically lead the user toward the SH database which promotes the idea that sustainable USWM projects are those integrating the widest range of competences and approaches (civil engineering, social and economical knowledge, landscape architecture). Of course this is not an exclusive condition but it is, to our opinion, a necessary one. Following the DayWater general aims at leading to sustainable USWM projects, a lot of effort was spent in linking the information included in these libraries to ADSS internal and external tools. These tools refer to various disciplines: some of them integrate biological knowledge (e.g. the biotests for toxicity characterisation of stormwater: Chapter 20), chemical knowledge (e.g. the risk assessment tool for selection and characterisation of priority pollutants – CHIAT: Chapter 11), hydrological and mathematical knowledge (Source and Flux Model: Chapter 14). Others include more social sciences (Flood Vulnerability Assessment Tool FVAT: Chapter 13), or geographical and sociological knowledge (Aspects of Water and the Ambition-Refection: Chapter 6).

Last but not least, as the ADSS is also a Decision Support System (DSS), the user can compare BMPs of his project in a Multi Criteria Comparator (when his project consist in implementing structural BMPs: Chapter 9) or alternative measures within a Matrix of Alternatives (when the variants of the project not only refer to source control but more widely to flood and water pollution control: Chapter 4). Libraries were designed in order to enable the user to discover and strengthen criteria used to compare possible BMPs and alternative measures for of his USWM project.

Part 2

Tools associated to the Adaptive Decision Support

Chapter 6

Urban dynamics

Roel Valkman, Pieter Lems and Govert Geldof

INTRODUCTION

Water managers and water policies generally support the idea that an integrated water approach leads to better solutions. In spite of this integration, new problems appear that cannot be solved with this approach. Climate change, desertification, peak discharges and floods in urban areas have forced the water manager to a more sustainable way of handling urban water problems. Instead of the traditional "end of pipe" measures, such as discharging stormwater by sewer overflows, source control measures are a new way to tackle these water problems. By local infiltration and day lighting stormwater flows, water becomes again a significant part of the urban environment, after decades of being hidden underground.

Because of such stormwater source control issues, the field work of the urban water manager has become more complex, dealing not only with the technical aspects of the system itself, but also with spatial, social and political aspects, within the context of the dynamic urban environment. In the urban environment, space is scarce and every square meter has to be won. Is the urban stormwater manager able to play his role in this space game? The finest water concepts may be suggested, but fine plans are more easily proposed than carried out. Once paper plans become reality and the costs of a water project become obvious, the support of other stakeholders cannot be taken for granted. Of course, the benefits of stormwater source control measures can be easily listed; however, in relation to other urban issues, such as parking space, road and social safety, they can rapidly be given lower priority. Thus some urban stormwater measures, initially planned, are never carried out (Vuurboom, 2001). The results of an enquiry, carried out at the beginning of the DayWater project, on the needs of urban water managers, demonstrated the same difficulty: such managers encounter difficulties positioning the urban stormwater management (USWM) within the urban environment issues. As it is hard to influence the behaviour of citizens and to stimulate water consciousness, finally although important, USWM issues are not presently a political priority (Valkman and Lems, 2004).

The problems mentioned by the urban water managers, when interviewed as DayWater end-users, cannot be solved solely by specifications of best management practises (BMPs) (Chapter 8), by tools for source and flux modelling (Chapter 14) or by multi criteria analyses (Chapters 4 and 9). Urban water managers need other information or knowledge in order to implement successfully BMPs, in the urban context. This chapter describes the results of our research for a better understanding of the problems related to the implementation of stormwater measures in the urban environment. Furthermore, this chapter presents one of the Adaptive Decision Support System (ADSS) component named "Aspects of water" as well as the three archetypes of water management that helps the water manager to interact with other stakeholders in the urban context and to reflect on his own ability to involve urban dynamics in USWM. Finally, this chapter presents how each of these new insights are incorporated in the prototype ADSS.

 ISBN: 9781843391609. Published by IWA Publishing, London, UK.

A DEFINITION OF URBAN DYNAMICS

In order to define Urban Dynamics, it is necessary to distinguish two concepts: system and context. The "system" comprises the urban water system, including groundwater, surface water, stormwater, sewer system, etc... The "context" consists of all events and processes, within the spatial and social environment, that interact with the system. This distinction between system and context is somewhat arbitrary. In reality, the system is interwoven with its context, they are inseparably bound to each other and there are strong interactions between them. All events and processes in the spatial and social environment, i.e. the urban context, that directly or indirectly influence the system are designated as "Urban Dynamics", referring to the altering and unpredictable nature of these processes (Geldof, 2005).

Figure 1 presents a commonly used strategy for decision-making in UWSM. In the narrow part of the sandglass, the urban context releases the stormwater problem, in order to find a sound solution. Once the solution to the problem is identified, this solution needs to be implemented in the urban context. Figure 1 shows how urban dynamics are located both at the top and the bottom of the sandglass, and are absent from the narrow part where the decision-making takes place. The enquiry on the needs of urban water managers, held at the beginning of the DayWater project, confirms this hypothesis: the real bottleneck of USWM is not the narrow part of the sandglass, but where the problem is emerging from the urban context and where solutions (BMPs) are implemented into it. Hence, the real bottlenecks are not at the centre but at the top and, especially, at the bottom of the sandglass.

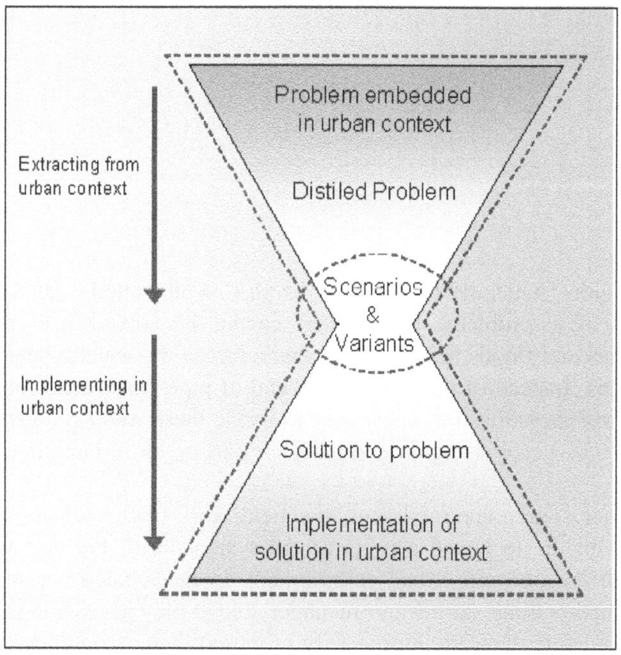

Figure 1: Sandglass scheme, placing urban dynamics at the top and the bottom of the traditional decision making process.

One of the reasons for extracting the water problem from its urban context comes from the assumption that the water problem can be solved within the system and needs to be isolated from the urban context. Unlike water problems embedded into the urban context, such isolated water problem fits into an mathematical model where, however, it looses its diversity of meaning and is reduced to a set of equations for water discharge, capacity, flow rates describing its physical, chemical and biotic aspects. Nevertheless, the solution, for stormwater source control, found by such optimisation process might not be the best one in the real urban context, where other aspects of water are contributing to the success of a scenario.

With respect to stormwater source control, the traditional sandglass strategy is far from being adequate (Figure 1). Source control is located at the border between the water system and the urban environment, where other stakeholders with other interest play an important role. In other words, source control scenarios cannot be "distilled" or extracted from the context since this context is an essential part of the water problem and of its solution. Successful cooperation with other stakeholders demands insights in the interests of these stakeholders. An ADSS component named "aspects of water" have been developed in order to map these interests in relation to water problems.

ASPECTS OF WATER

Mapping perceptions and valuation

Water has many different characteristics or aspects. Water first consists of a molecule made up of two hydrogen and one oxygen atom. In water live fishes, and trees and plants draw water from the ground. A dive in sea river provides cooling on a hot summer day, but the same water is life-threatening during floods. Thus there are different ways of looking at water and such perspective is the determining factor for how water experienced (e.g. wet in the swimming pool or wet under a rain shower). The perspective differs for each person and depends on the relationship between this person and water. Different characteristics of water are visible from different perspectives. The same characteristics are interpreted in different ways. The aspect theory helps to map the different facets of water that are important to cope with urban dynamics (Lems, 2005).

The aspect theory originates from Dooyeweerd, a Dutch philosopher at the beginning of the 20th century (Kalsbeek, 1983). Dooyeweerd uses the aspect theory to describe reality in all its richness, so to say, in all its aspects. He distinguishes fifteen aspects from the arithmetic up to the pistic aspect (Table 1). There is a difference between the lower six aspects and the upper nine ones. The lower six obey the laws of nature, determining the structure of natural life and matter. The upper nine aspects drive man in his way of dealing with natural live and matter, for example in USWM. Water partners deal with water according their moral beliefs with regard to good water management. These beliefs originate from a religious funded "weltanschauung" and influence the regulations on water. Water management also deals with the aesthetic, economic and other aspects of water. The different aspects light up different facets of USWM.

With respect to UWSM and water problems mentioned earlier, the aim of these aspects of water is threefold. Firstly, the aspects offer a framework for dealing with stormwater within the urban context, to draw a complete picture of stormwater related issues. Secondly, the aspects of water help to look at water problems, from different perspectives, in order to build a shared problem perception among stakeholders, a precondition for cooperation. Finally, the aspects of water "unlock" values of stormwater related to the "higher" aspects that are essential to a successful implementation of water measures.

Table 1: Aspects of water.

Aspect	Essence	Aspects of water
Human aspects		
Pistic	World opinion	Ritual washing - Moslems Washing in river the Ganges - Hindus Baptism - Christians
Moral	Views concerning good treatment	Views concerning good USWM Safety or the prevention of damage Sustainability
Legal	Law	Regulations for water Issue of permits for sewer overflow Authority and responsibilities of water partners
Aesthetic	Beauty	The beauty of water Reflecting water Sunset by the sea
Economic	Way of saving	Do the costs of water projects weigh up against the benefits? Treatment charged by the water company
Social	Dealing with people	Meeting by the water Discussion by the drinking water well, in Africa Resident's evening concerning roof runoff disconnection project
Linguistic	Symbolic significance	Writing about water Poems
Historical	Management by free forming	Intervention in the water system Land reclamation
Logical	Analytical distinction	Thinking about water Thales: "Everything is water".
Psychological	Perception	Water stimulates the senses Splashing water Delicious drinking water

Natural aspects		
Biotic	Life processes	Water as the first condition for life A person can survive for a maximum of 3 days without water Fish live in water
Physical/chemical	Matter	Water has a chemical composition (H_2O) and satisfies the physical laws of nature Water quality parameters Water flows downhill Navier-Stokes flow comparison
Kinetic	Uniform movement	Water flows A river carries snow melt and/or rain water to the sea Throughput in a water system
Spatial	Uninterrupted extendedness	Water occupies space The percentage of surface water in a district is 6% A pond contains 1000 m^3 water
Arithmetic	Discrete quantity	The expression of water in numbers 5 litres of water per day and per inhabitant

Aspects of water in Hydropolis ADSS

To be more practical the number of aspects in DayWater ADSS is reduced from fifteen to twelve (Table 1 and Figure 2): moral, legal, aesthetical, economic, social, linguistic, historical, logical, psychological, biotic, chemical and physical. The Aspects of Water component of Hydropolis presents a large database of mainly Dutch projects, highlighting these twelve aspects.

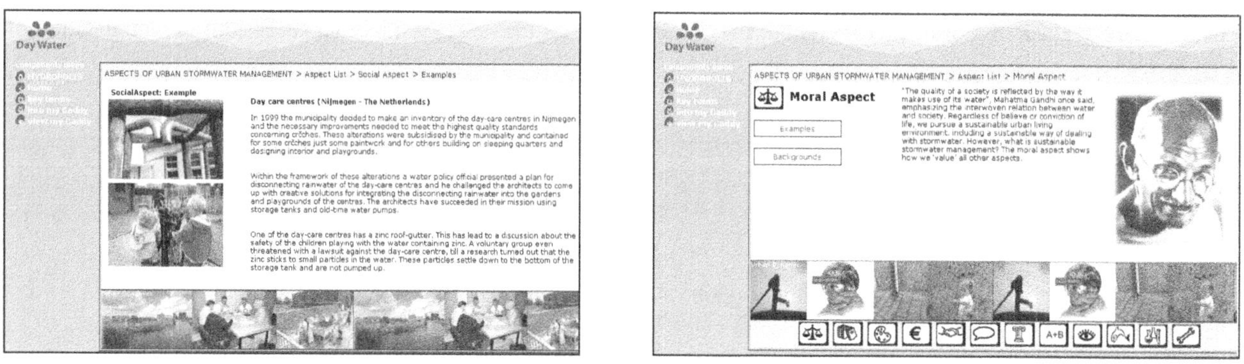

Figure 2: Two sample pages from the "Aspects of Water" component in Hydropolis ADSS: social and moral aspects.

COPING WITH URBAN DYNAMICS

Three archetypes in urban water management

Practically water management is also determined by the method of dealing with interaction between system and context. In other words, water management may be characterized by the way in which the complexity of the interaction between system and context is dealt with. In the framework of the DayWater project, three forms of water management are distinguished, all three being characterized as archetypes (Figure 3): basic, functional and contextual water management (Valkman and Lems, 2004).

In basic water management, the water management is directed inward: attempts are made to optimize the water system and the water chain, and to minimize any inconvenience at the boundaries of the system. In this way the interaction between system and context is minimized. Functional water management makes connections with the context controlling some functions, such as ecology, nature and recreation. The aim of functional water management is to integrate the system as well as possible within these previously defined functions. Then a question appears: the unilateral effect of context on the system. Finally contextual water management really confronts the context which is not frozen in some functions, but considers as a central issue the continuous interaction between system and allow the blossom of all values of water.

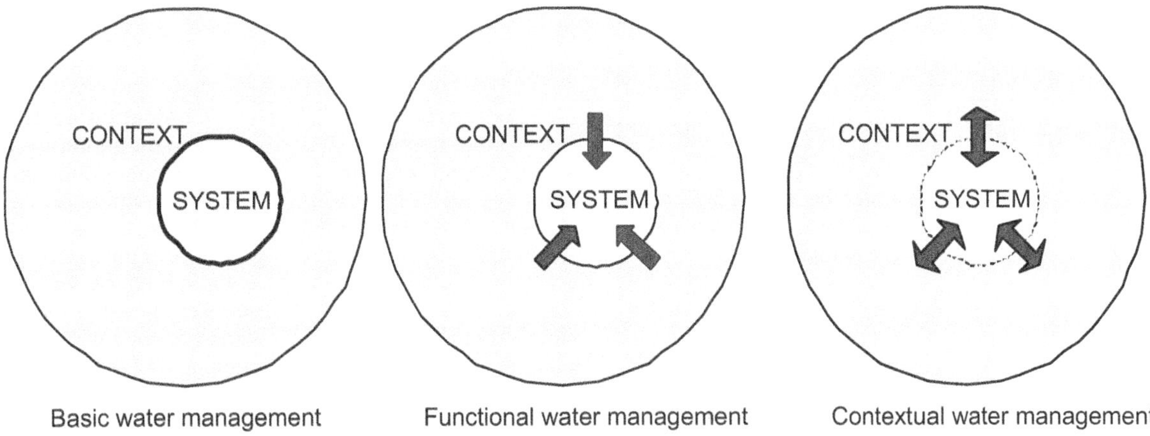

Figure 3: Interaction between the system and the context: (i) basic water management hardly experiences any interaction between system and context, (ii) functional water management allows a one-way traffic and (iii) contextual water management allows comprehensive interactions (Valkman and Lems, 2004).

Basic water management: static approach

Basic water management aims at optimizing the water system based on physical, chemical and economic criteria. In particular, the associated objective is the prevention of damage (Table 2). A core concept in basic water management is control: damage can be prevented by good control of the physical and chemical processes in the water system. The system is understood as independent: there is hardly any consideration given to the interactions between the system and the context. The context is only considered as the site imbedding the system. The advantage of this approach is that it results in relatively simple, normative water management. Optimization of the water system is based on standards. A disadvantage of this simplification is that alienation arises between the "optimized" system and its context: in a wider system, the optimum solution can become sub-optimal.

Functional water management: frozen dynamics

Just as for basic water management, functional one aims at the prevention of damage by controlling the system. However, in this approach, consideration is given to the context and possibilities are examined in order to fit the system (Table 2): this is a one-side influence of the system by the context, selecting some functions such as ecology, nature and recreation. Optimization of the system means here that the system is attuned, as well as possible, to these functions. The advantage of functional water management is that attuning the system to social need may also be considered, although such attuning takes generally place globaly. Functional water management still contains the core concept of control present in basic management, but the various functions result in a more differentiated, but still normative, management.

Contextual water management: from optimization to adaptation

For contextual water management, the interaction between the system and the context plays a central role (Table 2). Water partners play a social game in order to make the best possible use of the opportunities and possibilities of water. Contextual water management is not mainly aiming at the optimization of the system, but at contributing to the quality of living environment while exploiting the values of water as far as possible. The interactions between the system and the context are not neglected nor reduced, attempt to create an optimum living environment. However, such approach results in increasing management complexity. For example by considering the interactions with other municipal issues, uncertainty is introduced. Water management is no longer completely in the hands of the water partners, but must result from collaboration with third parties. These uncertainties are not only considered negative; they also offer the possibility to develop new creative solutions and to generate wide enthusiasm. In order to cope with this complexity, in a positive way, the management process of contextual water management considers the principle of Interactive Implementation.

Table 2: Major patterns of the three archetypes of water management.

Characterization	Basic water management	Functional water management	Contextual water management
Objectives	Aimed at optimizing the water system on the basis of physical, chemical and economic criteria	Aimed at optimizing the water system on the basis of the functions defined by the context	Aimed at optimizing the living environment by making use, as far as possible, of the values of water in both the system and the context
	Primarily aimed at preventing damage	Aimed at preventing damage and making use of the possibilities and opportunities for water	Aimed at preventing damage and making use of possibilities and opportunities for both water and other areas
Relationship between system and context	No interaction between system and context: water managers take no account of the context	Unilateral influencing of the system by the context: water partners are "sensitive" to the context	Interaction between system and context: water partners are concerned by the context
Dealing with complexity	Neglect complexity	Reduction of complexity	Takes complexity into account
Characterization of the management process	Normative management	Standard management, but standardization based on various functions	Adaptive management with interactive implementation as management process
Position of water partner in society	Autonomous water manager	Water partners driven by the province administration, which also determines the functions	Water manager taking part to the social play
Water partner involved in other policy areas (roads, traffic, environment)	Water partners have a compliant attitude	Water partners are reactive	Water partners are pro-active and anticipate issues in the living environment
Advantages	Management process is relatively simple	Internal integration results in more coherence between the various aspects within the water system	Unlocking a wide spectrum of values, whereby not only the costs but also the procedures are taken into account
	Easy communication between colleagues	Project may be attuned to the social needs	Society involved in water management and vice versa
	Clear organizational structure	The water partners receive more social recognition	Water manager obtains more diversified and more pleasurable work and a wider social recognition
	Normative character makes good legal anchoring possible		More opportunities and possibilities for water projects: water planning do vanish but are carried out More profit from water projects for both the water system and the living environment Enthusiasm for water measures originates in the interaction with practice
Disadvantages	Not very stimulating Water partner stands on the sidelines of the social discussion Gap between water and society Limits of possible solutions Limits of technician ability	Neither fish nor fowl Top-down approach planning achieved, but rarely carried out Theoretical coherence often appears not to result in integrative solutions Tackling diffuse sources is difficult to get off the ground	Greater uncertainty Basic water management threatens to disappear from sight because too much attention is paid to carrying out "attractive" projects

Ambition Reflection

In practise, what the water manager is able to achieve depends highly on his attitude towards urban dynamics. In order to support the water manager in becoming "contextual", the Ambition Reflection tool has been developed (Figure 4) including the three archetypes of water management. This tool considers the water manager's ability to implement successfully stormwater measures in the urban context: it is an introspection in order to reflect on its attitude (goals, philosophy, perception and position) and to help him to become a contextual water manager.

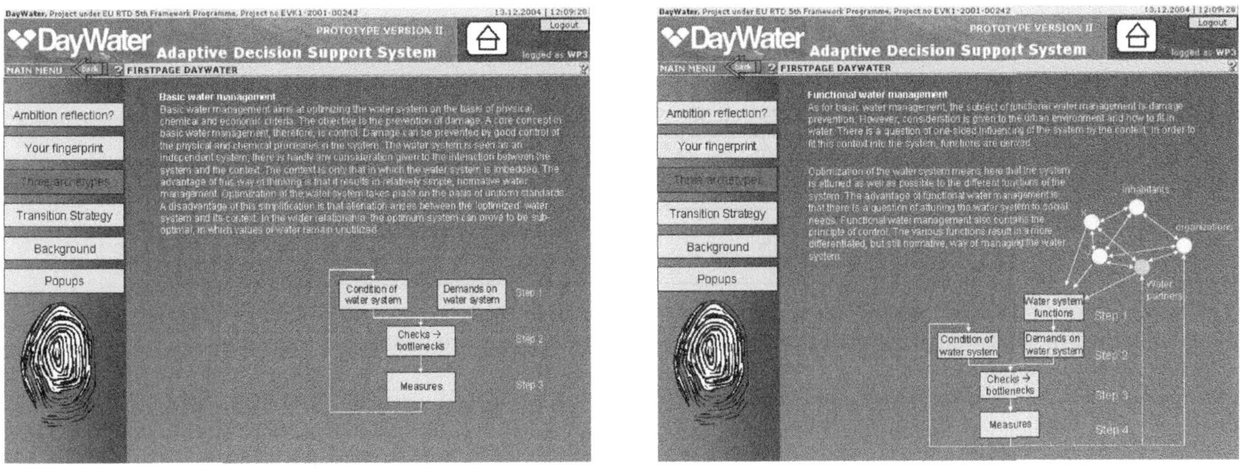

Figure 4: Sample pages from the Ambition Reflection component of the Hydropolis ADSS.

IMPLEMENTATION AS STARTING POINT

In the DayWater project, the archetype contextual water management has been given "flesh and blood" through the Interactive Implementation concept: it focuses attention on the implementation phase of water projects, using. three principles, parallel working, switching between scales, learning while doing (Geldof, 2005).

Working in parallel

In practical USWM projects, policy, planning, design, implementation and management tasks are commonly considered in series, as if it was a question of passing on a relay baton. When a problem is reported, policy makers make regulations. This policy will be incorporated in planning, on the basis of which designs will be made. These designs are then translated into a specification, after which actual operation or implementation will be carried out, the result being transferred to water managers. Such interactions between the various parties of this chain have been minimised: in some cases, there exist so-called "cold welds".

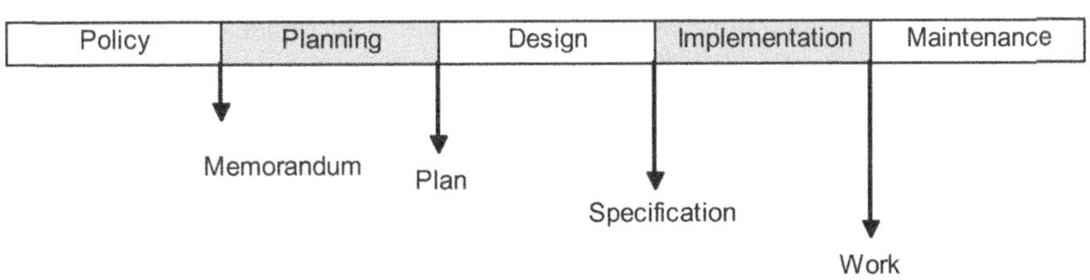

Figure 5: Static and linear chain, from policy to maintenance tasks, in an USWM project.

When the USWM project is isolated and a single implementation process is considered, such working method is logical. However, when many processes are concerned simultaneously and measures taken at different times and at different spacial scales, such sequence of operation appears too rigid. Therefore integrative and complex projects leads to failure with this linear method. In many interactive planning, many persons are involved for its implementation and their support is necessary. However, support for a plan does not automatically mean support for carrying out that plan. Plans often disappear in the drawer or are conducted in stripped-down format: this leads to frustration and is a great obstruction in the creative process.

For complex, integral projects, a management process is necessary that connects better with dynamic processes in reality. This management process is designated Interactive Implementation (Horst, 2003). The essence is that planning, design, implementation and management are not placed in series but are interwoven with each other (Figure 6). Immediately after setting up a view, the first experimental projects are defined and carried out. Policy makers participate actively in the implementation and managers enter into a dialogue with designers.

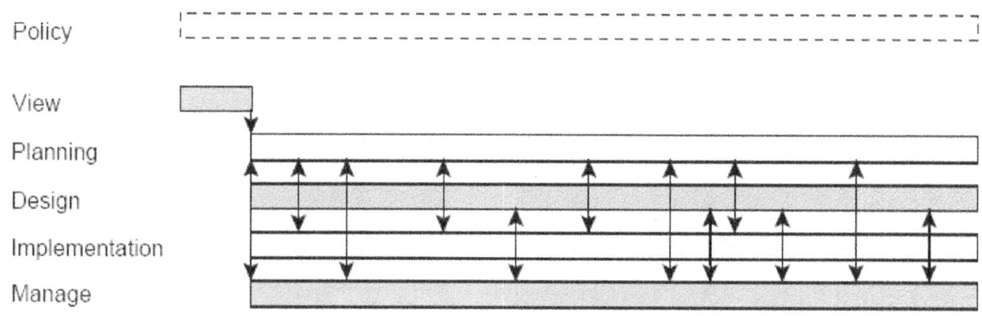

Figure 6: Interactive Implementation: phases in the implementation process are conducted in parallel.

Where storm water projects are interwoven with processes involving soil clean-up, traffic nuisance, renovation of dwellings, decontamination of industrial sites and programmes for education and integration, such as for municipal renewal, a similar parallel working method is a necessary condition, the water manager has to fit in. They must make their contribution at the right moment in order to be able to bring the values of the water project into the picture. This right moment cannot be determined beforehand, it is determined by the urban context.

Switching between scales

The second principle is based on the fact that USWM has to influence different levels of spatial scale (Table 3): for example the implementation of a BMP at a small scale, the discharge of storm water of a city at a large scale. Interactive Implementation suggests to switch between these levels, instead of translating the USWM to one specific scale.

Table 3: Spatial scales considered in USWM projects.

Small scale	Large scale
Concrete	Relatively vague
Short term dominates	Longer term dominates
Natural integration through intervention in the living environment (water, soil, traffic, social safety, etc.)	Focus on spatial coherence
Many differences in outlook	Possible consensus at an abstract level

Table 3 gives several characteristics of processes on large and small scales. On a large scale the water system can be pictured as a whole and a view of sustainability can be set up. However, it remains vague because of the scale level. This vagueness may result in "everyone agrees". Who does not want to have a durable water system? Viewing on the large scale level is necessary in order to discover the spatial coherence. However, it is impossible to bring about full integration with other environmental themes at this scale level. That is possible at the scale level of street or dwelling. At this level all the environmental themes come together. The radius of 100 metres around one project area is a suitable position to actually couple the themes and realize synergy. Reducing noise pollution, cleaning up contaminated ground, satisfying the Building Decree for the renovation of dwellings, setting up a centre for the homeless and disconnecting a paved surface, everything comes together at a small-scale level. This scale level is dominated by short-term thinking and differences in viewpoints must be resolved, because it is extremely concrete. In Interactive Implementation, a link is forged between the different scale levels. No pains are taken to describe all measures on one scale level (so that they can be compared with each other), but there is an alternating treatment and reflection at small and large scales. A considerable difference between the large and small-scale level is reflected in the involvement of inhabitants. In practice, water projects at neighbourhood level seem to arouse a great deal of enthusiasm amongst the inhabitants. This enthusiasm flows from the fact that water projects are carried out in their immediate living environment. It gives vitality to a neighbourhood and a street and it has a certain appeal. When inhabitants are given more responsibility for the design and the management of the water in the district, the water will be more highly valued. In this way the involvement of the inhabitants is increased. This can be a starting point for tackling many more deeply rooted problems.

The switch between scales can also be significant for the development of a new housing development plan. In the Netherlands, an urban development project is organised as follows. At first one looks on a regional scale, after which structures are set up at district level. Water can be introduced at this level. After global exploitation calculations have been made at district level, parcelization studies are carried out at neighbourhood level. Ideas for durable water management can then still be kept standing. Only after some time will – together with contractors and architects – the descent be made to the scale level of the street and the dwelling. The plans for making ready for development are then worked out and more accurate exploitation calculations are made. It then appears that many ideas for durable water management are put at risk, are scrapped or simply prove not to be possible. For example, a construction without a crawl space appears not to be possible because the contractor has a long-term contract with a flooring manufacturer. By switching between scales and already in the initial phase exploring at the scale level of street and dwelling, problems can be discovered early and agreements concerning the arrangement of the water systems can be better anchored.

Learning by doing

Where everything depends on everything, processes are surrounded by many uncertainties. "Do the measures have the desired effect?" "Will the inhabitants participate?" "Can we rely on political support?" These uncertainties are resolved by building experience by carrying out projects (Table 4). These projects – often designated as pilot projects – must be undertaken where there is a significant chance of success. The success of one project can lead to support for another project.

Table 4: Roadmap: conditions to consider before starting a pilot USWM project.

Shared sense of urgency This sounds evident, however at the start of a project many times the underlying problem is not known by the stakeholders. If communication takes place, it is about possible solutions and not about the problems. In particular, politicians demand a clear problem definition. In order to get approval of the local council an alderman the "sense of urgency" needs to be stated. For example: "If we do not take water management really seriously, once in 10 years the district will be under water with significant damage on houses." Or: "The chance that children will become ill by swimming in this water is unacceptably high". *Mutual dependency* When stakeholders are mutual dependent there is a need for cooperation. For example, if a water partner wants to disconnect storm water form a paved area on private property, he is unilaterally dependent on the voluntary cooperation of the landowner. If the landowner doesn't want to cooperate, the project has a great chance to fail, since the water partner does not have legal authority to force landowners to cooperate. However, for topics such as road safety and parking, inhabitants are dependent on the local authority. By taking all these topics together, lines of mutual dependence become visible and opportunities for carrying out water projects occur. For example a project aimed at disconnecting storm water can simultaneously contribute to an extension of green areas in a street. If there is mutual dependency, there is a basis for means that negotiations, since both stakeholders have an interest. To create mutual dependency, it helps urban water managers to have insight into the valuation of water by elaborating the aspects of water. *Trust* The condition for a successful project is the presence of trust between the parties concerned. Experience teaches that when inhabitants have little trust in the local government, they do not react to information, but to so-called peripheral stimuli. These stimuli are over-simplified messages and signals from all kinds of sources, often with an emotional component. Inhabitants distrusting the local government can exhibit a strong and violent reaction because of these stimuli. For example the peripheral stimulus that 'disconnected paved area will result in overflow of cellars', summoned by a reluctant inhabitant, can result in massive resistance of other inhabitants to the local government despite all kinds of rational arguments proving the message not to be true. Thus, only when there is trust, participation on honest arguments is possible. One of the ways to increase trust is to follow the principle of "do what you say and say what you do". A principle often neglected resulting in a lot of distrust. *Do at not avoid conflicts at all costs* Projects seldom result in win-win-situations, even though designers often create this impression. There are always winners and losers. It is fatal to a project if dormant conflicts are brushed aside. It is better to discuss points of conflict at an early stage of the project. Then there is still sufficient time to find solutions. By making ideas concrete quickly and by learning by doing, points of conflict soon become apparent. *No optimal but best solutions* For integral projects, optimum solutions are always sub-optimum. For the water partners, this understanding is of particular importance, because all too often the best solution for the present water problems in a large area is determined with the aid of model exercises. This solution is then applied generically to all projects. This will not work. The trick is to determine the best solution for each different project, a solution that reinforces the identity of an area. Water partners often are afraid of creating a precedent: "If we do that in one place, we have to do it on other places also". The more integrally an area is looked at, the stronger will be the justification for doing something different in each place. Optimum solutions are not determined in advance, but emerge in a surprising manner in an interactive process, by doing it.

Snowball effect

Only when plans are carried out most people become enthusiastic and room for water is created both literally and figuratively. Enthusiasm originates when a water manager with 'a gleam in the eye' is given the room to carry out his or her ideas and to involve others in the process. These may be colleagues in other departments, but may also be other water partners or administrators. The collaboration between municipal civil servants in different departments helps water partners to integrate different goals in a specific project. For example, a municipality can take the initiative in developing educational packages concerning the environment for use in elementary schools.

CONCLUSION

Source control takes place in the tentacles of the water system, stretched out into the urban environment. Therefore, source control forces the water manager to involve the urban environment as essential part of the water problem as well as the solution to this problem. Because of this, the work field of the water manager has become more complex, dealing not only with the technical aspects of the system itself, but also with spatial, social and political aspects within the context of the dynamic urban environment. In other words, source control cannot be distilled from the context. Successful cooperation with other stakeholders demands insights in the interests of these stakeholders. To map these interests with respect to water problems, the aspects of water are developed. The aspects of water offer insight into the different aspects of water as well as the valuation of these aspects by other stakeholders. Although Hydropolis offers a structured database with examples of projects, the aspects of water come to expression in practical use in water projects. The way in which water management is given form in practice is also determined by the way in which the interaction between system and context is dealt with. Put in another way, water management is characterized by the way in which the complexity of the interaction between system and context is dealt with. Table 2 presents three archetypes of water management. From these three archetypes, contextual water management gives the best chance of implementation a water project successfully. The Ambition Reflection tool incorporated in Hydropolis makes use of these archetypes and helps the water management to reflect on its attitude (goals, philosophy, perception and position). We do hope that the Urban Dynamics component in the ADSS makes the water manager more context-oriented.

REFERENCES

Geldof, G.D. (2005) *Coping with complexity in integral water management. On the road to Interactive Implementation.* Tauw bv, Deventer.

Horst, M. van der, et.al. (2003) Interactieve uitvoering geeft meer resultaat. [Interactive Implementation gives more result]. Tauw bv, Deventer.

Kalsbeek, L. (1983) De Wijsbegeerte der Wetsidee [The philosophy of Legal Concept]. Buijten & Schipperheijn, Amsterdam (in Dutch).

Lems, P. (2005) Towards a methodology to connect water and society. CPTS: Amersfoort.

Lems, P, Valkman, R. (2003a) New values of water – Theoretical framework. NIDO, Leeuwarden.

Lems, P, Valkman, R. (2003b) New values of water – Handbook for practise. NIDO, Leeuwarden.

Valkman, R. and Lems, P. (2004) Contextual water management. *Proceedings of the Novatech conference,* Lyon, 6-10 June 2004, 8 p.

Vuurboom, M. (2001) De doorwerking van integrale waterplannen, Graduation report, University of Twente, Department of Civil Engineering (in Dutch).

Chapter 7

Assessment criteria for sustainable urban drainage systems

D. Michael Revitt, Lian Scholes and J. Bryan Ellis

INTRODUCTION

The use of source control techniques within urban drainage systems for the treatment of stormwater runoff arising from impermeable surfaces is becoming increasingly popular within Europe. The adoption of such approaches presents particular issues for developers, regulatory agencies and water service companies given the increasing pressure to achieve sustainable drainage solutions. Best Management Practices (BMPs) can offer flow control and pollution removal, as well as the secondary benefits of amenity/ecology improvements. They can also be cost-effective, at least in terms of investment, if not operation and maintenance (O&M), when compared with conventional systems. The decision-making process to select both new and retrofitted urban drainage systems is necessarily complex, involving a variety of stakeholders within public and private sectors, holding differing opinions regarding the significance they attribute to differing technical, environmental, economic and social criteria. The involvement of wider stakeholder groups in sustainable urban land use planning is of particular importance in Europe given the diffuse pollution drivers and requirements for stakeholder engagement set out in the EU Water Framework Directive, WFD (2000). This is further strengthened by changes in planning regulations proposed for some European countries which will push regulators, planning authorities and developers to ensure public participation in the decision-making processes on new urban developments.

A significant challenge to European water companies will be to convince stakeholders, particularly customers and financial regulators, that full-cost water and drainage services will provide clear and sustainable whole-life investment. Decisions in the field of urban stormwater runoff are no longer entirely in the domain of drainage engineers but must represent the needs of all user and interest groups. Identification of appropriate stormwater management options within this dynamic and multi-partner setting can be assisted by the development of robust quantifiable sustainability criteria and indicators which cover a range of scenarios including long term performance, lifetime costing and receiving water impacts. Such an approach was adopted within the DayWater EU 5th Framework Programme project to contribute to the development of a web-based Adaptive Decision Support System (ADSS) (Chapter 2) aimed at assisting end-users to identify possible BMP solutions for urban stormwater management appropriate to their particular development location and community situations. Within this project the identification of appropriate criteria has represented the critical first stage in the development of a matrix-based, multi-criteria analysis approach for the comparison of the benefits and disadvantages of different source control techniques.

A basic structured approach to a multi-criteria, multi-objective methodology for the assessment of water resource systems was proposed within the 1998 UNESCO International Hydrology Programme (American Society of Civil

Engineers and UNESCO, 1998). This study recommended the incorporation of sustainability criteria and guidelines covering economic, ecological, social and environmental factors into a decision-making process relating to the investment in and construction of water resource systems. Within the field of urban water management, several studies have reported on the use of relevant generic sustainability criteria to support stormwater managers in identifying appropriate sustainable urban drainage systems (Crabtree, 2000; Everard and Street, 2001). The breadth and depth of the selected criteria can vary substantially. Thus, Scholz (2005) devised a decision support tool to assist planners in their choice of BMPs based almost entirely on site characteristic criteria, including land costs, whereas Quigley *et al.* (2005) have developed a range of urban runoff management 'objectives' which they identify as being important when considering source control options. The Sustainable Water industry Asset Resource Decisions (SWARD) project represents a European based approach which has been primarily developed for assessing the options available for water supply and sewerage assets (Ashley *et al.*, 2004). The assessment criteria promoted are the economic, environmental, social and technical dimensions of proposed projects and these are discussed in more detail in Chapter 25 where the continuation of this work into the UK based WaND (Water cycle management for New Developments) project is also fully described. In Australia, a methodology based on the 'triple-bottom-line' approach has been employed in which financial, ecological and social criteria are used to help urban stormwater managers bring about improvements to urban waterway health (Taylor, 2005).

In this paper, key criteria which can assist stakeholders in determining the most appropriate stormwater BMP for a specific catchment area are identified. The selected criteria are fully described in terms of how they will support users to select the most appropriate stormwater BMPs in the face of the conflicting demands posed by factors such as system performance, environmental impact, adoption, community benefits and life cycle costs. More detail on the characteristics of individual BMPs can be found in the comprehensive on-line information resource and database which constitutes the DayWater BMP catalogue (Chapter 8). The developed criteria, and associated indicators, form the basis of the performance matrix which has been developed for evaluating and prioritising stormwater source control systems within the DayWater Multi-Criteria Comparator (MCC) described in Chapter 9. In situations where it is particularly important that stakeholders consider how their selection of suitable BMPs impacts on urban social structures, a broader consideration of the urban dynamics may be relevant as described in Chapter 6.

SELECTION OF STORMWATER CONTROL OPTIONS

Identification of criteria

The main factors which influence the identification of the appropriate strategies for the treatment of urban storm drainage are outlined in Figure 1 (Ellis *et al.*, 2004a). Located in the centre of the diagram is a triangle with apices representing the major benefits associated with using BMPs in terms of improving water quality, achieving attenuation of water volumes and flow rates and enhancing the ecology and amenity characteristics of the local environment. At the corners of the surrounding square are four of the most important aspects (technical, environmental, economic and social/urban) which need to be taken into account when comparing possible drainage options with regard to performance, impact, cost and acceptability. Figure 1 also shows the variety of stakeholders who can be involved within the illustrated regulatory and planning frameworks and in the evaluation of the sustainability of different source control techniques (Chapter 5). It is expected that responsible developers will take into account all four indicated criteria whereas for others there will be a clear emphasis with, for example, ecologists, landscape architects and regulatory bodies having a specific interest in the application of the environmental criterion. The general public along with local politicians and community pressure groups would be expected to be more focussed on the social/urban criterion.

The primary components in the structure of a decision-making process must identify all those aspects which can influence the choice of alternatives from all different possible perspectives. The information shown in Figure 1 has commenced this process in the case of BMPs for the control of urban storm drainage by identifying four relevant basic criteria. For the purposes of this paper, criteria are defined as the major established factors on which the final judgement, evaluation or decision is made. Three further criteria were added to facilitate a full assessment of the factors influencing the choice of a BMP. These involve consideration of the site characteristics, operation and maintenance requirements, and legal and urban planning aspects. Site characteristics are necessary to establish any factors which make the installation of a particular BMP inappropriate because of a conflict of requirements. For example, the presence of a high water table can act as an effective barrier to downward percolation (or exfiltration) of stormwater and thus reduce the capability of an infiltration BMP to drain properly. The reduction in the depth of the unsaturated zone as a result of an elevated water table will also reduce the BMP purification capability. In addition, an elevated water table could leak groundwater into retention/detention basins and swales. The infiltration process is critically dependent on the hydraulic conductivity (or the infiltration rate) of the prevalent soil with values of 10^{-6} cm/sec (or 5 mm/hr) being advocated as threshold levels below which the use of BMP infiltration devices are not recommended. The types of soils not supporting infiltration are those predominantly based on silt or clay characteristics. A third important site property influencing the feasibility of 'stand-alone' BMP options is the proportion of catchment area which contributes to surface runoff during a storm event. Detention basins and constructed wetlands require a significant drainage area (normally greater than 8 hectares) to ensure an effective operation within a typical urban development and to counter drying-out and wilting effects caused by prolonged dry periods. A lower limit of suitability (6 to 7 hectares) is set by the orifice size for extended detention basins and a similar limit might be adopted to maintain water level capacity in retention basins and wetlands. By contrast, individual infiltration and grass filter facilities are generally only applicable on sites smaller than 2 to 3 hectares due to flow velocity constraints (as well as space and cost) and their low storage volume/storm runoff (Vs/Rv) ratios.

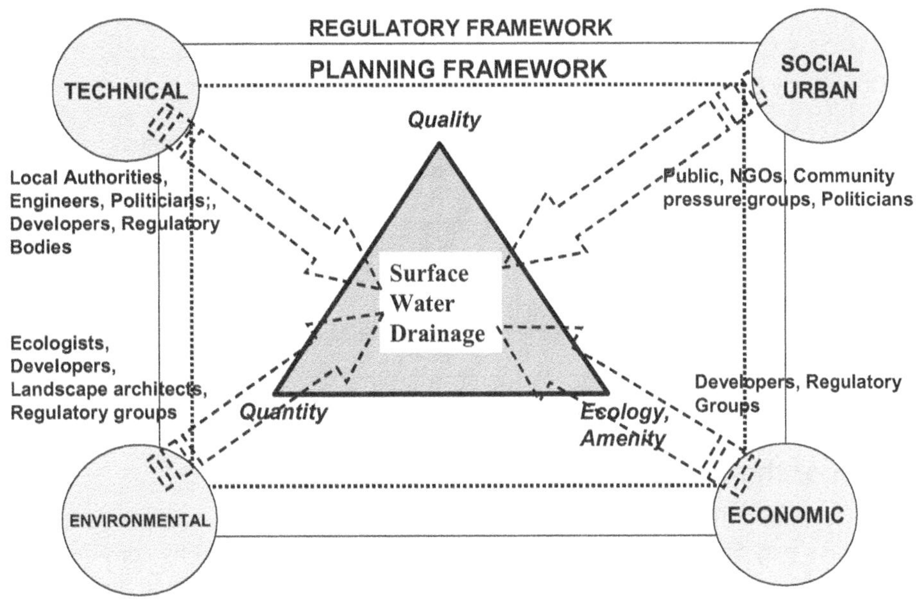

Figure 1: Diagrammatic representation of the interactions between source control techniques for surface water drainage and the relevant stakeholders.

Identification of indicators and benchmarks

The six identified criteria, excluding site characteristics (as discussed earlier), are shown in Table 1. Each of these criteria has the flexibility to be adapted to meet changing circumstances and constraints within differing organisations, regulations and customers. Each criterion can be sub-divided into indicators, which represent the diagnostic states or conditions required to describe the relevant and appropriate properties of the established criteria. The choice of indicators has been derived following comprehensive discussion with a variety of users and the pragmatic need to satisfy a range of concerns across Europe. However, it is recognised that, in some specific cases, alternative and/or additional indicators might be equally appropriate depending on local conditions and circumstances. This situation is addressed within the DayWater ADSS through the provision of a Matrix-of-Alternatives approach (Chapter 4) which supports users in evaluating alternative drainage scenarios. It is also possible to have more detailed descriptions of the selected indicators through the use of 'primary' and 'secondary' categories.

To further characterise the allocated indicators, each of these has been benchmarked and a threshold value or condition (qualitative or quantitative) identified which allows the criteria to be compared against national sustainability targets (DETR, 2001a, b; Water UK, 2000), or other standards, in a consistent, impartial and transparent manner. This process is illustrated in Tables 2 to 7 which show a range of possible benchmarks that are proposed to provide points-of-reference for the criteria and their associated indicators and to support robust, auditable decision-making. In principle it is possible to have a wider range of indicators and benchmarks for all the criteria but this could prove discouraging to the user as well as lead to an overlap of the influence of each variable across the identified criteria thus negating the validity of using a multi-criteria analysis approach. The mix of objective (quantifiable) and subjective (qualitative) information is seen as an advantage as the use of subjective input can demonstrate the flexibility of the process to users.

Table 1: Criteria and indicators relevant to the selection of BMPs for source control of urban drainage.

Criteria	Indicators
Technical	Storage and flood control Pollution control System adaptability to urban growth
Environmental	Receiving water volume impact Receiving water quality impact Receiving water ecological impact
Operation and Maintenance	Maintenance and servicing requirements System reliability and durability
Social and Urban Community Benefits	Public heath and safety risks Sustainable development Public/community information and awareness Amenity and aesthetics
Economic	Life cycle costs Long term affordability
Legal and Urban Planning	Adoption status Building development issues and stormwater regulations

Short description of the Technical criterion

The contributions of the three different indicators to the Technical criterion are considered independently but the impacts of the benchmarks may be treated in different ways. This can be illustrated by considering the four different benchmarks which could be used to assess BMP performance in relation to the storage and flood control indicator (Table 2). To determine the relevance of this indicator to the choice of a particular BMP, it is possible to amalgamate the impact of all four benchmarks or to use a specific benchmark. For example, peak runoff control has been selected for use in the DayWater project as it relates to a major objective of drainage design which is to retain the pre-development hydrograph or flow regime of a site, using appropriate BMPs such as those possessing infiltration capability or storage volume potential. In contrast, for the pollution control indicator it is not realistically possible to separate the dissolved and solids associated pollutant capture benchmarks as both are equally important in assessing the performance of a BMP. Therefore, a combined approach needs to be adopted in which removal is considered for specific pollutants taking into account their physico-chemical characteristics (e.g. solubility, adsorption tendency) and the relative importance of the different removal processes within the different BMPs for particulate and dissolved pollutants (Scholes *et al.*, 2007; Ellis *et al.*, 2005). The system adaptability indicator relates to the viability of retrofitting a BMP into an existing urban situation where flooding and/or pollution problems are occurring as a result of the expansion or intensification of land use activities within an urban catchment. A relevant benchmark would relate to the hydraulic robustness of performance following increased impermeability within a catchment due to expansion or change in land use activities. As demonstrated by Bardin *et al.* (2002), such increased urbanisation would modify the runoff coefficient (R_c) as a result of changes in the surface impermeability factor. In this context, adaptability can be represented by the change in ultimate treatable volume (V_t) arising from the ratio of the increase in impermeability to the ultimate or final R_c value at which overflow from the system would occur. Further details on the identified indicators and benchmarks can be directly accessed by clicking on the appropriate indicator in the MCC component of the DayWater ADSS (Chapter 9).

Table 2: Indicators and benchmarking within the Technical criterion.

Indicators	Benchmarks	Units
Storage and flood control	Overflow frequency	1...n
	Design storm return interval	RI, years
	Peak runoff and storage volume	volume (m^3); (effective contributing area, ha)
	Extreme event control	H/M/L
Pollution control: water quality improvement	Dissolved pollutant capture	%; H/M/L
	Solid(s) pollutant capture	%; H/M/L
System adaptability to urban growth	Ease of retrofitting	H/M/L
	Design freeboard	% ; volume, m^3
	Changes in impermeability/runoff coefficient	% ; fractional value

Key: RI = return interval; H/M/L = high/medium/low.

Detailed consideration of the Environmental criterion

The three indicators, all relating to receiving water impacts, and the associated benchmarks allocated to this criterion are identified in Table 3. The preferred benchmarks for assessing the impact of using different BMPs to control urban runoff, together with the treatment approaches involved are discussed in the following sections.

Table 3: Indicators and benchmarking within the Environmental criterion.

Indicators	Benchmarks	Units
Impact on receiving water volume	Draw-down times	(hours)
	Downstream erosion	H/M/L; RI (year)
	Groundwater recharge/low frequency maintenance	Hydraulic conductivity (cm/s)
	Thermal impacts	Temperature change (°C)
Impact on receiving water quality	Threshold concentrations for general water determinands	mg/l
	Compliance with EQS values for priority pollutants	mg/l
	Sediment metal toxicity levels	mg/kg
Impact on receiving water ecology	Biotic diversity score relative to water use	Very good/good/fairly good/fair/poor/bad

Key: RI = return interval; H/M/L = high/medium/low.

Receiving water volume impact

The major environmental considerations selected as benchmarks to describe how the impact of installing a BMP affects receiving water volume are the potential for downstream channel and bank erosion, the thermal impacts of discharges and the potential for groundwater recharge and maintenance of minimum flows. Those BMPs capable of controlling peak discharges for the 0.5 to 2 year return storm event will provide some degree of downstream bank erosion control by reducing the occurrence of channel-forming bankfull discharge. Extended detention basins and some infiltration BMPs, if properly sized, can effectively reduce the frequency with which such high frequency bankfull flooding occurs and thus have potential for downstream erosion control. Retention and wetland basins however, have a lower capability in this regard.

Aquatic organisms are susceptible to thermal impacts due to having characteristic temperature preferences and tolerance limits with game fish, for example, requiring a maximum water temperature of 13°C for egg incubation and spawning whereas higher sustained temperatures, in excess of 21°C, are either extremely stressful or lethal. A rise in temperature of only a few degrees over ambient conditions can reduce or eliminate sensitive receiving water species such as stoneflies and mayflies. In prolonged summer periods, urban runoff can warm receiving waters to the detriment of these organisms not only in terms of direct thermal impact but also because warm discharge waters can reduce the available dissolved oxygen. Average stream temperatures are strongly influenced by the impervious area percentage and urban BMPs can exacerbate this thermal effect when they retain water for an extended time in hot weather. Temperature stratification is possible in water bodies as shallow as 0.5 m and runoff leaving shallow wetlands and wet retention ponds can be 5 to 10°C warmer than the runoff entering the BMP structure. The release of impounded waters, which have been heated during retention between storm events, can therefore be detrimental to the receiving water ecology.

The recharge of groundwater and hence a contribution to the long term maintenance of minimum receiving water flows is feasible for infiltration BMPs. In contrast, retention and wetland facilities are frequently lined or have compacted beds which prevent seepage to groundwater. However, it should be noted that whilst the volume effect of infiltration percolation is beneficial to the groundwater resource, the quality impacts may be detrimental.

In determining the comparative benefits to BMP volume controls which can be gained from the three potential benchmarks, it is necessary to decide whether they merit equal consideration or if an appropriate weighting should be applied. A careful assessment of each of the contributing factors suggests that downstream erosion will have a much greater environmental impact than would be caused by thermal changes. The impacts on groundwater are likely to represent an intermediate situation and therefore a weighting distribution of 3:2:1 for the downstream erosion potential, minimisation of groundwater recharge and thermal impact is proposed. Although this may be appropriate for temperate climates, change to the applied weighting would be required in order to take account of local circumstances such as the higher ambient temperature conditions encounter within southern Europe.

Receiving water quality impact

Where the physical structure of a watercourse has been irrevocably and heavily modified, as may well be the case for many urban receiving waters, the EU WFD (2000) requires that further deterioration must be prevented and objectives set for the achievement of "good chemical and ecological potential". Chemical status can be related to Environmental Quality Standards (EQSs) for a range of water pollutants chosen on the basis of the risk that they present to the aquatic environment. These standards typically incorporate general physico-chemical parameters (e.g. dissolved oxygen (DO), biochemical oxygen demand (BOD) and total suspended solids (TSS)) as well as substances which are increasingly toxic to aquatic species (e.g. ammonia). Chemicals identified as presenting a significant risk to or via the aquatic environment are specified in Annex 10 of the WFD (2000) as Priority Pollutants (PPs) for which EQSs will be set according to EU-wide protocols. At the present time there are no equivalent standards relating to the concentration of PPs in biota or sediment although the WFD does contain provision for their future introduction. Eriksson *et al.* (2007) have recently proposed a list of 25 pollutants which are identified as selected stormwater priority pollutants.

Under the new Groundwater Directive proposed in relation to Article 17 of the WFD, the locations and conditions under which direct discharges to ground are allowed are either restricted or prohibited. The direct or indirect discharge of diffuse urban surface runoff which is known to contain toxic substances, and particularly species in a soluble form, to ground via soakaways, infiltration trenches/basins etc, can present a potential threat to underlying ground waters. The transfer to groundwater of toxic species weakly adsorbed to organic colloids can lead to their release and mobilisation on encountering anaerobic conditions.

TSS, DO, BOD and ammonia are prime water quality determinands and are widely recognized indicators of pollution that can be applied to all watercourses including urban receiving waters. In the UK, the Chemical General Quality Assessment (GQA) scheme (Environment Agency, 1998) utilises these parameters to establish receiving water quality class grades in which threshold concentrations are set for categories varying from very good to bad. These classifications can be matched to the available data on BMP effluent pollutant concentrations (e.g. US EPA National BMP Database, 2004) to evaluate the exceedance potentials of individual BMPs in respect of the indicated benchmark threshold values for each receiving water grade. This is feasible for BOD and TSS for which monitoring data is relatively abundant but is not the case for DO although the limited available data suggest that surface outflows to receiving waters are unlikely to cause significant oxygen depletion.

Enhanced effluent BOD levels, compared to other BMPs, in the effluent from constructed wetlands has been associated with the mobilization and release of oxygen-demanding materials e.g. Fe^{2+}. Elevated BOD levels can also occur in infiltration facilities such as infiltration basins and soakaways as the downward-percolating water passes through the reducing sub-surface conditions of the benthal sediment and underlying soils. Such reducing conditions are not evident for porous paving (with reservoir structures) which appear to maintain a favourable oxygen balance to the receiving surface or ground waters. The situation is somewhat reversed for TSS impacts where below-ground BMPs e.g. infiltration basins, soakaways, are likely to provide the best receiving water protection. Wetlands, swales and retention basins are normally equally efficient although they occasionally exhibit elevated TSS concentrations due to overturning and release of suspended solids from the BMP facility. Whilst some BMPs behave comparatively poorly with regard to both BOD and TSS exceedances e.g. porous asphalt and settlement tanks, other BMPs, such as swales, exhibit different exceedance potentials for the two pollutant groups. In such instances, the emphasis should be on the "worst-case" condition as it is this lower performing parameter that is most likely to prejudice receiving water quality.

Receiving water ecological impact

Natural ecological variability does not allow absolute biological standards to be established for implementation across the EU. The WFD proposes that biological quality (or ecological status) be judged on the basis of the degree of deviation of the observed condition from that which would be expected in the absence of significant anthropogenic influence. The biological elements for most water bodies will include aquatic flora, macro invertebrates and fish. In the UK, a biological General Quality Assessment (GQA) scheme for receiving waters based on a prediction and classification system for macro invertebrates (Wright, 2000) is used to correlate ecological status with associated water use. The classification grades (varying from very good to bad) and associated descriptions are consonant with those included in the WFD guidelines.

Benthic macro invertebrate communities can be substantially altered in many ways as a result of the impact of acute polluted discharges as well as the lower level intermittent events which are characteristic of urban surface runoff. They are also vulnerable to the *in-situ* chronic accumulation of sediment toxicity due to the gradual release of bioavailable contaminants. The sedentary nature, ubiquity, responsiveness to disturbance, ease of sampling and importance to other ecosystem components make the monitoring of benthic macro invertebrate communities highly relevant to aquatic systems. Decreases in taxa richness and total abundance of macro invertebrate taxa provide clear indications of detrimental effects and therefore are ideal benchmarks for the receiving water ecology indicator. Biotic indices provide a simple and understandable presentation of biological data for management purposes and are the foundation of European ecological indices. In a polluted stream, the more sensitive Plecoptera, Ephemeroptera, Trichoptera and Gammarid species will be replaced by Asellus, Chironomid and Tubificid worms. The biology associated with the highest biotic scores ($\geq$8) will yield a high diversity of families, usually with several species in each, whilst the lowest biotic scores ($\leq$3) are defined by a small number of very pollution tolerant families (e.g. worms, midge larvae, leeches). These may be present in high numbers although the presence of a toxic sediment might substantially reduce the total numbers present. This biotic index approach is most suitable for organic pollution as toxicity arising from metals can complicate the biotic relationship. Nevertheless, it has been shown that an ecological approach based on biotic diversity can be successfully applied to urban receiving waters which are subject to a combination of toxic, chemical and physical impacts as well as loss of habitat (Bascombe *et al.*, 1988).

Individual BMPs (and BMP combinations) introduced into a single development site are unlikely to offer a substantial improvement in receiving surface water ecology, if the remainder of the urbanised catchment (or the upstream catchment) has no extensive BMP source controls and possesses imperviousness levels above 20 – 25%. Quantification of cumulative BMP performance can be obscured by benefits arising from other upstream sources such as riparian buffers, intact forest, wetlands etc. Constructed wetlands score highly using this benchmark as they provide compensatory habitat for the normal paucity and deficiencies associated with urban watercourses. Such wetland habitats provide additional opportunity for emergence of the adult terrestrial stages of organisms such as the caddis fly and dragon fly. Thus wetlands enable a variety of aquatic species to complete their life cycle and enhance the adjacent receiving water habitat. Such species will also have high aesthetic visibility and community value. A retention basin with vegetated margins might approach constructed wetlands in terms of ecological efficiency whilst detention basins having no permanent water pool, would exhibit a lower efficiency. Infiltration basins and swales have more limited ecological opportunities for receiving water habitats whilst porous paving/asphalt systems, filter drains, roof storage, settlement tanks and soakaways offer few if any direct ecological benefits.

Overview of remaining identified criteria and indicators

In addition to the 'Technical' and 'Environmental' criteria, four further criteria with their supporting indicators have been identified as contributing to the decision-making process for the selection of BMPs (Table 1). In this section, each of these is discussed briefly with an emphasis on the benchmarks specifically chosen as being of prime importance in influencing the choice of BMP. For each of the criteria, more detailed information explaining the development of the associated indicators and benchmarks can be obtained by accessing the MCC part of the ADSS on the DayWater web-site (www.daywater.cz ; also see Chapter 9).

Operation and maintenance criterion

The identified indicators for this criterion are 'System reliability and durability' and 'Maintenance and servicing requirements' (Table 4).

Table 4: Indicators and benchmarking within the Operation and Maintenance criterion.

Indicators	Benchmarks	Units
System reliability and durability	Risk of system failure e.g. clogging in infiltration systems	% probability
	Consequences of storm event exceeding design storm RI	Flooding depth (m)
	Climate change scenario impacts	H/M/L
	Hydraulic retention time	hours
	In-basin quality condition and health hazards	Trophic state; smell; stagnant water; bacteriology etc…
	Alarm/intervention procedures	yes/no
	Safety level/provision for accidental pollution etc	H/M/L
	Number of in-basin/receiving water pollution complaints	Number/year
Maintenance & servicing requirements	Need and frequency for O & M servicing to maintain: technical/environmental/amenity/habitat objectives	H/M/L; frequency/year

Key: RI = return interval; H/M/L = high/medium/low.

Maintenance and servicing requirements

'Operation and maintenance (O&M) servicing' has been identified as the critical benchmark which strongly influences the ability of BMPs to function at their optimum design efficiency throughout their operational lifetimes. Regular and effective O&M strategies, linked to a clear line of responsibility, ensuring their continued operational benefits to the adopting authorities, the receiving environment and/or the local community are often lacking, and relatively little guidance is available. The flood, pollution, amenity and ecological component functions of BMPs need to have O&M servicing that minimise performance risks in terms of:

- regular and effective maintenance of BMP ancillary structures (inlets, outlets, silt traps, flow controls and storage structures, headwalls, low flow channels, bypass etc...),
- management of litter, sediment, vegetation and herbicide/pesticide applications etc…
- protection and exploitation of amenity, recreational and habitat benefits with associated landscape management.

An O&M management plan should specify schedules of work relating to both regular and occasional maintenance together with monitoring activities linked to potential remedial requirements. It is possible to assign qualitative values to each of these O&M activities although it is difficult to foresee the nature, type and extent of remedial maintenance which will be required during a BMP lifetime, as this will depend on factors such as vandalism, extreme storm event damage, inadequate design etc… Given the fundamental importance of regular O&M activities, it is considered that this component should be more heavily weighted than the occasional O&M and monitoring components, and a 3:2:1 weighting ratio has been adopted in the DayWater project to assess the relevance to specific BMPs.

System reliability and durability

The long-term reliability and durability of BMPs can be evaluated against a range of possible benchmarks, each of which can provide important information with regard to identifying the most appropriate BMP for use within a particular catchment area. However, several of these overlap with other criteria and indicators previously considered, such as 'adaptability to urban growth' which has been considered within the 'Technical' criterion, and their use within an MCC methodology should not be duplicated. The current range of identified indicators do not consider how the onset of global warming, with its implications for water quantity and water quality, could influence the reliability and durability of BMPs and therefore this aspect has been selected as an appropriate benchmark for evaluation under the system reliability and durability indicator. Predicted climate changes include reductions in soil moisture content by up to 40% for the UK by 2080 (UK Foresight Project, 2003) and the resulting drier and more compacted soils during the summer months would favour rapid runoff during convective storms. Combined with the predicted higher storm intensities this would lead to the need for BMP systems to be able to accommodate higher runoff volumes within short delivery times and would render infiltration – based treatment systems susceptible to clogging by the higher sediment loads carried by the intense summer storms. There is also the probability of an increased clustering of storm events introducing the need for BMPs to be able to quickly treat large volumes of received water. The extreme situation is that BMPs would regularly need to cope with superimposed storm events requiring shorter drawdown times and hence reduced retention times for high magnitude events resulting in lower treatment efficiencies. In the UK, there are expected to be longer wetter periods during the winter months leading to a reduced soil capacity for infiltration although this will be for a shorter winter period. Hence BMPs will progressively be required to be able to deal more efficiently with more extreme runoff events ranging from the rapid delivery of large runoff volumes (initially with high sediment and associated pollutant loads) in summer to the reduced ability of infiltration systems to function effectively in winter.

These predictions have been used to assess the reliability and durability of BMPs particularly with respect to the anticipated changing patterns of runoff volume and quality.

Social and urban community benefits

Four indicators, which are closely linked to this criterion, have been identified in Table 1 and are further described in Table 5 and the subsequent discussion below.

Table 5: Indicators and benchmarking within the Social and Urban Community Benefits criterion.

Indicators	Benchmarks	Units
Public health and safety risks	Local community concerns (injury, infection, drowning etc)	% user survey
	Formal technical risk exposure audit (flood risk, health risk, safety risk)	H/M/L
Sustainable development potential	Contribution to urban sustainable development policies	H/M/L
	Role in Agenda 21	H/M/L
	Role in Biological Action Plans (BAPs)	H/M/L
	Additional benefits offered by different BMPs	yes/no
	Material use: aggregate/concrete/top-soil use and costs	H/M/L; Euro
	Energy use: construction, operation and maintenance energy consumption	kW; kW/m^3 storage
Public/community information and awareness	Interpretation boards, signage, brochures/literature, visitors centre etc	H/M/L
	Awareness in local/regional community	% awareness survey
Amenity and Aesthetics	Level of amenity and recreation provision	H/M/L
	Aesthetic contribution to the local environment	H/M/L
	Social inclusion and multi-functional use	H/M/L

Key: H/M/L = high/medium/low.

Public health and safety risks

Consideration of the potential level of risk presented by individual BMPs during their operational lifetimes can be used to benchmark this indicator based on the five threshold conditions listed below:

- the presence of an open water body in or adjacent to urban developments can potentially provide a safety hazard, especially for small children,
- malarial vectors may be encouraged by shallow, vegetated standing waters particularly during hot, humid weather conditions,
- bacterial contamination from bird and pet excreta and vermin infestation where high discharges of litter and gross solids occur,
- flooding of impermeable surfaces during wet weather events leading to possible aquaplaning with freezing conditions causing ice hazards for both motorists and pedestrians,
- surface heave and obstructions caused by surface water draining over paved areas leading to frost heave during winter with prolonged dry periods causing surface cracking and subsidence.

By treating each of these five threshold conditions as being equivalent in terms of the potential level of risk, a combined assessment for each BMP can be obtained. As expected, those systems possessing a permanent water pool pose the highest health and safety risks, whereas porous surfacing and infiltration BMPs present least likelihood of risk.

Sustainable development potential

This indicator can be benchmarked using a screening approach associated with three threshold conditions relating to biodiversity potential, community participation potential and energy/resource consumption. The relevance of each of these to the evaluation of a BMP is usefully based on their compatibility with, and support for, the guiding principles and attributes contained in the EU Sustainable Development Strategy (e.g. HMSO, 2005). This strategy, adopted in June 2001, is intended to reconcile environmental protection with social cohesion/equity, and economic development with the protection and enhancement of natural resources and the environment. BMPs which enhance ecological habitats and control hydrological regimes will maximise the biodiversity potential benchmark. Thus, wetlands score very highly compared to infiltration facilities although trenches and basins do have biodiversity potential by virtue of reducing discharge and pollutant impacts on receiving watercourse habitats. BMPs which are likely to require or support involvement from a range of organisations and which benefit from the fostering of long term partnerships will contribute significantly to the community participation benchmark. Ponds and basins possess high potential for community participation and partnerships but most other BMPs have relatively little to offer as far as this social community benefit benchmark is concerned.

Energy and resource use are prime sustainable development benchmarks which seek to minimise the future impact of environmental development as well as to reduce or prevent the impact of development upon additional environmental compartments. BMPs which utilise minimum energy and material consumption in their implementation, operation and maintenance will have high sustainable development potentials and are unlikely to exceed the "carrying-capacity" of the local environment. BMPs requiring heavy material use during construction as well as material transport represent the reverse situation, particularly if there is a continued energy demand during their O&M lifetimes. However, it is important to recognise that infiltration facilities have potential for resource re-cycling through groundwater recharge whilst surface water BMPs can also offer some potential for additional use as irrigation water within the local community if gravity-fed systems can be utilized.

Public/community information and awareness

An assessment of a BMP in terms of public perception and acceptability as primarily judged by aesthetic appeal and potential for enhancing community participation and environmental education awareness will contribute significantly to this indicator. Factors which contribute towards the public perception potential benchmark include perceived aesthetic enhancement, elevated property values, provision of educational opportunities and existence of "green" and "blue" space for community involvement. BMP facilities having a permanent pool together with enhanced habitat provision (wetlands, extended detention basins, retention basins) clearly present the highest potential in terms of public perception and acceptability as they can be aesthetically pleasing and offer a variety of amenity and recreational opportunities in addition to their primary flood and/or pollution control functions. Detention basins and lagoons offer more limited opportunities but do have the advantage of green open space for games, informal leisure and pet-walking activities with the latter having the advantage of an open water body. Filter strips, swales and porous surfacing have less potential and without the correct maintenance can be aesthetically non-pleasing whilst the public perception and awareness potential of below-ground BMP facilities is inevitably much more restricted and requires major educational effort to enhance and maintain local community awareness of recharge and harvesting potential. Green roofs may enhance the 'rural' feel of a development and thus have potential for increasing local perception.

Amenity and aesthetics

This indicator can be benchmarked in terms of a general quality assessment of BMPs based on the integration of their aesthetic, amenity and recreational opportunity potentials. Evaluation of the aesthetic condition can be primarily based on the occurrence of litter (gross litter, general litter, dog faeces and sewage litter), oil scums, foams and fungus as well as prevailing colour and odour. The occurrence of sewage litter and odour are the most frequent sources of public complaint. Amenity and recreational potential can be evaluated in terms of the site and landscape opportunity potential for a range of community activities and the provision of adequate access and management facilities. This potential takes into consideration both aquatic and terrestrial ecosystems as well as the aesthetic condition of the drainage system and includes both active and passive forms of recreational opportunities. A well designed, maintained and managed constructed wetland has the potential to be of high ecological and chemical status, aesthetically pleasing and offering a considerable variety of both amenity and recreational potentials. However, many constructed wetlands are subject to oil sheens and some litter, with the vegetation (particularly *Typha* species) during die-back periods being aesthetically unattractive.

Economic costs

The two indicators identified as being important to this criterion relate to 'Life cycle costs' and 'Long term affordability' respectively (Table 6).

Table 6: Indicators and benchmarking within the Economic Cost criterion.

indicators	Benchmarks	Units
Life cycle costs	Design and capital costs	Euro
	Operational & maintenance costs	Euro/year
	Sediment disposal costs	Euro/year (or lifetime)
	Community costs including stormwater fees	Increase/decrease
	Refurbishment and replacement costs	Euro/year
Long term affordability	Adoption and financial viability	H/M/L
	Long term insurability	H/M/L
	Economic add-on value (enhanced land/property values)	Euro/ha
	Amenity income streams	Euro/year
	Long term management provision	H/M/L

Key: H/M/L = high/medium/low.

Life cycle costs

The quantification of BMP lifecycle costs needs to be based not only on a consideration of design, project management, capital and operation and maintenance (O&M) costs but also on a range of other costs which include those relating to replacement, risk and disposal charges. Set against these should be the benefits arising from environmental/amenity and residual values. However, most of the costs and benefits vary according to local circumstances and national construction costs and this needs to be remembered when developing a lifecycle costing approach. Benchmark quantification for the life cycle costs indicator has therefore been based on a combined consideration of design, project management, capital and O&M costs using financial data from Greece, France and the UK. The costing approach used for individual BMPs is described in more detail in Chapter 10. For comparative purposes, all BMP capital and O&M costs within the lifecycle benchmark have been converted to a 50 year lifetime equivalent and a standard 6% discount rate applied with an initial design and project management overhead of 25% being assumed. No allowance has been made for land-take costs, risk, and residual value or for off-set environmental benefits. Costs have been determined in terms of the unit cost per storage volume provided (€/m^3) and the BMP lifetime cost distributions have been determined to illustrate the effect of local circumstances and national costing regimes.

Long term affordability

The long term viability of a BMP can be considered in terms of economic affordability and insurability, although lifetime affordability is probably best achieved by managing different storage volumes at different spatial locations within an urban development and by the integration of conventional and alternative source control drainage systems. However, the inclusion of source control BMPs as an integral element of the core building infrastructure requires much greater attention to flood proofing e.g. raising floor and service outlet levels, use of water resistant foundation bricks, wet proofing of electrical cables and outlets etc. Long term affordability must also consider insurance indemnity with flood risks exceeding the drainage system capacity being increasingly liable to excess premiums or even to non-insurance. Hence, the benchmark for long term affordability needs to be based on a consideration of the flood risks arising from future climate change in terms of the likelihood of increased runoff volumes and rising groundwater levels combined with the long term potential for insurance liabilities posed by individual types of BMPs. Few BMPs apart from porous surfaces, filter strips, swales and green roofs, rate highly on the affordability indicator.

Legal and urban planning

The two indicators identified in Table 7 for this criterion are 'Adoption status' and 'Building development issues and stormwater regulations'.

Adoption status

A major barrier to the introduction of BMP drainage systems is the unwillingness of authorities to formally adopt such source control approaches and the perceived risks associated with their long term funding as well as operation and maintenance. In the ideal situation, legally binding obligations in respect of the adoption of BMPs should be in place which have been negotiated and agreed formally between the relevant stakeholders e.g. local or highway authority, developer, water and/or sewerage utilities, environmental regulatory agency, private land owner(s) etc. For an adopting authority, a major consideration is the potential and liability of 'failure' which can be interpreted in the context of BMP adoption as the probability that a particular drainage component will not perform as intended with respect to a range of engineering, technical, hydrological, economic, legal and social aspects. A risk assessment approach to the issue of adoption requires that the potential for failure is identified and that this probability of failure is assessed separately from consequence. A consequence of failure may be flooding or non-achievement of a performance objective such as the additional amenity or recreation value provided by a wetland. The consequences of failure are likely to be much greater in a BMP which has multi-functional uses e.g. a retention basin compared to porous paving. The assessment of probability of failure can be determined through modelling, historic evidence or by expert judgement. Sand filters have a high probability of failure in comparison to infiltration trenches which in turn have a higher probability of failure than a detention basin. Risk can be assessed by constructing a matrix to combine probability with consequence and this provides the basis of a benchmark approach for BMP adoption.

Table 7: Indicators and benchmarking within the Legal and Urban Planning criterion.

Indicators	Benchmarks	Units
Adoption status	Balance of level of acceptability against liability of risk.	H/M/L
	Legally binding contract	yes/no
	Health and safety	yes/no
	Operation and maintenance	yes/no
	Legislation to enforce use of BMPs on private property	yes/no
Building development issues and stormwater regulations	Compatibility with planning and development requirements	H/M/L
	Achievement of WFD regulations	H/M/L
	Housing/urban density	H/M/L
	Design flexibility and structural safety of buildings	H/M/L

Key: H/M/L = high/medium/low.

Building development issues and stormwater regulations

A benchmark relating to 'compatibility with planning and development requirements and stormwater regulations' has been developed in the light of key issues posed by current and proposed urban land use planning and legislative pressures arising from major EU Directives and national strategic environmental policy such as stormwater taxes, sustainable drainage requirements etc. The benchmarking procedure applied in this study involves an evaluation of three discrete parameters relating to increased housing/urban densities, use of BMPs in relation to the structural safety of buildings and the potential of BMPs to contribute to EU WFD regulations. Increasing housing/urban densities, particularly around areas having good transport and utility accessibility and infrastructure, is seen as a way to contribute to a variety of sustainability initiatives and policies such as sustaining local services and public transport systems, leading to a reduction in social exclusion and enhancing quality-of-life indices. For example in the UK, housing densities of less than 20 homes/hectare are now seen as unsustainable, with 25-50 dwellings/hectare currently being recommended in official planning guidance documents (Planning Policy Statement (PPS) 3, 2005). The evaluation of BMPs with respect to increasing housing/urban densities is based on their ability to be incorporated into high-density design approaches such as those proposed under 'Low Impact Development' (USA) (Coffman, 2000) and 'Sustainable Development Design' (UK) (Ellis *et al.*, 2004b) scenarios. Ensuring the long-term structural stability and safety of buildings is clearly a fundamental requirement under any development management plan. However, the potential for infiltration, particularly when concentrated in close proximity to buildings, to adversely affect ground stability is an issue that has been raised recently (e.g. in the UK through Planning Policy Statement (PPS) 25 (2005). In addition, hydraulic loading and waterproofing is a key major issue for large scale green roof developments. Implementation of the EU WFD (2000) will have a major impact on urban water management practices in urban areas with respect to water quality, quantity and socio-amenity targets. With regard to the anticipated introduction of and compliance with a range of new and modified legislation and regulations in the urban planning arena, the potential for BMPs to make a contribution to WFD requirements has been evaluated with respect to their ease of application/installation i.e. the need to undertake compliance assessments and/or obtain planning permission.

The overall analysis suggests that BMPs such as filter drains and filter strips, together with permeable surfacing materials (porous asphalt and porous paving), have the greatest overall potential to jointly contribute to meeting increasing housing density, building structural safety and WFD requirements. Filter drains score comparably well against all 3 requirements, whereas both permeable surfacing materials score highly in relation to high housing density requirements (great potential for multi-functional use in paving and car parking areas etc.) but less for structural building safety (infiltration systems located close to buildings may impact on foundation stability) and WFD requirements (relatively complex installation procedure). At the other end of the scale, BMPs such as retention ponds and detention basins have comparatively less potential to meet typical local planning requirements as their larger size limits their ability to fit into highly urbanised environments as well as leading to the need to obtain planning permission and/or demonstrate compliance with health and safety legislation.

CONCLUSION

As part of a broader move to enhance the quality-of-life of local communities, an increasing number of EU Directives require that regulatory authorities fully engage stakeholders within all decision-making processes. This offers a particular challenge in urban areas where there can be a diversity of stakeholders with a wide variety of backgrounds, concerns and priorities. The EU WFD prioritises the control of diffuse pollution and therefore requires the development of approaches which can inform and support discussions involving both specialists and non-specialists in identifying appropriate stormwater management options. This paper has contributed to the existing 'information gap' by identifying key sustainability, technological and socio-economic issues (i.e. criteria) against which various BMPs may be evaluated and, through the use of indicators and benchmark thresholds/values, provides guidance on how BMPs may be assessed for their suitability in a particular location. The identification of the described criteria together with the supporting information and guidance have made a significant contribution to the development of the MCC (Chapter 9) in the DayWater ADSS, which provides a consistent and transparent process for assisting the wide profile of relevant stakeholders in their selection of appropriate BMPs for a particular stormwater treatment option.

REFERENCES

American Society of Civil Engineers and UNESCO (1998) *Sustainability criteria for water resource systems* Report of Joint ASCE/UNESCO Task Committee on Sustainability Criteria, Project M-4.3, IHP IV. American Society of Civil Engineers, Reston, VA, USA.

Ashley, R., Blackwood, D., Butler, D. and Jowitt, P. (2004) *Sustainable water services: a procedural guide.* IWPA Publishing, London, UK.

Bardin, J-P., Barraud, S., Alfakih, E and Dechesne, M. (2002) Performance assessment of stormwater infiltration strategies; A multi-indicator approach. In *Proceedings 9th International Urban Drainage Conference* (ed. E.W. Strecker and W.C. Huber), Portland, Oregon. US. American Society of Civil Engineers, Reston, Va. USA. ISBN 0 7844 0644 8.

Bascombe, A.D., Ellis, J.B., Revitt, D.M. and Shutes, R.B.E. (1988) The role of invertebrate biomonitoring for water quality management within urban catchments. In *Proceedings of International Symposium on Hydrological Processes and Water Management in Urban Areas* (ed. J.C. Hooghart). IHP/UNESCO, The Netherlands, 403-412.

Coffman, L. (2000) Low impact development design strategies: An integrated design approach. *Report EPA 84LB-00-003.* Department of Environmental Resources, Planning Division, Prince George's County, Maryland, USA.

Crabtree, R.W. (2000) A UK perspective on institutional constraints limiting advances in stormwater management. *Paper given at NATO Workshop*, St, Mariethal, Germany,.

DETR. (2001a) *Quality of life counts: indicators for a strategy for sustainable development for the UK.* Department of the Environment, Transport and the Regions, HMSO, London.

DETR. (2001b) *Achieving a better quality of life: review of progress towards sustainable development.* Department of the Environment,Transport and the Regions, HMSO, London.

Ellis, J. B., Deutsch, J-C., Mouchel, J-M., Scholes, L. and Revitt, D. M. (2004a) Multicriteria decision approaches to support sustainable drainage options for the treatment of highway and urban runoff. *Sci. Tot. Environ.*, **334-335**, 251-260.

Ellis, J.B., Scholes, L., Revitt, D.M. and Oldham, J. (2004b) Sustainable approaches for urban development and drainage in the twenty-first century. *Proceedings of the Institution of Civil Engineers, Municipal Engineer*, **157**, 245-250.

Ellis, J. B., Deutsch J-C., Legret M., Martin, C., Revitt, D. M., Scholes, L., Seiker, H. and Zimmerman, U. (2005) The DayWater Decision Support Approach to the Selection of Sustainable Drainage Systems: A Multi-criteria Methodology for BMP Decision Makers. In *Proceedings of the 10th International Conference on Urban Drainage*, (eds. E. Eriksson, H. Genc-Fuhrman, J. Vollertsen, A. Ledin, T. Hvitved-Jacobsen, and P. Steen Mikkelsen), Copenhagen, Denmark, 21-26 August.

Environment Agency, (1998) *The State of the Environment of England and Wales: Fresh Waters.* The Stationery Office, London, (ISBN 0 11 310148 1).

Eriksson, E., Baun, A., Scholes, L., Ledin, A., Ahlman, S., Noutsopoulos, K. and Mikkelsen, P. S. (2007) Selected stormwater priority pollutants-a European perspective. *Sci. Tot. Environ.*, **383**, 41-51.

EU Water Framework Directive (2000) Directive 2000/60/EC of the European Parliament and of the Council of 23 October 2000 establishing a framework for Community action in the field of water policy.

Everard M. and Street P. (2001) *Sustainable urban drainage systems: an evaluation using the natural step framework.* Draft report, January 2001, The Natural Step Group, London.

HMSO. (2005) *Sustainable Development; The UK Government's Approach.* March 2005. HMSO, London.

Planning and Policy Statement 3 (2005) *Consultation Paper on a New Statement for Housing,* Office of the Deputy Prime Minister Publications, Wetherby, UK.

Planning and Policy Statement 25 (2005) *Consultation Paper on a New Statement for Development and Flood Risk,* Office of the Deputy Prime Minister Publications, Wetherby, UK.

Quigley, M. M., Strecker, E. W., Leisenring, M., Huber, W. C., Heaney, J., Weinstein, N., Sansalone, J. and Bodine, D. (2005) The integrated unit process design approach for source control and best management practice design. In: *Proceedings of 10th International Conference on Urban Drainage*, (ed. E. Eriksson, H. Genc-Fuhrman, J. Vollertsen, A. Ledin, T. Hvitved-Jacobsen, and P. Steen Mikkelsen), Copenhagen, Denmark, 21-26 August.

Scholes, L., Revitt, D. M. and Ellis, J. B. (2007) A systematic approach for the comparative assessment of stormwater pollutant removal potentials. *J. Environ. Manage.*, (in Press) (doi:10.1016/j.envman 2007.03.003).

Scholz, M. (2005) The Glasgow sustainable urban drainage system management project. In: *Proceedings of 10th International Conference on Urban Drainage*, (ed. E. Eriksson, H. Genc-Fuhrman, J. Vollertsen, A. Ledin, T. Hvitved-Jacobsen, and P. Steen Mikkelsen), Copenhagen, Denmark, 21-26 August.

Taylor, A. C. (2005) *Guidelines for evaluating the financial, ecological and social aspects of urban stormwater measures to improve waterway health.* CRC for Catchment Hydrology, Melbourne, Victoria. Available at www.catchment.crc.org.au (October 2005).

UK Foresight Project (2003) Impacts of future flood risk *Phase 2 Report of the UK Foresight Project on Flood and Coastal Defence* http://www.foresight.gov.uk/Previous_Projects/Flood_and_Coastal_Defence/Reports_and_Publications/index.html (verified June 2007)

US EPA (2004) Internet web site. National Menu of Best Management Practices for NPDES Stormwater Phase II. http://cfpub.epa.gov/npdes/stormwater/menuofbmps/menu.cfm (Verified June 2007)

Water UK (2000) Towards environmental sustainability; UK water industry environmental sustainability indicators. Water UK, London.

Wright, J. F. (2000) An introduction to RIVPACS. In *Assessing the biological quality of fresh waters: RIVPACS and other techniques* (ed. J.F. Wright, D.W. Sutcliffe and M.T. Furse), Freshwater Biological Association, Cumbria, pp 1-24.

Chapter 8

The Best Management Practice catalogue developed within the DayWater project

D. Michael Revitt, Lian Scholes and J. Bryan Ellis

INTRODUCTION

The term Best Management Practices (BMPs) refers to a wide range of techniques, measures or structural controls which enable the management of stormwater to be achieved from hydrological, environmental, economic and public amenity perspectives (CIRIA, 2001). The adopted approaches may involve both structural (e.g. infiltration systems, ponds and permeable surfaces) and non-structural (e.g. snow management and street cleaning practices). BMP options may be used as an alternative to, or in combination with, conventional stormwater drainage systems. The BMP catalogue, contained within the DayWater ADSS (Chapter 2), is a comprehensive on-line information resource which provides detailed information on 15 structural and 5 non-structural BMPs, which have been selected as being broadly representative of the main types of stormwater BMPs currently in operation within Europe (see Table 1). The catalogue synthesises information and data provided by several DayWater partners (see Scholes *et al.,* 2003 for specific references) together with a variety of peer-reviewed and on-line publications and databases which are fully cited in each down-loadable PDF file, as and where appropriate. Examples of typical structural and non-structural BMPs which have been reported for the USA can be found in a recent Water Environment Federation /ASCE report (WEF/ASCE, 1998). It should also be noted that different terminologies are commonly used in different countries with Sustainable Drainage Systems (SUDS) being the accepted descriptive term in the UK and Water Sensitive Urban Designs (WSUD) being the all embracing descriptor in Australia and the Far East. Another frequently used term in the USA to describe the incorporation of these drainage practices is Low Impact Development (LID).

The BMP catalogue has been designed as an urban stormwater decision support tool to assist end-users according to their existing level of expertise in the use of BMPs for stormwater management (Chapter 2). Thus, for those practitioners with a limited experience of using stormwater BMPs, it provides a comprehensive overview of the wide variety of system types, designs and modes of operation of structural BMPs, as well as providing background and information on non-structural stormwater BMP approaches. As an on-line facility, it also acts as an accessible reference point for urban drainage practitioners who require specialist information on the use of BMPs. For example, in the case of BMP designers (such as landscape architects) or those responsible for approving the inclusion of BMPs within new or existing developments (local authorities and regulators), the catalogue provides clear information on both BMP performance and operation and maintenance requirements. Comparisons of the costings associated with the different BMPs are provided within the catalogue through both a review of available data (accessed through each BMP type) and an interactive format (accessible through the "BMP costings tool" link; see Chapter 10).

An important objective relating to the preparation of the BMP catalogue was to provide a general description for each BMP, supplemented by information on performance, operation and maintenance, costings, photographs and example types (overviews providing information on actual examples of a particular BMP system in use within Europe). In some instances it has not been feasible to include information under a specific heading for a BMP (e.g. "performance" and "operation and maintenance" for the non structural BMP "Control of impervious area development"). In these cases, the reason for the absence of information is explained. The catalogue also contains information on the sources and loads of a range of stormwater pollutants, as well as BMP dimensioning and costings tools. Where it is considered relevant, the BMP catalogue also provides links to a range of external BMP design manuals and websites.

Table 1: List of non-structural and structural stormwater source control measures presented in the BMP catalogue.

Non-structural BMPs	Structural BMPs
Control of impervious area development	Filter drains
Educational aspects	Porous asphalt
Reduction in pollutant usage	Porous paving
Snow management practices	Sedimentation tank
Street cleaning	Filter strip
	Swales
	Soakaways
	Infiltration trench
	Infiltration basin
	Retention ponds
	Detention basins
	Extended detention basin
	Lagoons
	Constructed wetlands

USING THE BMP CATALOGUE

On entering the "introduction" page of the BMP catalogue through the HYDROPOLIS tile "BMP" (Chapter 3), the user can access information on each of the BMPs listed in Table 1 by clicking on the "list of BMPs" hyperlink (see Figure 1). This home page also provides links to information on typical urban pollutant sources, loads and concentrations (by clicking on "sources and loads"), generic BMP design (by clicking on "design information") and comparison of different BMP costings (by clicking on "costings"). Examples of sizing procedures and costing requirements for some individual BMPs, which will provide further support to end-users designing BMPs, are provided within the "Dimensioning Tables" and "Cost Estimate Tool", respectively. A full description of the development and use of the "Cost Estimate Tool" is provided in Chapter 10.

Selecting an individual BMP from the list generated by clicking the "list of BMPs" hyperlink enables the user to access a page on which there is a short description and photograph of the selected BMP. Each BMP-specific page is arranged as shown in Figure 2, where "constructed wetlands" have been selected as the illustrative example. A more detailed description of this structural BMP is provided as a PDF file by clicking on the hyperlink entitled "characteristics". Similarly, extensive information in the form of PDF files is available on the "performance", "operation and maintenance" and "example types" (short case-study descriptions). Photographs of specific BMPs installed throughout Europe can be accessed by clicking on the appropriate heading listed either in the lower left-hand column of the page or at the bottom of the page. The generic pages on urban pollutant "sources and loadings" and "costings" can also be accessed from each BMP specific page. To return to the full list of BMPs, the user simply clicks on the "BMP list" hyperlink or, alternatively, returns to the Hydropolis front page by clicking on the hyperlink "HYDROPOLIS" in the top left-hand corner of each page (Figure 2).

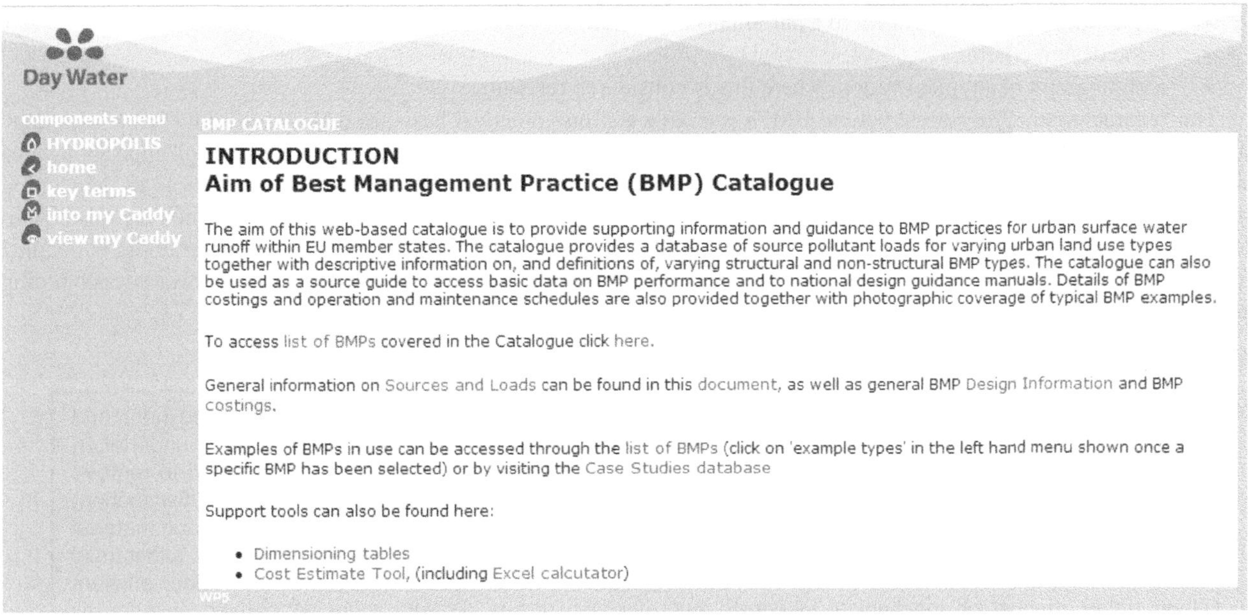

Figure 1: Introduction page of the BMP catalogue.

Figure 2: Example of the layout of a typical BMP specific page: constructed wetland.

BMP CHARACTERISTICS

The objective of the "Characteristics file" is to provide the reader with a substantial amount of general information about each BMP. The format will vary from BMP to BMP and particular differences exist between structural and non-structural systems. The information given on structural BMPs can typically be categorised as follows:

- different system types which have been developed for the treatment of urban drainage,
- pertinent pollutant removal processes,
- representative diagrams of idealised systems emphasising the relevance of the different treatment components,

- the applicability of the system to a particular urban area,
- basic design criteria,
- a photograph of a typical system where this is considered relevant.

The "characteristics" of non-structural BMPs can have a strong practical basis, as is the case with "Street cleaning" and "Snow management practices", or may be more theoretically based as is the case for "Control of impervious area development", "Educational aspects" and "Reduction in pollutant usage". The consequent differences in developing appropriate sections for the BMP catalogue are illustrated by the extracts shown in Boxes 1 and 2 which are taken from the "Characteristics files" for "Street cleaning" and "Educational aspects", respectively. The latter example is detailed and illustrated in the "Urban Dynamics" component of the ADSS, especially in the "Water Aspects" awareness building tool (see Chapter 6 for further information on this component).

Box 1: Extract from street cleaning characteristics file.

Street cleaning is a non-structural BMP technique which helps to reduce the amount of sediment and associated pollutants reaching the stormwater or combined sewer system. Cleaning is normally undertaken by manual, mechanical or vacuum cleaning methods as well as by hydrojetting and is intended to remove litter, refuse, detritus and dog fouling materials. Whilst hydrojetting (sometimes combined with disinfection) can be highly effective in removing solids from the impermeable surface, it only removes the surface material to the adjacent kerbside gutter channel or to the gullypot, and therefore does not serve any substantial pollution control function; it is a purely cosmetic technique. Vacuum cleaning appears to be more efficient than either manual or mechanical sweeping but efficiency drops off with reducing particle size for all methods. However, such cleaning is again not really effective in terms of stormwater pollution control as sweeping only removes the larger solids leaving behind the bulk of the fine particles with which the majority of the pollutants are associated; only some 10% - 20% of particles less than 43 μm will be removed. Field studies suggest that as much as 5 to 6 passes by a mechanical sweeper would be required to achieve anywhere near a 100% removal efficiency. The remaining fine particles are readily flushed from the street surface and roadside-gutter channel to the gully pot chamber and below-ground sewer during the next rainfall event.

Box 2: Extract from educational aspects characteristics file.

The public's perception of the use of structural BMPs within their local environment will be influenced by a range of factors such as their aesthetic impact, their ability to attenuate water flow rates and volumes (flood abatement) and their contribution to water quality improvement. Other issues to be considered when BMPs, incorporating open water areas, are located within residential areas, include public safety, enhanced amenity opportunities, increased biodiversity and the ability to contribute to broader sustainability concepts such as the quality of life in urban areas and optimal 'urban space management'. The provision of high quality amenity opportunities can involve a variety of aspects including visual and habitat enhancement, wildlife and biodiversity benefits, and the provision of facilities for recreation, relaxation and leisure. There are potential educational opportunities and benefits associated with all of these aspects and their contributions to supporting sustainable stormwater management initiatives as a 'non-structural' BMP are addressed in this document.

BMP PERFORMANCE

By clicking on the "performance" hyperlink in any BMP specific page (e.g. see Figure 2), the user will be directed to a PDF document containing current information on the pollutant removal efficiencies of specific BMPs. This performance data has been obtained from a range of different international sources but, because of the different criteria which have been used to determine the removal efficiencies, the data is subject to different levels of reliability. Boxes 3 and 4 contain extracts from "Performance" files which can be accessed within the BMP catalogue using "Snow management practices" (see also Chapter 22) and "Swales" as examples of non-structural and structural BMPs.

Box 3: Extract from snow management practices performance file.

Development of an optimal snow handling system involves evaluating the whole process to allow the best practice possible to be selected with the lowest environmental impact at the most reasonable cost. A snow handling strategy that takes the environment into account involves separation into different areas on the basis of the snow quality, which varies with location and time. The strategy involves dividing the city into white, grey and black zones, where the white zones represent areas with clean snow during the whole winter, the black zones represent dirty snow throughout the whole winter and the grey zones represent snow that is clean at the beginning of the winter but dirty at the end. This zoning must be carried out for each city individually since conditions, and therefore snow quality, will obviously vary according to local conditions. The snow that is most polluted should be transported to snow deposits that are appropriately located, designed and operated to minimise any negative environmental effects. In contrast, clean snow can be dumped at stormwater outlets. The use of such a snow handling strategy results in a significant reduction in the volume of snow dumped at depots in comparison to those deposits currently made.

Box 4: Extract from swale performance file.

The Table below shows the range of pollutant removal efficiencies, based on event mean concentrations and pollutant loads that have been noted by various European workers for grass swales. It is clear that whilst in general, good removal rates can be achieved by such systems, there is still considerable variability in performance. Investigations of soluble metal species, nutrients or bacteria suggest that direct removal of these contaminants is limited and it may be that swales can only provide an efficient performance for solids and heavy organics such as leaf litter (and their attached pollutants) as well as oils. Solids removal performance increases as total suspended solids (TSS) concentrations increase, and with inflow concentrations below 30-40 mg/l, little reduction can be expected.

Swale pollutant event mean concentrations (EMC), loadings and removal efficiencies.

Pollutant parameter	EMC and range (mg/l)	Load (kg/ha/yr)	Removal efficiency (%)
Total suspended solids	25.0 (7.0 - 47.0)	–	86 (55 - 91)
Total Zn	0.032 (0.011 - 0.143)	7.05 (1.85 - 9.2)	83 (63 - 93)
Total Pb	0.079 (0.014 - 0.144)	0.78 (0.25 - 2.61)	54 (17 - 76)
Chemical oxygen demand	39.0 (2.0 - 76.0)	–	62 (53 - 74)

BMP OPERATION AND MAINTENANCE

Users of the BMP catalogue can access a PDF file detailing the operation and maintenance needs of a specific BMP by clicking on the "Operation and maintenance" hyperlink located on the BMP specific page (e.g. see Figure 2). Currently, many of the major concerns about the applicability of BMPs as source control facilities in urban areas are related to aspects of operation and maintenance (O&M). Therefore, it is important that potential end-users are fully conversant with the O&M requirements associated with a particular BMP. Representative extracts of the O&M details given in the BMP catalogue are provided in Boxes 5 and 6 for "Gully pot cleaning" (part of the street cleaning process) and "Porous paving", respectively. These represent selected examples for non-structural and structural BMP systems.

Box 5: Extract from gully pot cleaning operation and maintenance file.

The roadside gullypot chamber or sump gradually fills up with sediment and debris and regular emptying is necessary to maintain a flow through the outlet trap and to prevent flushing of sediment from the chamber. Gullypot cleansing is normally undertaken on a cyclic basis with recommended frequencies being within a minimum frequency of 2-3 times per year (reducing to once per year for rural locations). Commercial centres and market areas subject to heavy littering, will often receive cleaning on a weekly basis.
Gullypot emptying is normally undertaken by vacuum/suction tankers (eductors) having a capacity to collect up to 5000 L of waste sludge "arisings". The gully cleansing operation usually incorporates a "flush test" to ensure that the outlet trap is not blocked. The "black-water" arisings are either discharged to sewer or taken for disposal at a registered landfill site. On-road parking presents severe problems to gully cleansing O&M operations and in the UK, up to 85% of authorities organize temporary car park restrictions to facilitate cleansing schedules. The efficiency of gully cleansing O&M can vary considerably from 20% to 95%, but averages around 70%. The residual material left after cleansing is mainly comprised of large (25 mm) aggregates and debris with typically some 10% of basal chamber sediment being washed into the sewer by the flush test.

Box 6: Extract from porous paving operation and maintenance file.

The different types of porous paving are relatively high maintenance BMPs, requiring regular attention, as lack of maintenance can result in severe clogging of the void/pore spaces and loss of pollutant removal capacity as well as a failure to infiltrate stormwater from the surface. The following lists some general maintenance recommendations:

- The pavement surface should be protected from fine sediment wash-in during construction.
- Oil and grease spillages should be cleaned as soon as possible from the pavement surface.
- The surface should be vacuumed or brushed approximately every 4 to 6 months after the initial 6 month bedding-down period, preferably followed by pressure washing of the entire surface. The frequency of vacuum/pressure hosing should be increased if the surface is subject to high sediment loadings.
- If clogging occurs, drilling the pavement surface may restore some capacity; if this fails, then replacement of the entire surface is required.

The table below identifies both regular and occasional maintenance as well as monitoring requirements that can be recommended for porous paving.

Regular O&M	Occasional O&M	Monitoring
• Check and repair block paving/alignment (annually) • Surface brush/vacuum cleaning and high pressure hosing, especially following ice/snow (annually or 2 per year) • Check/clean major oil spills.	• Renewal of filter course (as required) • Inlet/outlet cleaning (1 per 3 - 4 years) • Repair cracks / potholes / ruts and missing/loose blocks (as required)	• Surface ponding inspection (annually) • Check/clean overflow (1 per 3 - 4 years) • Check drain time (1 per 4 - 5 years)

BMP PHOTOGRAPHS

The "photographs" section includes images from across Europe chosen to represent the typical applications/installations of a particular BMP. The use of pictorial representations indicates to prospective end-users of BMPs the situations in which different BMPs can be and have been used. The stored images are categorised by their country of origin and are supported by descriptive captions. This part of the BMP catalogue is continually being added to as more photographic evidence becomes available. Figure 3 presents an overview of the type of photographic information available using the relevant page within the BMP catalogue for retention ponds, as an example. Figure 4 presents a photograph of a retention pond as a typical example of the type of image available within the BMP catalogue photographic library.

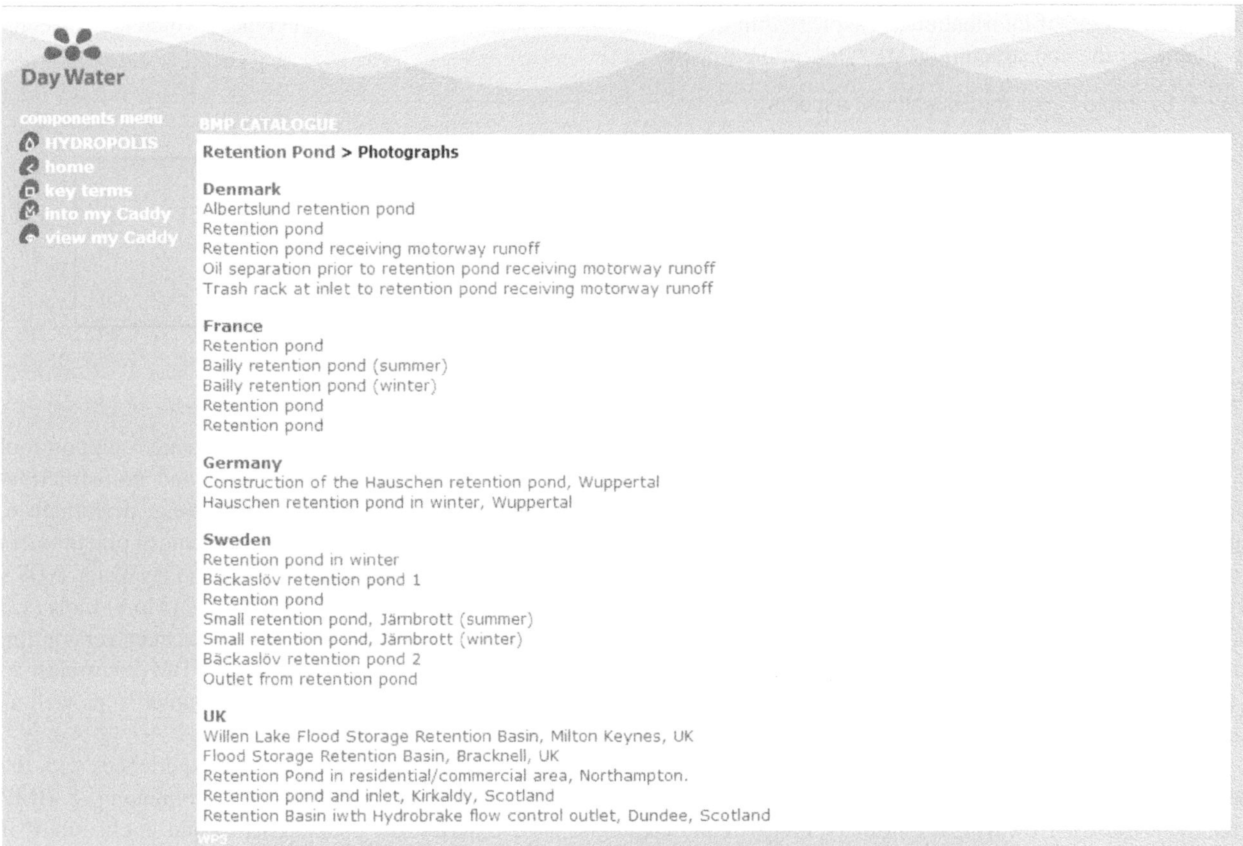

Figure 3: Photographic examples of retention ponds currently contained within the BMP catalogue.

Figure 4: Retention pond: typical example of the type of image available within the BMP catalogue photographic library.

BMP EXAMPLE TYPES

This section of the BMP catalogue effectively complements the "photograph" section by enabling information regarding the actual use of a specific BMP in the field to be described. The information is contained in PDF files which are accessed by clicking on the hyperlink "Example types" in the appropriate BMP specific page (e.g. see Figure 2). Where available, examples of both successful and unsuccessful BMP systems/approaches have been included, as information on both positive and negative experiences are obviously equally valuable as an information resource. To illustrate the type of information available within this section, Box 7 identifies the "Example types" which are currently available for the non structural BMP "Education aspects".

Box 7: Example Types for Educational aspects.

- Camley Street Natural Park, Kings Cross, London, UK
- Sandwell Valley nature Reserve, Birmingham, UK
- Communication on urban water from Seine-Saint Denis county (Paris suburb), France
- Scottish Oil Care Campaign, UK
- Ouzel Valley Park, Milton Keynes, UK

CONCLUSIONS

As an on-line freely available information resource, the BMP catalogue is an easily accessible management support tool which can be used to illustrate to those not familiar with BMPs the wide range of structural and non-structural approaches available. The inclusion of BMP-specific information on a range of technical issues (e.g. performance, operation and maintenance requirements and costs) also enables the catalogue to support more experienced practitioners in identifying potential stormwater management solutions, as well as facilitating the use of other DayWater ADSS components such as the Multi-Criteria Comparator (MCC) and the Guided tour/Matrix of Alternative (MoA) tools (see Chapters 9 and 4, respectively). The BMP catalogue is also designed to provide support to stormwater managers during the project design and construction phase through the provision of information on actual examples of BMPs currently in operation (further case studies are also contained within the ADSS "Case studies" database; see Chapter 5) as well as providing links to a range of general design documents and data resources.

Within the "Example types" component of the BMP catalogue, there is information on current experiences with the use of stormwater management approaches which incorporate the use of two or more BMPs in combination (i.e. BMP treatment trains). However, the primary focus of the catalogue is to describe the characteristics and applications of individual BMPs. This approach was adopted to enable a broad overview of the main treatment systems to be provided from which users could select combinations which would be appropriate to their specific requirements. Further support in developing BMP treatment train options is contained within the Guided tour/Matrix of Alternative (MoA) tool (see Chapter 4).

REFERENCES

CIRIA (2001) Sustainable urban drainage systems: Best Practice Manual. Report C523, Construction Industry Research & Information Association, London.

Scholes, L., Revitt, D.M. and Ellis, J.B. (2003) Review of the use of BMPs in Europe. DayWater Deliverable 5.1. www.daywater.org Verified June 2007, 98 p.

WEF and ASCE. (1998) Urban runoff Quality management. Water Environment Federation, Alexandria, Virginia, USA. ISBN 1-57278-039-8.

Chapter 9

The DayWater Multi-Criteria Comparator

J. Bryan Ellis, D. Michael Revitt and Lian Scholes

INTRODUCTION

Multi-criteria analysis (MCA) is a decision-making tool originally developed for the evaluation of options where not all of the considerations have an easily assignable value. It may be considered as an extension of the more traditional decision-making techniques (e.g. cost-benefit analysis) in that, as well as financial information, it also enables a range of other factors, such as the environmental and social impacts of the possible alternatives to be taken into account. Wenzel (2001) has argued that the resolution of best practice source strategies require an integrated evaluation of such environmental, technological, economic and social values which are normally set in a local planning and/or site master-scenario context, and which is inevitably inter-disciplinary and inter-institutional in character. Any resolving optimisation procedure must be capable of addressing dimensions of multi-criteria complexity in a manner which is both understandable and auditable to the variety of stakeholders and, if possible, expressed in a simple but objectively quantified methodology. MCA provides such an optimisation approach that facilitates transparent and auditable decision-making, enabling the views and concerns of all stakeholders to be taken into account in combination with scientific and technological data as argued by Brunner and Starkl (2004). However, the variety of stakeholders (or "deciders") will normally also hold different opinions (or strategies) regarding the significance (or 'weight') that might be attributed to various influencing criteria and Key Performance Indicators (KPIs), although weighting procedures within MCA methodologies can be contentious (Figueira *et al.*, 2005). There are several types of MCA but all approaches involve the development of a performance matrix containing data/information on how a range of possible solutions perform in relation to a variety of criteria and/or indicators.

MULTI-CRITERIA APPROACHES FOR WATER MANAGEMENT

A basic structured approach to a multi-criteria, multi-objective methodology for the evaluation of sustainable water resources systems was proposed within the 1998 UNESCO International Hydrological Programme (ASCE and UNESCO, 1998). The report recommended a holistic conception of decision-making processes incorporating sustainability criteria and guidelines covering economic, ecological, social and environmental factors influencing investment and construction of water resource systems. Other workers have recently presented approaches for urban runoff management which also utilise generic sustainability criteria or system conditions defining limiting constraints on best practice drainage control and treatment technologies (Hellstrom *et al.*, 2000). The procedure provided by the **S**ustainable **W**ater Industry **A**sset **R**esource **D**ecisions (SWARD) project provides a multi-criteria approach which has been widely applied within Europe for assessing the relative sustainability of options involving urban water supply and sewerage assets (Ashley *et al.*, 2004). The guideline procedure is based on a multi-objective support system for the

 DayWater: an Adaptive Decision Support System for Urban Stormwater Management. Edited by Daniel R. Thévenot. ISBN: 9781843391609. Published by IWA Publishing, London, UK.

assessment of technical, environmental, economic and social criteria with a range of tools and techniques supporting each assessment step (Ashley *et al.*, 2002). Various workers have utilised criteria/indicator values from the SWARD procedure in their own multi-criteria applications, such as Gilmour *et al.* (2005) in their development of Information Flow Diagrams (IFDs) for the assessment of stakeholder involvement in wastewater treatment design. Martin *et al* (2007) have adopted the criteria listing developed within the current DayWater Multi-Criteria Comparator (MCC) workpackage for their multi-criteria decision analysis for urban stormwater BMP selection. However, they utilise the ELECTRE III fuzzy logic, outranking methodology to determine pair preferences, which ranks BMP options based on a derivation of strict, weak and indifference thresholds. It is argued that this pseudo-criteria approach takes into account uncertainty in the data information, but the outranking relationships are complex and the allocation of threshold values can be subjective. Ellis *et al.* (2004) developed a generic, matrix based, multi-criteria methodology as an initial screening assessment of highway drainage systems, and the approach forms the basic framework for the current MCC structure.

The 'triple-bottom-line' (TBL) approach that has been widely adopted in Australia for the assessment of urban stormwater projects uses financial, ecological and social dimension criteria but follows a very similar methodology to SWARD (Taylor, 2005). However, the TBL approach requires considerable resources to undertake the assessment process, can generate substantial complexity if large numbers of criteria and/or conflicting stakeholders are involved and there is no guarantee of a "sustainable" outcome. The **Wa**ter cycle management for **N**ew **D**evelopments (WaND) project also contains within its complex, fuzzy logic, decision-support workbench, multi-objective screening and "optioneering" tools for guiding stakeholders to the selection of sustainable water management options (Makropoulos *et al.*, 2005). Further background to the WaND and SWARD projects, and their context to sustainability principles, is given in Chapter 25. Parikh *et al* (2005) have also utilised 'market' mechanisms and incentives as a basis for a multi-disciplinary decision-making approach to stormwater management which attempts to integrate hydrologic, economic and legal criteria.

THE DAYWATER MCC APPROACH

The multi-criteria comparator (MCC), contained within the DayWater ADSS, is an electronic version of MCA specifically designed to support stakeholders in identifying the most appropriate type of structural Best Management Practice (BMP) for a particular site (www.daywater.org and access to ADSS). However, the multi-criteria approach does not include non-structural measures and is only applicable to individual "stand-alone" BMP devices i.e. it does not consider hybrid or treatment train applications and it assumes a mixed land use catchment. An overview of the steps in the assessment methodology is given in Figure 1 which also illustrates the stages at which stakeholder involvement and feedback in the procedure can take place; the methodological framework adopted is deliberately similar to that of SWARD in order to provide companion compatibility (see Chapter 25).

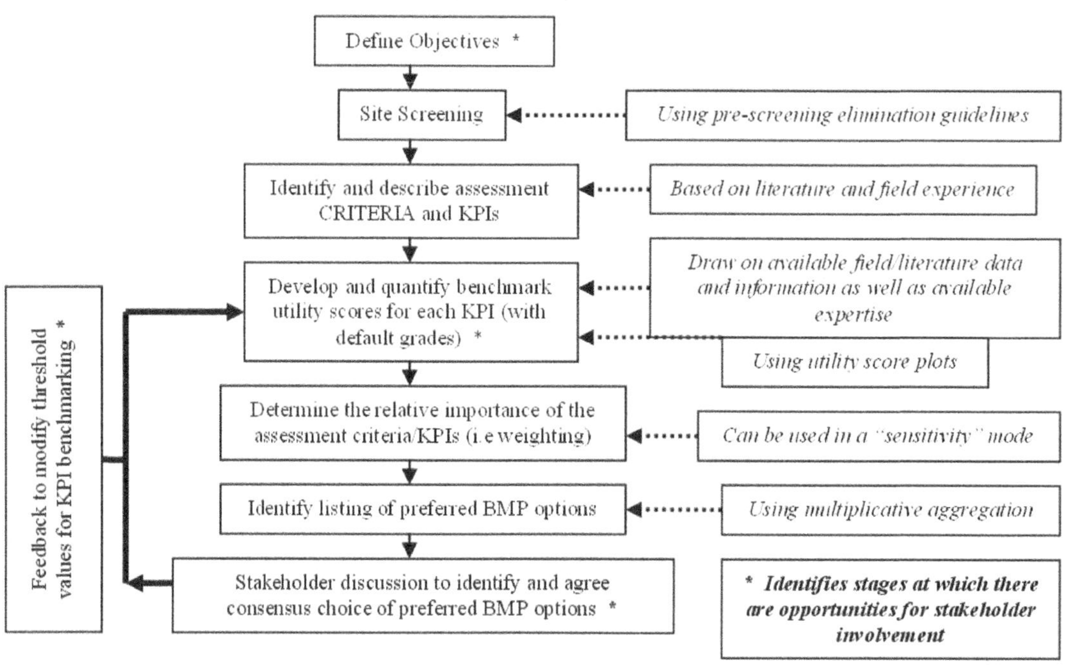

Figure 1: Flow chart of the MCC methodology.

In the MCC performance matrix, the possible solutions are a range of BMPs (the potential use of 15 different structural BMP systems may be evaluated), and the six criteria against which each of the BMPs are evaluated are categorised as Technical, Environmental, Operation and Maintenance, Social and Urban Community Benefits, Economic Costs and Legal and Urban Planning (Table 1). Criteria are considered to be primary, high-level established factors (or Areas of Concern) on which the final judgement, evaluation or decision is made. Each criterion is supported by a series of Key Performance Indicators (KPIs) comprising diagnostic states or conditions which describe relevant and appropriate properties of the given criteria. Indicators essentially account for the changes in a "driver-pressure, impact-response" system as described by Ashley *et al.* in Chapter 25, and provide a comparative basis to decide when and whether the criteria have been satisfactorily addressed. The KPIs provide more detailed discriminators (e.g. flood storage, pollution control and system flexibility for the Technical criteria), and are benchmarked by threshold values (qualitative or quantitative) which can comprise a point of reference for decision-making. The criteria and KPIs were selected following extended DayWater end-user consultation and are additionally supported by the extensive workshop validation undertaken within the SWARD project, as described in Chapter 25.

Table 1: MCC performance matrix for 15 different structural BMPs and 6 criteria.

Criteria	**Key Performance Indicators**	**Performance scores**															**Weighting**	
		Filter drain	Porous asphalt	Porous paving	Green roofs	Settlement tank	Filter strip	Swales	Soakaways	Infiltration trench	Infiltration basin	Retention ponds	Detention basins	Extended detention basin	Lagoon	Constructed wetland	**Indicators**	**Criteria**
Technical	Flood control																	
	Pollution control																	
	Adaptability to urban growth																	
Environmental	Impact on receiving water volume																	
	Impact on receiving water quality																	
	Ecological impact																	
Operation & Maintenance	Maintenance & servicing requirements																	
	System reliability and durability																	
Social and Urban Community Benefits	Public H&S risks																	
	Sustainable development																	
	Public/community information & awareness																	
	Amenity & aesthetics																	
Economic Costs	Life Cycle Costs																	
	Long term affordability																	
Legal & Urban Planning	Adoption Status																	
	Building development issues and stormwater regulations																	
TOTAL	(Sum of score x weight: %)																100	100

Performance benchmarking serves to consider the position of an optional source control BMP in relation to specified key performance characteristics and provides standardised information against which individual performance claims can be judged and so can comprise a framework for technical evaluation at a generic level. There can be difficulties getting stakeholder agreement on benchmarking values and thresholds and in defining the data to derive an appropriate performance score. Thus it is essential that high level performance does in fact measure down to those aspects (or key drivers) that have the greatest impact on performance. The first step in setting benchmark scores is to establish an appropriate interval scale with justified references or endpoints; normally the worst performance is allocated a zero score and the best 1 (or 10 or 100). Some workers recommend an interval scale spanning ranges of 0 – 33, 33 – 66 and 66 – 100 to represent low, medium and high descriptors respectively. The approach therefore develops standardised merit ratings having a general functional relationship of x = f (where f is the user utility score input or a default grade value), such that $0 \leq x \leq 1$ (or 10 or 100). The relationship can also be used as 0 = No and 1 (or 10) = Yes or alternatively as 0 = Very Low and 1 (or 10) = Very High, if the benchmark is expressed in a qualitative manner. The various BMP options are then benchmarked (or scored) in terms of their performance against each KPI based on field/laboratory, literature, questionnaire, expert systems or other proven performance data/information. The utility (or performance) scores for the various BMP options can then be recorded as direct entries to the performance matrix. However, given that there is undoubtedly uncertainty associated with the allocation of utility scores to individual BMP KPI performance (Figueira *et al.*, 2005), another alternative is to use coarser grade bands (1 – 5) to cover the appropriate interval scale used for the utility scores, and it is this option that has been used in the current DayWater MCC methodology. The score allocation using an interval scale means that the differences in scores are consistent within each criterion and KPI, although it does not allow the conclusion that a utility score of 8 is twice as good as a score of 4; this would require a ratio scale of measurement.

An example of the benchmarking procedure used to produce a graphical plot of the utility scores and default grade values is given in Figure 2 which illustrates the utility plot for the 'Impact on Receiving Water Quality' KPI within the Environmental criteria category. Similar utility score procedures were adopted within the decision matrix for urban water resource management developed by Sieker *et al* (2005), based on four principal criteria and over 28 KPI indicators. The utility scores for the 'Impact on Receiving Water Quality' KPI have been benchmarked against the exeedance potential of individual BMPs as expressed in terms of their average outflow performance for dissolved oxygen (DO), biochemical oxygen demand (BOD), total suspended solids (TSS), ammonium-nitrogen (NH_4-N) and toxic substances, as defined within the water quality guidelines of the Water Framework Directive (WFD).

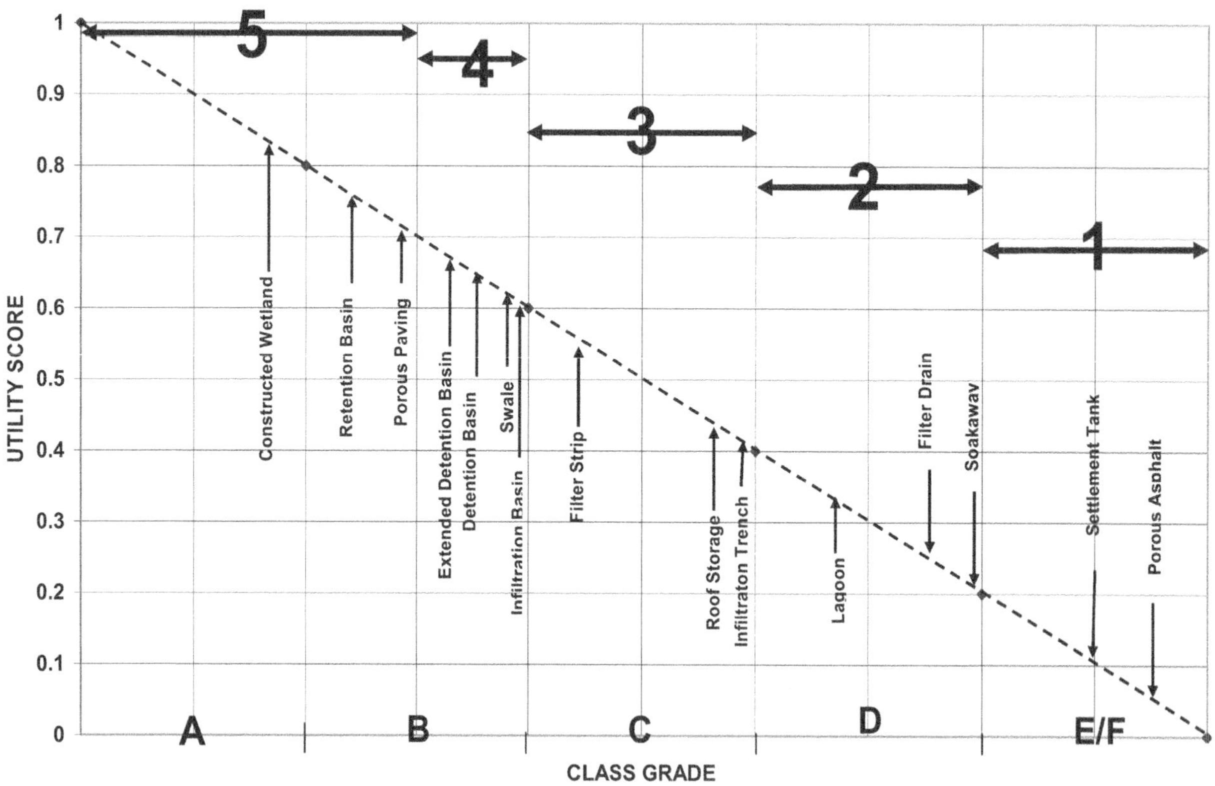

Figure 2: Performance/utility score and grade plot for the 'Impact on Receiving Water Quality' Key Performance Indicator.

The Class Grades (A – F) in Figure 2 refer to the water quality assessment classification contained within the WFD for which further information can be obtained by accessing www.daywater.org. A brief summary of the KPI benchmark can be obtained by clicking on the KPI in the performance matrix and then following the 'detailed description' link at the end of the summary. The exceedance concentrations serve as target outflow threshold values to measure the performance of individual BMPs against WFD receiving water standards and are based on storm event data recorded in various UK, European and North American databases. These databases can be viewed through the Hydropolis BMP Catalogue (www.daywater.org) accessed via the 'Sources and Loads' link or via the Hydropolis Pollutants window and accessed via the 'Source characterisation' link to the same 'list of sources' given in the BMP Catalogue (Chapter 8). This same 'Source characterisation' entry is included in the CHIAT tool accessed via the Hydropolis Risk and Vulnerability or Tools windows (Chapter 11). The input required for analysis of the pollutant build-up and washoff process within the STORM/SEWSYS model (Chapter 14), accessed through the Hydropolis Tools window (Chapter 3) and the 'SFM' (Sources & Flux Model) link, can also use default values presented in the BMP Catalogue 'Sources and Loads' database (Chapter 8). This database presents urban runoff quality data expressed as event mean concentrations (EMCs) (with medians, ranges, standard deviations or quartiles), or as loads (per annum, per unit area, per traffic densities etc.) and drawn from field studies in the UK, Europe, USA, Canada and Australia/New Zealand. The STORM/SEWSYS model also contains an option to use Scandinavian and German data accessed from a Microsoft® Access database. Chapter 14 provides further details of this database and additional information is given in Ahlman and Svensson (2002). However, it should be noted that this database and the US EPA national BMP database (www.bmpdatabase.org) are essentially concerned with inflow concentrations to BMP devices, whereas the 'Impact on Receiving Water Quality' KPI is based on outflow performance data. The STORM module of the SFM modelling procedure can also directly generate flow volumes for varying rainfall depths/intensities within a catchment, which could be used for benchmarking the Flood Control KPI. Further examples of the KPI benchmarking procedure are given in Ellis *et al* (2005), which also provides an overview of the MCC methodological approach.

USING THE DAYWATER MCC PERFORMANCE MATRIX

The structural stormwater BMPs included in the MCC are those described in the DayWater ADSS BMP catalogue and which can be separately referred to in Chapter 8. On accessing the MCC tool via the DayWater website (Chapter 3), users enter a webpage which provides a short introduction to the use of the multi-criteria approach. There is also an explanation of how the MCC can support users in the decision-making process and a brief overview is provided of the various modes in which the MCC may be operated.

MCC Instructions

From the MCC opening page, users can access simple instructions on how to use the MCC by clicking on 'instructions' in the menu of options or when prompted to do so in the main body of the introductory text. This page explains that the MCC can be used in 3 modes of operation, as follows:

- Mode 1: Use of default scores,
- Mode 2: Development of scores by the end-user,
- Mode 3: Refinement of an initial order of preference by introducing new scores for the initially generated BMP options.

The different modes of operation support stakeholders who are new to urban stormwater management through the provision of default scores derived from the utility plot procedure (i.e. Mode 1), as well as offering a useful tool to those with more experience by enabling them to generate their own values using site-specific characteristics (i.e. Mode 2). This Mode 2 approach is similar to that adopted in the Matrix of Alternatives (MoA) which is explained later and for which further information is given in Chapter 4. The use of Mode 3 is a combination of Modes 1 and 2, enabling a robust shortlist of BMP options to be generated for which further site-specific information can be gathered as required, effectively enabling data collection which needs to be identified and completed in a targeted manner. Detailed step-by-step instructions on the completion of the MCC procedure using each mode are sequentially presented as shown in Figure 3 for a section of the instructions outlining the use of the MCC in Mode 1.

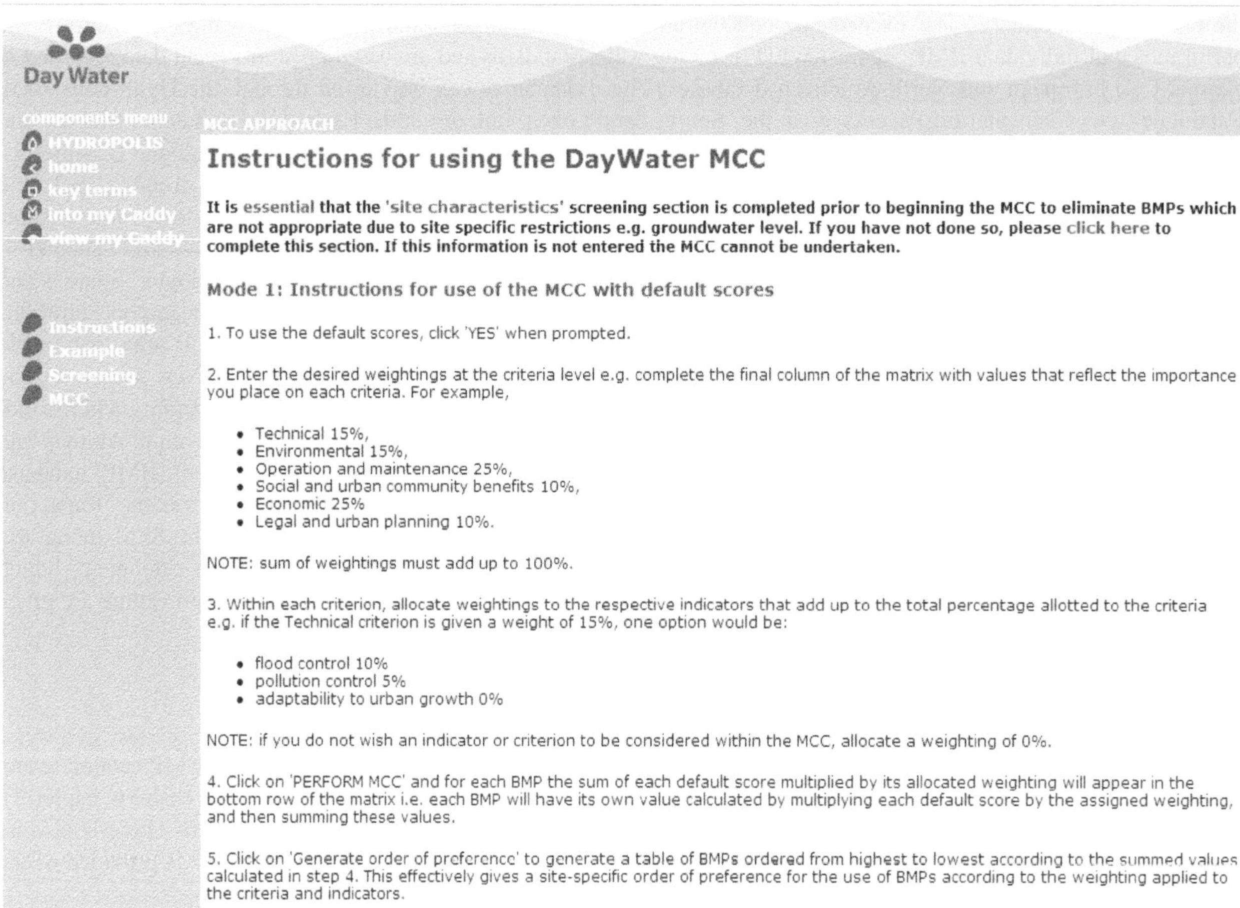

Figure 3: Section of the instructions webpage outlining how to use the MCC in Mode 1.

Site screening characteristics

On accessing the 'instructions' webpage after opening the Hydropolis MCC window on the DayWater website, users are initially directed towards the 'site characteristics" screening section of the MCC where they are requested to enter site specific data. These include depth to groundwater and information on either the soil hydraulic conductivity, soil type or infiltration rate, together with more general information on the size of the contributing catchment area (Figure 4). Users are strongly recommended to complete this section to enable BMPs which are not appropriate for use due to site-specific conditions to be identified at an early stage by being highlighted in red in the performance matrix (in relation to water table depth and soil characteristics) or orange (with respect to the contributing drainage area information; Figure 5). The appearance of a red colour for a particular BMP indicates that it should not be used e.g. infiltration systems are not recommended at sites with a predominantly clay soil or where the water table is close to the surface. An orange colouration advises the user that a particular BMP should not be used in isolation because of the size of the contributing catchment area e.g. a single, "stand-alone" swale system may be inadequate to manage the volume of runoff generated by a large catchment area. An important difference between orange and red highlighting is that the former acts as a warning regarding the use of a specific BMP whereas red advises that a BMP should not be used.

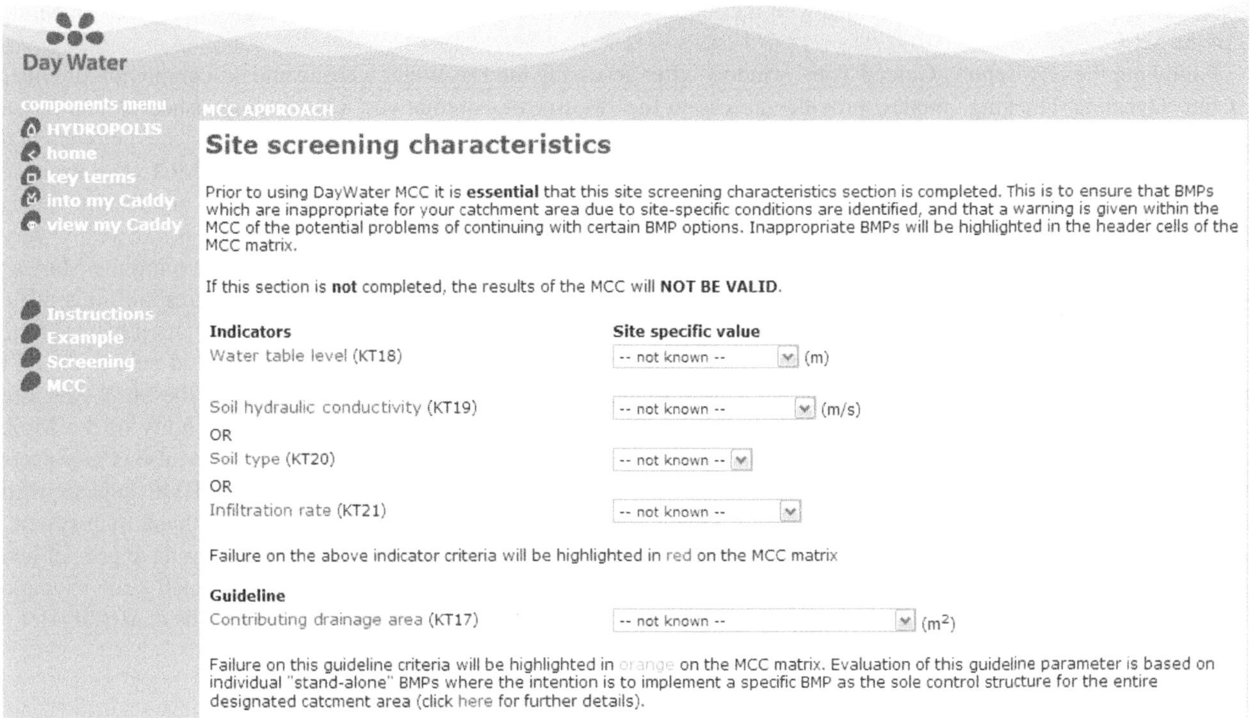

Figure 4: The 'site screening characteristics' web page.

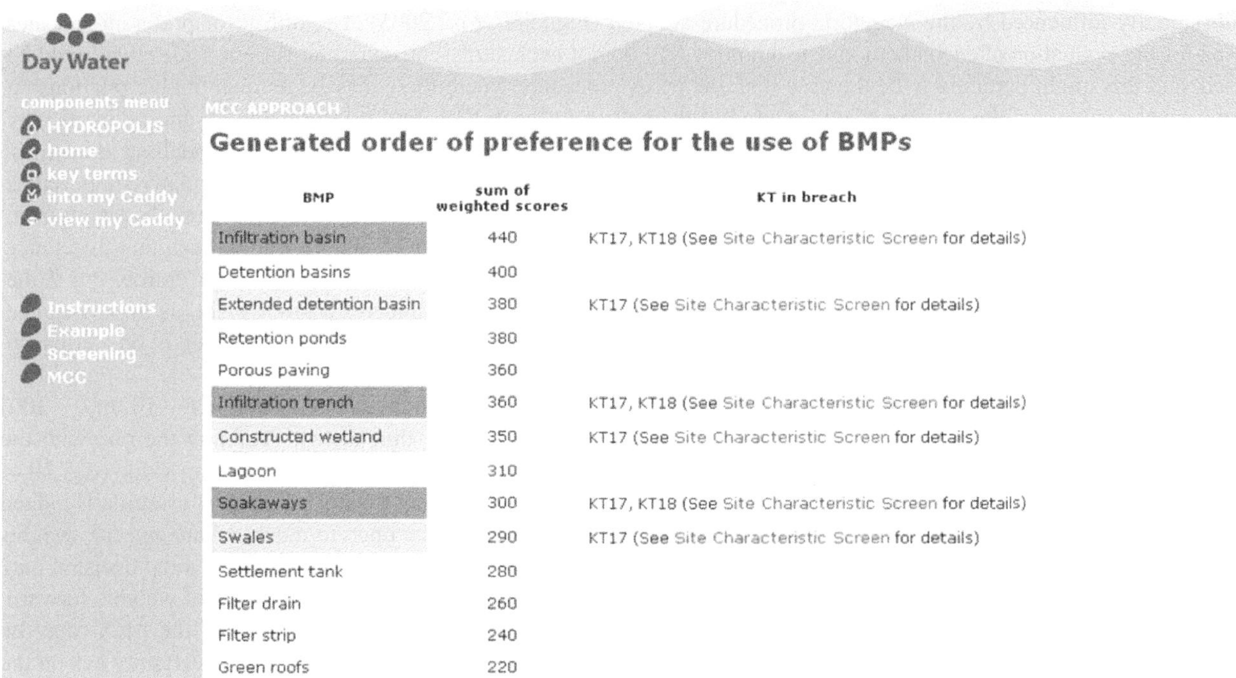

Figure 5: Example of an order of preference generated for the use of BMP options at a particular site.

MCC performance matrix

Following completion of the site characteristics section, users are directed to the 'MCC performance matrix' which represents a structured illustration of the multi-attribute decision space. The user is required to decide whether to use the default scores (i.e. the MCC in Mode 1) or whether to develop and enter their own scores (i.e. use the MCC in Mode 2).

Users can apply the information provided with the Mode 1 approach as a guide to developing their own scores for Mode 2 of the MCC.

Following the 'Hydropolis Guided Tour' window, after accessing the DayWater website and on completion of either of the 'Dynamic Thinking' modes, provides access to the 'Matrix of Alternatives' (MoA) spreadsheet which has a similar format to the MCC performance matrix although the Economic Costs criteria and KPIs are not included (Chapter 4). A larger set of benchmarks are however included in the matrix than appears in the MCC performance matrix. Users can enter their own benchmark values and scores into the matrix and there are no default values presented. Further information on the procedure can be obtained from Chapter 4. In effect this is not really very different from the Mode 2 approach in the MCC procedure. However, there is no guidance accompanying the MoA as to what might constitute an appropriate benchmark value or score for a specific BMP or what threshold quantities, properties or conditions might comprise a low or high value or grade. This must be agreed in consultation between the stakeholders in relation to specific site conditions and individual projects, or alternatively the user could refer to the KPI benchmarking descriptions contained in the MCC to provide some general guidance. Nevertheless, the MoA structure recognises that certain criteria and KPIs will be more relevant in differing locations and the flexibility of the MoA procedure allows for a unique set of concerns and values to be considered in the matrix. The flexibility also allows users to enter additional criteria and KPIs to those listed, and the MoA also allows a comparison of BMP devices with conventional drainage systems as well as enabling a comparison of hybrid combinations within a treatment train system. Indeed, the UK WaND project has suggested that there may be circumstances when it is better not to have pre-defined criteria and/or KPIs to meet certain types of decisions and stakeholders (Sakellari *et al.*, 2005). A small-scale example of the application of the MoA methodology to the treatment options for highway runoff is given in Ellis *et al.* (2004).

Applying weights

The final stage in the MCC process is for users to allocate weightings at the criterion and/or indicator level to reflect their particular concerns or to address site-specific issues. The assigning of weights to criteria (and/or KPIs) is a highly contentious issue within multi-criteria analysis and undoubtedly the outcomes of the performance matrix can be substantially influenced by the weighting procedure adopted (Figueira *et al*, 2005). The utility (or preference) scales used for the evaluation of each alternative technology KPI do not necessarily equate on the various scales that might be used, and this might particularly be the case with the MoA procedure where there may be no quantitative relationship between the value and the grading scores entered in the matrix for each KPI. End-user preferences and opinions may often be fuzzy, conflicting, open to change during the decision process or influenced by the modelling outcomes. Weighting is therefore required to equate the scales, and effectively this is equivalent to judging the relative importance of the scales used. Weights can be assigned using a "swing-weighting" method which determines how the swing from 0 – 10 (or 0 – 100) on one scale compares to 0 - 10 (or 0 – 100) on another scale, and also how much that difference matters (ODPM, 2003). For example, cost might be an important factor in choice of a technology option, but if the difference in cost between all the options is low, then the cost criterion could receive a low weight, as the difference between highest and lowest costs is so small that it would not really affect the decision. Hence the weight on a criterion reflects both the range of difference between the options, and how much that difference matters.

To apply the "swing" method, it is first necessary to identify the criteria with the greatest relative 0 – 10 (or 0 – 100) "swing" in preference. A pair-wise comparison may be necessary to identify this, always retaining the one with the biggest "swing". The criteria which emerges as having the biggest "swing" is assigned the maximum value (e.g. 10 or 100), and this becomes the standard to which all others are compared. Thus if the "swing" of the next criterion is judged to be half that of the first, it is given a weight of 5 (or 50). This process can be open to much debate and the weights selected should be those which best represent the views of the stakeholders, as a whole, although the final decision may well be political in nature. If consensus cannot be reached, it may be best to take two or more sets of weights forward, as agreement on choice of options can sometimes be reached without agreement on weights, or the MCC may be repeatedly run using different values which reflect differing views to ascertain the effect that this uncertainty has on the generated BMP order of preference. Operated in this mode, the re-iteration can also serve as a sensitivity test. If a user does not wish a particular indicator or criterion to be considered within the MCC, a weighting of 0% should be applied, but the sum of all the weighting values is required to equal a total value 100%. To calculate the final weighted scores, the BMP option's score on a criterion (or KPI) is multiplied by its weight and this is done for each criterion and the products are added together to give the overall preference score for that option.

Outcomes

The actual MCC process involves combining the allocated performance scores and weightings (Table 1). This procedure can be achieved in a variety of ways and, following a consideration of alternative techniques, the approach of multiplying scores and weightings together was selected as this method highlights the 'extremes' and enables a higher level of discriminatory power to be achieved. The procedure is carried out automatically when the user clicks on the 'Perform MCC button'. The sum of each 'score x weighting' value for each performance indicator for each BMP option appears in the bottom row of the matrix. Having completed the MCC procedure, a button entitled 'Generate Order of Preference' appears leading to a table in which the BMP-specific summed values are ranked from highest to lowest, effectively generating an order of preference for the use of BMPs at a particular site. As an example, Figure 5 illustrates the format of an order of preference for the use of BMP options generated on completion of the MCC procedure. Where the use of a KPI has led to a breach of the site characteristics screening conditions, the relevant Key Term (KT) number is noted in the comment column (see Figure 5). The KT numbers refer the user to the particular technical KPI benchmark which has not been satisfied in the screening conditions e.g. KTs 16 to 21 cover design flow volume (KT16), available land area (KT17), water table level (KT18), soil hydraulic conditions (KT19), soil type (KT20) and infiltration rate (KT21). Further information on the Key Term list is given in Chapter 5.

If default scores have been used to generate an initial order of preference, an end-user may further refine this ranked list of BMPs by developing site-specific scores for the top four or five BMP options (or as many as are required) and re-running the MCC analysis as outlined above (i.e. completing the MCC procedure using Mode 3). It is important to note at this stage that the results generated through this approach are relative rather than directly comparable in terms of the values produced i.e. the MCC can determine that option one is more preferable than option two, but it cannot assess how much better option one is in comparison to option two. However, it might be argued that this deficiency is not specific to the DayWater MCC approach. The problem of evaluating the relative preference and sustainability of environmental alternatives, in the presence of cross-boundary institutions, agencies and vested interest groups, inherently possesses contradictions and contentions as argued by Ashley *et al.* in Chapter 25. Whilst it may be reasonably feasible to set up a quantitative benchmarking of KPIs for technical, environmental, economic and operation & maintenance criteria, it is much more difficult to arrive at such KPI benchmarking for socio-cultural, administrative and legal criteria, which will be readily acceptable to all stakeholders.

CONCLUSION: APPLICATION OF THE DAYWATER MCC AS AN URBAN STORMWATER DECISION SUPPORT TOOL

The MCC can support urban stormwater stakeholders, both directly and indirectly, in a variety of different ways. The performance matrix (Table 1) provides a decision-making framework within which the application of benchmarked performance scores enables a large amount of complex information to be handled in a consistent way. In addition, the inclusion of descriptions relating to the development of the default performance scores (both as short simplified summaries and more detailed PDF documents), provides stakeholders with a comprehensive source of information and explanation about each indicator and facilitates discussions in areas in which not all stakeholders may be familiar. Hence, decision-making is supported by enabling stakeholders to see the "bigger picture": a vital requirement in ensuring the identification of decisions which may be more sustainable over both the short and long term. The opportunity of selecting and implementing alternative weighting schemes enables a wide variety of views and concerns to be explored in a meaningful way, allowing the identification of a solution which best meets the needs of the stakeholder group as a whole. Flexibility and adaptability are major considerations in both the MCC and MoA structures. Adaptability is particularly important as it facilitates the inclusion of both expert and public lay opinion and participation in the decision process, and thus provides a genuine cross-boundary decision support tool for multi-attribute evaluation of best available stormwater practice (Chapter 2).

The MCC itself can be used to support a number of objectives, including identifying the single most preferred BMP option, producing a shortlist of possible BMP options and distinguishing acceptable from unacceptable possibilities. The latter two situations could both be regarded as initial 'filtering steps' in which the least favourable options are eliminated thereby reducing the number of possible solutions requiring detailed consideration and data gathering. This can be invaluable with regards to both focusing stakeholder discussions as well as preventing unnecessary data collection on BMPs which are unlikely to be selected, saving both time and associated costs.

The multi-criteria methodology provides a simple, robust and readily-understandable procedure that can be appropriately applied to the evaluation of selection criteria of alternative stormwater source control technologies as a means of identifying preferred solutions. The procedure is flexible and amenable to a variety of stakeholder and institutional interests with the methodology primarily intended to provide guidance for stakeholders within the context of a participatory negotiation and learning process.

REFERENCES

Ahlman, S and Svensson, G. (2002) Modelling substance flow in urban sewer systems using MATLAB/Simulink. In: Strecker, E.W and Huber, W.C (eds); *Global Solutions for Urban Drainage. Proc. 9th Int.Conf on Urban Drainage*, 8-13 September 2002. Portland, Oregon. CD-ROM, ASCE Publications, Reston, Virginia, USA. ISBN 0-7844-0644-8.

ASCE and UNESCO (1998) *Sustainability Criteria for Water Resource Systems*. Report of Joint ASCE/UNESCO Task Committee on Sustainability Criteria, Project M-4.3, IHP IV, American Society of Civil Engineers, Reston, Virginia, USA.

Ashley, R.A., Blackwood, D., Butler, D., Jowitt, P., Oltean-Dumbrava, C., Davies, J., Meilkenny, G., Foxon, T., Gilmour, D., Smith, H., Cavill, S., Leach, M., Pearson, P., Gouda, H., Samson, W., Souter, N., Henry, S., Moir, J and Bouchart, F. (2002) Making more sustainable decisions for asset investment in the water industry: Sustainable Water industry Asset Resource Decisions – The SWARD project. In: Strecker, E.W and Huber, W.C (eds); *Global Solutions for Urban Drainage. Proc. 9th Int. Conf on Urban Drainage*, 8-13 September 2002. Portland, Oregon. CD-ROM, ASCE Publications, Reston, Virginia, USA. ISBN 0-7844-0644-8.

Ashley, R.A., Blackwood, D., Butler, D and Jowitt, P. (eds) (2004) *Sustainable Water Services: A Procedural Guide*. IWA Publishing, London. UK. ISBN 1-843390-65-5.

Brunner, N and Starkl, M. (2004) Decision aid systems for evaluating sustainability: A critical review. *Environ. Impact Asst. Review*, **24**, 441 – 469.

Ellis, J.B., Deutsch, J-C., Mouchel, J-M., Scholes, L and Revitt, D.M. (2004) Multi-criteria decision approaches to support sustainable drainage options for the treatment of highway and urban runoff. *Science Total Environ.*, **334/335**, 251 – 260.

Ellis, J.B., Deutsch, J-C., Legret, M., Martin, C., Revitt, D.M., Scholes, L., Sieker, H and Zimmerman, U. (2005) The DayWater decision support approach to the selection of sustainable drainage systems: A multi-criteria methodology for BMP decision makers. In: *Proc 10th Int. Conf on Urban Drainage*. 21-16 August 2005, CD-ROM. Technical University of Denmark, Copenhagen, Denmark. ISBN 87-89220-80-3.

Figueira, J., Greco, S and Ehrgott, M. (eds) (2005) *Multiple-Criteria Decision Analysis: State-of-the-Art Survey*. Springer-Verlag, Berlin, Germany. ISBN 0-387-23081-51045.

Gilmour, D.J., Blackwood, D.J and Picken, K. (2005) Stakeholder involvement in wastewater treatment design. In: *Proc 10th Int.Conf on Urban Drainage*. 21-16 August 2005, CD-ROM. Technical University of Denmark, Copenhagen, Denmark. ISBN 87-89220-80-3.

Hellstrom, D., Jeppsson, U and Karrman, E. (2000) A framework for systems analysis in urban water management. *Environ. Impact Asst. Review*, **20**(3), 311 – 321.

Makropoulos, C.K., Morley, M., Memon, F.A., Butler, D., Savic, D and Ashley, R.A. (2005) A decision support framework for sustainable urban water planning and management in new urban areas. In: *Proc 10th Int.Conf on Urban Drainage*. 21-16 August 2005, CD-ROM. Technical University of Denmark, Copenhagen, Denmark. ISBN 87-89220-80-3.

Martin, C., Ruperd, Y. and Legret, M. (2007) Urban stormwater drainage management: The development of a multi-critera decision aid approach for best management practices. *European Journ. of Operational Research*, **181**(1) 338-349.

ODPM (2003) Multi-Criteria Analysis Manual. Office of the Deputy Prime Minister, London .

Parikh, P., Taylor, M.A., Hoagland, T., Thurston, H and Shuster, W. (2005) Application of market mechanisms and incentives to reduce stormwater runoff: An integrated hydrologic, economic and legal approach. *Environ. Sci. Policy.*, **8**, 133 – 144.

Sakellari, I., Makropoulos, C.K., Butler, D and Memon, F.A. (2005) Modelling sustainable urban water management options. *Proc. Inst Civil Eng: Engineering Sustainability*, **158**, ES3, 143 – 153.

Sieker, H., Bandermann, S., Schroter, K., Ostrowski, M., Leichtfuss, A., Schmidt, W., Thiel, E., Peters, C and Uhleck, M. (2005) Development of a decision support system for integrated water resources management in intensively used small watersheds. In: *Proc 10th Int.Conf on Urban Drainage*. 21-16 August 2005, CD-ROM. Technical University of Denmark, Copenhagen, Denmark. ISBN 87-89220-80-3.

Taylor, A.C. (2005) Triple-bottom-line assessment of proposed urban stormwater measures to improve waterway health. In: Proc 29th Hydrology & Water Resources Symposium, Engineers Australia: Water Capital, 21-23 February 2005. Canberra. CD-ROM, Engineers Australia, Canberra, Australia (available at www.catchment.crc.org.au verified August 2007).

Wenzel, V. (2001) Integrated assessment and multi-criteria analysis. *Phys. Chem. Earth (B)*, **26**(7/8), 541 – 545.

Chapter 10

Best Management Practice costing: methodologies for lifecycle cost assessment

J. Bryan Ellis and Emmanuel Aftias

INTRODUCTION

The cost of constructing and maintaining any stormwater source control drainage Best Management Practice (BMP) device will inevitably be a variable parameter depending primarily upon site location and catchment/drainage conditions. BMP costing data is also expressed in various forms with most literature sources providing only base capital construction costs with relatively limited information on front-end design/project and back-end operation and maintenance (O&M) costs. An appreciation of the long term financial implications are important when considering the overall costs of individual BMP devices given that, whilst capital costs may be lower than conventional piped systems, operational and maintenance costs may be higher. In addition, very few studies report on initial land costs which are possibly the most significant variable influencing overall BMP cost. In addition, there are costs related to risk, residual value and on-going environmental benefits which need to be considered in any long term cost comparison. This paper presents simplified lifecycle methodologies for the assessment of structural BMP drainage systems (see Chapter 8 for full listing and detail of various BMP devices) utilizing Capital / Operation and Maintenance costs (CAPEX/OPEX) principles with the balance of initial and future BMP expenditure evaluated at net present value (NPV) to a standard 50 year equivalent lifetime.

LIFECYCLE COST ANALYSIS

A whole life cost methodology is summarised in Figure 1 which identifies the various data components needed to undertake a meaningful appraisal. 'Capital costs' (CAPEX) include: (i) planning and site investigation costs, (ii) design and project management/site supervision costs, (iii) clearance and land preparation costs, (iv) material costs, (v) construction (labour and equipment) costs, (vi) planting and post-construction landscaping costs, and (vii) land-take costs. 'Operation and Maintenance' (O&M) costs include monitoring, planned (regular) operational maintenance and periodic unplanned remedial maintenance. 'Refurbishment and Replacement' costs include rehabilitation (resolving problems such as blocked culverts and screens) and/or replacement (such as plant replacement in wetlands). 'Risk costs' cover residual public and community costs not directly covered by any adoption or agreement clauses which might include exceptional legal costs associated with insurance liability action or household flood damage. 'Environmental' costs/benefits are "off-set" costs to recognise the direct (and indirect) local community and societal benefits that can accrue from the introduction of BMPs in an urban development including amenity, recreation, ecological and aesthetic enhancements, aquifer recharge as well as flood control and

Edited by Daniel R. Thévenot. ISBN: 9781843391609. Published by IWA Publishing, London, UK.

water quality improvements. 'Disposal' costs include requirements for disposal of vegetation (grass mowing, turfing, vegetation etc), sediment (which may need de-watering and safe disposal), contaminated geotextile/backfill or failed permeable block materials. 'Residual' costs (or more appropriately residual value) recognize the final NPV of the land, following the assumed operational lifetime of the site used for drainage control and treatment, and which are very likely to exceed the original land-take costs.

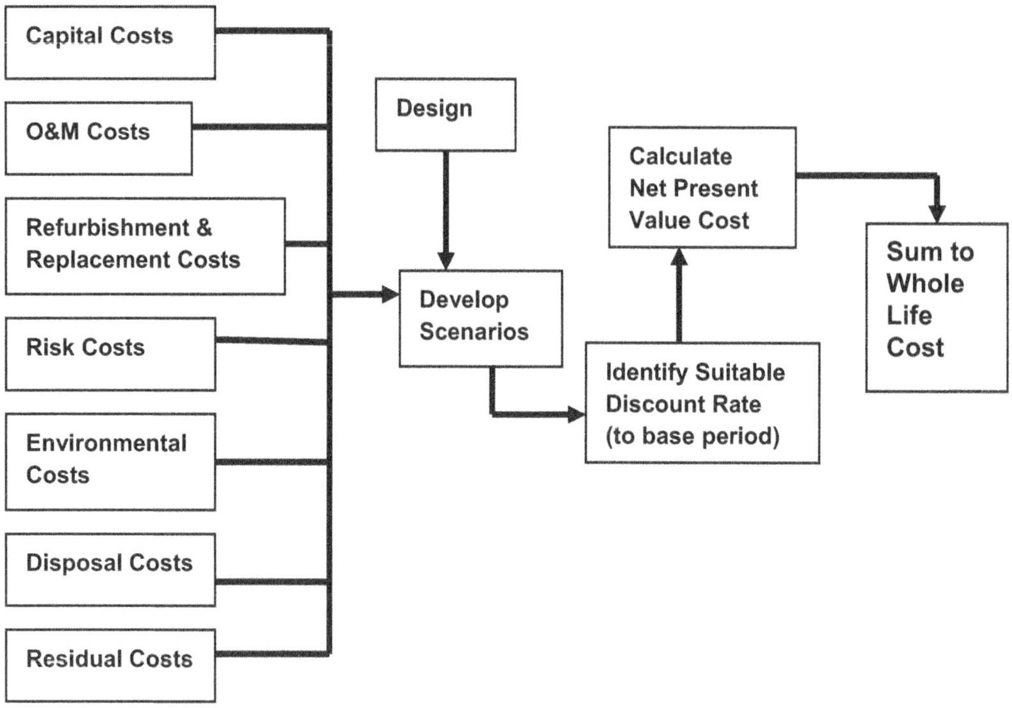

Figure 1: Whole life cost BMP appraisal.

The flow chart depicted in Figure 1 indicates the procedural steps in deriving a whole life costing based on the various data components. Scenarios refer to differing time frames and conditions relating to the costing period, to which an appropriate discount rate to allow for future inflation must be applied in order to derive NPV. It must be recognized that identifying overall average costs for each of the above components will be difficult as BMP costs are inherently variable even when conducted on a unit (volume or area) basis as they must depend on local conditions as well as size of the urban development. In addition, indirect factors such as public awareness and education, style and effectiveness of management and supervision practices, rigour of O&M schedules, as well as scale of development, will all affect the costing regime. As BMPs are relatively new source control drainage systems, many developers, contractors and adopting authorities are unfamiliar with costing schedules for many of the whole life cycle components shown in Figure 1, such as O&M or disposal costs. Quotations provided for BMP O&M for example, can vary considerably as evidenced by the UK M40 Oxford Motorway Service Area BMP treatment train and associated landscaping which received whole site quotations ranging from €32,000 to €64,000 per annum (Bray, 2001). Weigand *et al* (1986) consider that O&M costs can be simply considered as ranging between 5% - 10% of construction costs. There is therefore considerable uncertainty and lack of experience in appreciating the requirements of these systems and a resulting costing variability is to be expected even where specifications and schedules have been clearly provided. Most lifetime costing approaches therefore exclude some of the components shown in Figure 1, particularly risk and disposal costs, as well as environmental and residual benefit values, with many workers assuming that life cycle cost to be the sum of the net present value of the annual O&M cost incurred over time plus the initial capital costs (Sample *et al.*, 2003). Capital costs in this regard include land-take and easements, project planning and design, construction and any other environmental mitigation costs. Capital costs (C_{cap}) are frequently expressed as a power function of the explanatory variable(s) as $C_{cap} = a.x^b$, where x is a measure of size such as volume or area drained with the exponent b representing the economy of scale factor. The two parameters a and b are obtained from regression analysis of either specific site or literature data (Heaney *et al.*, 1999; Sample *et al.*, 2003). Most regard the inclusion of risk, environmental and residual cost components as being highly site specific, and as such likely to render meaningless any BMP comparison. With the exception of disposal costs, which have been incorporated into O&M costs, these components have also been omitted from the current cost methodology.

COST FUNCTIONS

Some sub-elements of the individual whole life cost components do have useful data and background information particularly in respect of capital costs per unit storage volume. Such cost functions have been well developed for retention basins, settlement tanks and infiltration trenches/basins e.g. see DayWater Deliverable 6.2 (Tarnaras *et al.*, 2004). However such cost function analysis rarely extends to biofilters, wetlands, roof storage or porous paving and also normally excludes consideration of post-construction landscaping costs or land-take costs. Figure 2 illustrates the inherent variability and underlying uncertainty in cost function methodology based on capital costing data for wet retention basin storage derived from the UK (Woods Ballard *et al.*, 2003), Europe (Tarnaras *et al.*, 2004), Australia (Tucker *et al.*, 2000) and the USA (Schueler, 1997; Harlow *et al.*, 2000). The continental European data are based on theoretical values derived from unit cost prices primarily related to the Greek market, whilst all other data are derived from real turn-out construction costs and costed operations collated from contractors and industry. The UK field data is the nearest to the theoretical European data and illustrates economy of scale up to about 8000 m^3 storage volumes which is also reflected to some extent in the USA data. The USA Texan data provided by Harlow *et al* (2000) are based on cost-performance information using a pollutant removal effectiveness index which can standardize the effects of BMP size.

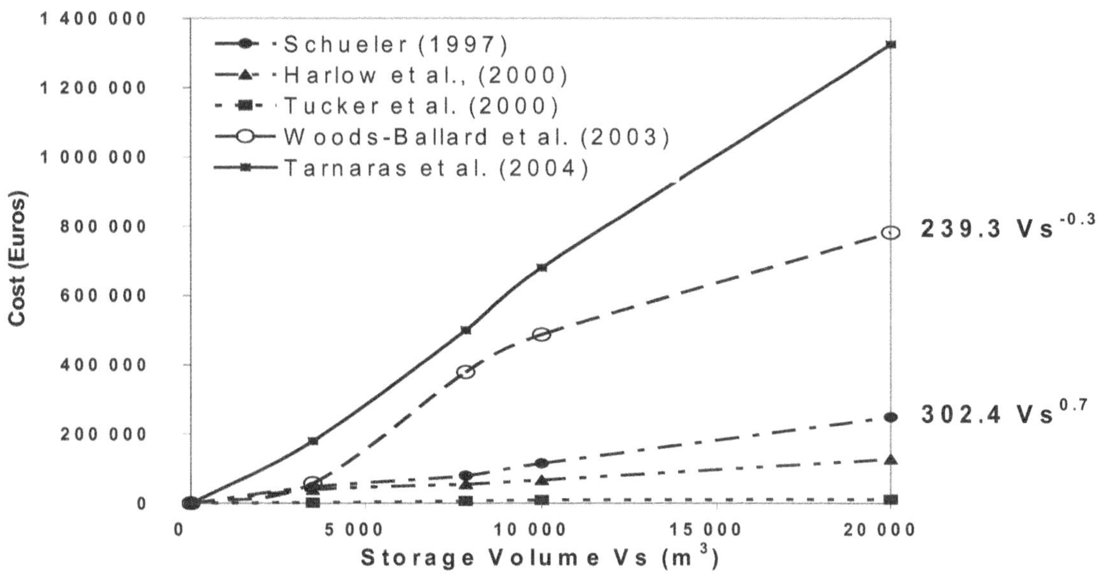

Figure 2: Capital cost functions (€) for storage in wet retention basin according to their storage volume (Vs m^3)

Despite the clear deficiencies exhibited by Figure 2 in terms of inherent variability, it is still nevertheless possible to utilise the cost function approach for at least the Capital Costing (CAPEX) component of the whole life cost index approach by selecting an appropriate curve from Figure 2 and scaling the cost in terms of performance/utility scores (e.g. 0 – 1.0, 0 – 10 or 0 – 100) (Chapter 9). The cost function equations (€ Vs m^{-3}) for the UK and USA data sets are given in Figure 2, although both of these are very different from that ($C = 61{,}000\ Vs^{0.8}$) quoted by Young *et al.* (1996) or by Wossink and Hunt (2003) ($C = 13{,}900\ Vs^{0.7}$) for highway retention ponds. Cost functions for a number of other BMPs can be obtained from US EPA (1995) and Tarnaras *et al* (2004). There is no doubt that cost functions are the most desirable method for providing this cost information, with capital and O&M infrastructure costs being defined as mathematical functions of one or more variables relating to the physical size, nature, throughput, capacity or other type of infrastructure item. Such cost functions should be as simple and generic as practically possible and it would be ideal if one function could be developed for each cost item. Efficiency of modelling and data entry require as few cost functions as possible and for some items it is difficult, if not impossible, to develop clear unit cost functions. The other essential problem with this approach is that there is very little similar cost data for BMPs other than retention and settlement basins. Nevertheless, despite these reservations, cost-function approaches do provide a logical basis for comparing whole life costs for conventional and alternative BMP devices, as part of a sustainable infrastructure analysis.

DERIVING EQUIVALENT LIFETIME COSTS

Stormwater source control BMPs have different design life spans and thus their economic comparison has to be estimated for the same life duration. Estimation of life time capital costs uses a timeframe equal at least to the longest lasting option: Table 1 illustrates net present value (NPV) coefficients which have been used as a basis for a 50 year reference with the coefficient (f) derived as:

$f = 1 / (1 + i)^{n}$

where i is the rate of interest (taken as a standard historic 6% discount rate rather than a declining long term discount rate), and the exponent n is the life span duration of each BMP.

Table 1: BMP Present Value Coefficients for a Lifetime Duration of 50 years (see text for coefficient calculation).

BMP	Life span (years)	Net present value coefficient
Swale	10	2.14
Green Roof	10	2.14
Porous Pavement	10	2.14
Porous Asphalt	10	2.14
Filter Drain	15	1.66
Soakaway	15	1.66
Infiltration Trench	15	1.66
Settlement Tank	50	1.00
Detention Basin	50	1.00
Extended Detention Basin	30	1.17
Infiltration Basin	30	1.17
Retention Basin	50	1.00
Constructed Wetland	30	1.17

Thus the present value capital cost (CAPEX) over 50 years for a swale having an average 10 years lifetime duration, i.e. replaced every 10 years, would be equal to:

C_{total} (50 years) = C_{total} (10 years) x f

where:

$f = 1 + [1/(1 + i)^{10}] + [1/(1 + i)^{20}] + [1/(1 + i)^{30}] + [1/(1 + i)^{40}] = 2.14$

It is evident from the Greek examples (Tarnaras *et al.*, 2004) that reduced capital CAPEX cost is associated with increasing storage volume as indicated in Figure 3 which shows the 50 year lifetime cost distribution for settlement tanks of varying storage volumes.

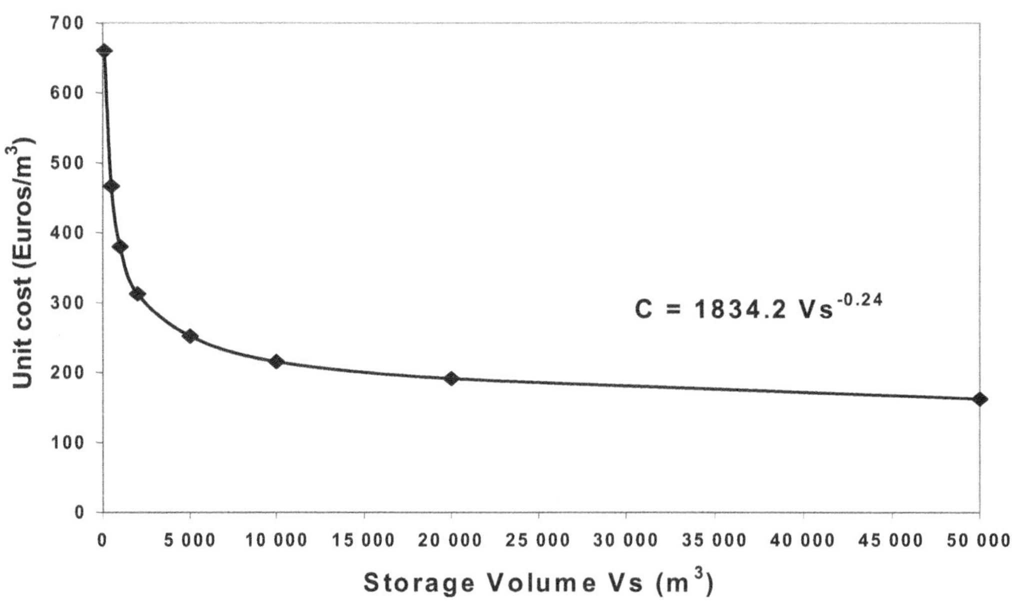

Figure 3: Variation of unit storage volume costs (€ m-3) for settlement tanks according to their storage volume Vs (m3).

Table 2: Estimated UK BMP capital construction unit costs.

Component	Unit	Mean UK cost value (€; 2003)
Site Preparation		
Light	Lump sum	408
Heavy	Lump sum	680
Site Clearing (including material removal/transport	Hectare	1629
Grubbing Site (including material removal/transport)	Hectare	5430
Excavation		
General	m^3	8.1
Trench excavation	m^3	8.25
Wetland excavation	m^3	7.05
Rock excavation	m^3	70.5
Placing and compact fill	m^3	5.4
Levelling	m^2	20.4
Grading	m^2	2.61
Formwork		
Concrete	m^3	150
Steelwork	kg	1.5
Wood shuttering/formwork	m^2	4.5
Site Development		
Crushed stone fill	m^3	39
Riprap	m^3	61.5
Polyethylene sheeting	m^2	6
Geotextile fabric	m^2	15
Grade-control structure	Each	10599
Basin inlet	Each	9768
Basin outlet	Each	13575
Observation well	m	814.5
Porous paving	m^2	87
Landscaping		
Salvage topsoil	m^2	2.28
New topsoil	m^2	5.43
Grass seed	m^2	1.5
Seed, mulch and	m^2	5.43
Fertiliser	m^2	4.89
Turf	m^2, hectare	
General landscaping etc..		8422.5
Design & Project Management	25% of total cost	

Table 3: Estimated UK BMP operation and maintenance unit costs

Component	Estimated UK unit cost (€; 2003)
Landscape Maintenance	
General grass/turf care	€16.5 / 100 m^2
Mowing	€1.5 / 100 m^2
Grass re-seeding (with mulch and fertiliser)	€0.65 / m^2
Re-turfing	€4.95 / m^2
Structural Maintenance	
Basin inlet maintenance	3% capital cost per annum
Basin outlet maintenance	5% capital cost per annum
Periodic inspection	€44 per inspection
Infiltration Maintenance	
Minor trench rehabilitation	€1.13 – 2.76 / m^3
Major trench rehabilitation	€2.42 - 3.6 / m^3
Soil levelling	€0.71 / m^2
Observation well monitoring	€171
General O & M	
Wet basin sediment removal	1% capital cost per annum
Infiltration basin sediment removal	1770 € / ha
Nuisance control (graffiti, vandalism etc.)	842 € / ha
Basin debris and litter removal	168 € per basin per annum
Swale debris and litter removal	1.0 € / m^2 per annum
Porous paving vacuum sweeping/High pressure hosing	633 € / ha per annum
O&M project management and inspection	84 € / basin or trench per annum

LIFETIME COSTING

Tables 2 and 3 identify the unit costings applied to various BMP capital and O&M components based on UK experience, as derived from questionnaire analysis of five leading consulting engineering companies. Such unit costs can form the basis for the development of a lifetime cost analysis for individual BMPs and this approach has been used in the working-up of the methodology for the DayWater project. Further detail can be obtained from the

DayWater Hydropolis Modelling Tools window under BMP Costing and Dimensioning Tools where users can work through the tables and spreadsheets to quantify specific BMP costs for particular site situations and dimensions (Tarnaras *et al.,* 2004). The 50 year cost formula provides a generic basis for the calculation of capital costs for differing settlement tank volumes and similar cost function graphs have been constructed for retention basins and wetlands. It is also evident that storage basin BMPs having extended operational lifetimes are likely to have near equivalent unit costs to those BMPs possessing shorter life spans given the higher NPVs of the smaller, short-lived source control devices such as swales or porous paving. The 'Operation & Maintenance' costs (OPEX) are based on the 15.76 NPV value of 1€ per year capitalized over 50 years at 6% discount which is the multiplier used to derive the equivalent 50 year lifetime O&M cost (see Box 1 for further detail of this calculation).

Figure 4 summarises BMP unit lifetime (CAPEX and OPEX) costs (€ m^{-3}) based on a combination of Greek, French, German and UK data which have been compiled within the DayWater project and which illustrates the considerable national variation in costing that can occur for some BMP devices: this is particularly noticeable in the case of green roofs. The Greek unit costs are primarily derived from commercial Bills of Quantity with the French and German data based on field examples whilst the UK costing data (Ellis, 2005) is averaged from data supplied by five consulting engineer and development companies (and originally expressed in terms of m^3 storage or m^2 surface area). The full data set can be inspected in the 'BMP Catalogue' accessed via the DayWater Hydropolis home page and then opening the 'Cost' window of the 'BMP Catalogue' (Chapter 8). All BMP capital costs have been referred to the equivalent timeframe for the longest lasting option i.e. over 50 years and the UK 'Capital Cost' data includes an initial design and project management overhead of 25%. Whilst the majority of storage devices in Figure 4 show a generally similar unit costing regime, some of the infiltration systems as well as green roofs and settlement tanks indicate that there can be a considerable variation in national unit costing between member states.

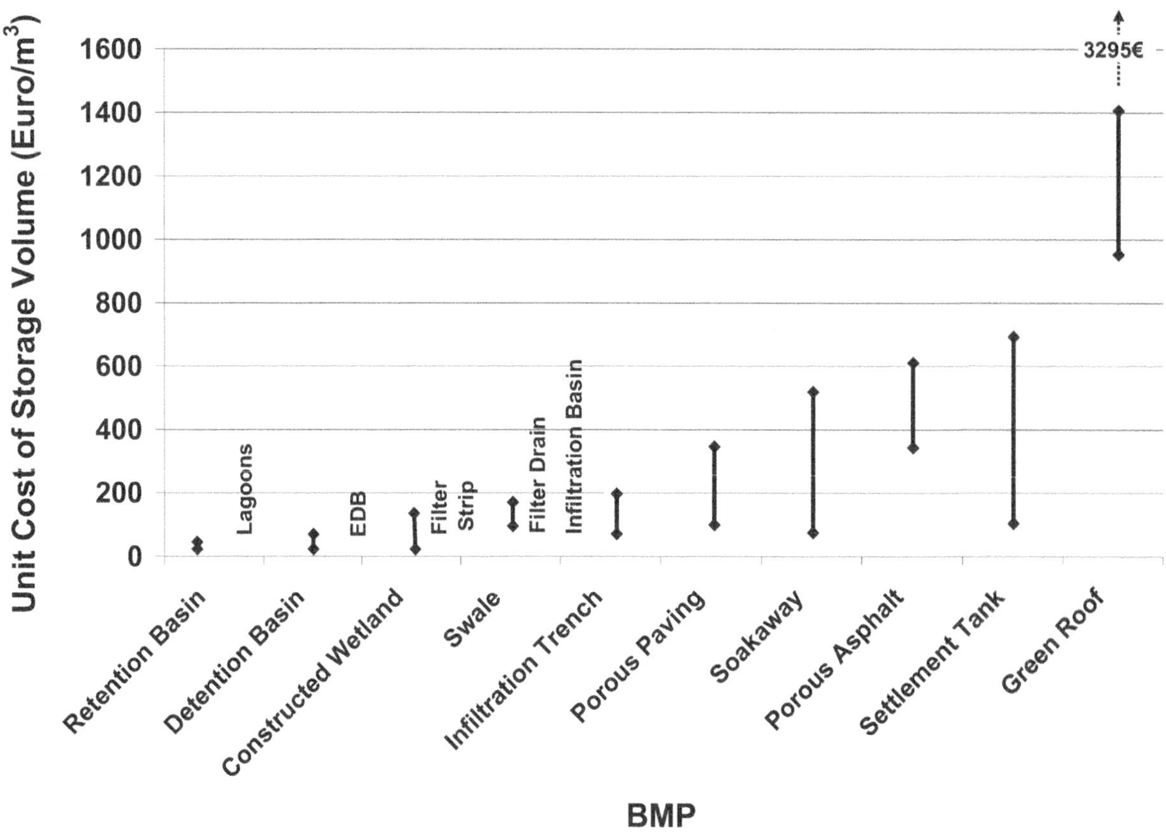

Figure 4: European BMP lifetime cost distribution range including unit CAPEX and OPEX costs (€ m^{-3}).

No allowance has been made for land-take costs, risk, and residual value or for off-set environmental benefits. A recent USA/UK joint project has suggested that land-take costs (BMP plan area, access/easement and any amenity/recreation area) can be as much as 40% of the total BMP whole life cost (Garden *et al.*, 2005). However, this is likely to be favourably balanced by the residual value at the end of the BMP lifetime. It has not been possible to identify sufficient information on costing regimes for a number of BMP units including filter drains, extended detention basins, lagoons and infiltration basins which have been inserted into Figure 4 on the basis of assumed functional comparability with other BMPs.

The data ranges shown in Figure 4 reflect not only differing EU national costing regimes but also the effects of BMP scale, as the larger the unit in terms of storage capacity, the more economical the overall costing. In the case of green roof storage based on the Greek data, the capital costs per storage volume have been obtained from the original 121.98 € m^{-2} cost by assuming an aggregate porosity of 40% for a depth of 70 mm and a corresponding 20% value for a soil/vegetation depth of 50 mm:

$$C_{total} / m^3_{storage\ volume} = [(121.98\ €\ m^{-2}) / (0.40 \times 0.07\ m + 0.20 \times 0.05\ m)]$$
$$= 3{,}210\ €\ m^{-3}_{storage\ volume}$$

This translates to an initial capital cost (CAPEX) estimate of 121.98 Euro per m^2 roof surface (see Box 1), which shows that green roofs cannot be considered economically efficient if considered purely in terms of storage volume, although they may retain events of up to 40 mm of rainfall. This approach has also been used for BMP infiltration facilities, with for example, the cost of an infiltration trench having an estimated initial cost of 46.00 € m^{-2} and a 40% aggregate porosity value being:

$$C_{total} / m^3_{storage\ volume} = [(46\ €\ m^{-2}) / (0.40)] = 115\ €\ m^{-3}_{storage\ volume}$$

In the case of porous paving, the porosity of the sub-surface aggregate has been assumed to be 40% for a depth of 0.5m with the 0.10m sand layer having a porosity of 30% and thus the cost per storage volume, based on an initial unit cost of 86.00 € m^{-2} is:

$$C_{total} / m^3_{storage\ volume} = [(86.00\ €\ m^{-2}) / (0.4 \times 0.5\ m + 0.3 \times 0.1\ m)]$$
$$= 374\ €\ m^{-3}_{storage\ volume}$$

The range of unit costing values shown in Figure 4 are primarily associated with variation in BMP storage capacity and where no cost information is available for a BMP, the assumed position of the facility has been inserted on the plot based on working experience.

Box 1 illustrates the steps in deriving the lifetime cost estimate for a green roof using the Greek commercial Bills of Quantity data for capital (CAPEX) costs and UK data for O&M (OPEX) costs.

Box 1: Life time costs of a green roof.

The estimated capital costs (CAPEX) for the transformation of a flat roof into a green roof are given below based on Greek Bill of Quantities data for 2004. The costing includes materials and labour but does not include any taxes, overheads and financial charges, bonds or design and project management costs, which have been additionally estimated at 100% of the total direct cost and thus the total cost has been multiplied by a factor of 2.

Component	Unit	Unit Cost (€)	Quantity	Total Cost (€ m^{-2})
Site Development				
Waterproofing,	m^2	4.20	1.00	4.20
30 mm Insulation barrier.	m^2	6.70	1.00	6.70
Geotextile fabric layers (above water protection layer and between soil layers).	m^2	9.00	1.20	6.60
Crushed stone (70 mm)	m^3	11.00	0.07	0.77
Landscaping				
Soil (50 mm).	m^3	3.50	0.05	0.18
Planting e.g. Sedum etc.	m^2	10.00	1.00	10.00
TOTAL				€ 28.45

Total cost for the construction of a green roof would therefore be 2 x 28.45 € m^{-2} = 57 € m^{-2}. Based on a 10 year lifetime, the capitalising coefficient would be 2.14 (see Table 1) and thus the equivalent 50 year lifetime cost would be 2.14 x 57 = 122 € m^{-2}.
Total O&M costs (OPEX) for a 100 m^{-2} green roof (based on UK data as no information is available for Greek conditions) are estimated at 1.47 € m^{-2} per year. Given the NPV of 1€ per year for 50 years is 15.76 €, the OPEX NPV capitalised over 50 years is (1.47 x 15.76) = 23 € m^{-2}.
Thus the total 50 year lifetime costs of a green roof would be 122 + 23 = 145 € m^{-2}.
As a comparison, based on UK costing data, the equivalent 50 year capital cost (CAPEX) would be 2.14 x 110 € m^{-2} = 235.4 € m^{-2} and therefore the total lifetime cost would be 235.4 + 23 = 258 € m^{-2}. In the calculation of the coefficients a standard 6% discount rate has been applied.

It should be noted that the data outcomes as seen in Figure 4 and Box 1 have been based on a standard 6% discount rate applied to the 50 year lifetime duration period. However, the use of alternative rates (including a variable rate) would significantly affect the outcomes and this is important given, for example, that the UK Treasury has recently recommended a shift to a lower discount rate of 3.5%. At lower discount rates, a high CAPEX/low OPEX BMP model would be preferred as a higher weighting is placed on future costs. At the 12.5% discount rate generally applied to developing countries, the reverse would be true with low CAPEX/high OPEX BMP applications being favoured. The aim of shifting to a lower discount rate is to encourage longer term, more sustainable development, which is entirely consonant with the functional objectives of BMP drainage solutions.

A similar approach by Tucker *et al* (2000) as applied to a hypothetical BMP design for a 26 hectare residential development in Brisbane, Australia showed that at low interest rates (5 - 9%), the BMP lifecycle costs were higher than traditional kerb-gully-pipe drainage systems. At a 5% interest rate, BMP lifetime costs were 11% higher than traditional methods but, at approximately 10% interest rates, cross-over was reached and above this, BMP costs became progressively cheaper.

COST-EFFECTIVENESS APPROACH

Based on the methodology developed by Harlow *et al* (2000), a cost-effectiveness approach relating BMP removal efficiency to lifecycle costs has also been undertaken. Most literature studies would confirm the findings of Barrett (1999) that grass swales and filter strips are the cheapest treatment options to construct, although detention basins and extended detention basins are probably the least expensive BMPs in common usage with constructed wetlands costings being some 25 to 30% more expensive than wet retention basins.

A cost-effective index (CE_i) can be derived as:

CE_i = [(Lifetime Cost; € / Unit of stormwater treated; m^{-3}) / (Pollutant Load Removed; kg)]

where cost and performance efficiency as well as stormwater volume processed are taken into account. At a cruder level of analysis, the denominator in the equation can be expressed as percentage pollutant removed, for which literature values can be readily obtained for a variety of BMPs e.g. see the BMP Performance data in the DayWater BMP Catalogue (Chapter 8). Even the most efficient BMPs can be limited in their capacity and this can result in a much higher cost per unit of stormwater treated. The index approach can be applied to individual pollutants or as a weighted average developed to cover a number of combined pollutants e.g. heavy metals or hydrocarbons. Given the level of uncertainty associated with the long term performance of BMPs, the index approach can only provide an initial screening analysis.

Figures 5a and 5b show results based on an analysis of total suspended solids (TSS) removal efficiencies for a range of BMPs with performance data obtained from several UK studies involving a variety of urban runoff sites. The swale, filter strip and porous paving studies (Figure 5a) represent results from a series of 2 year return interval storm events (n >4) providing greater than 65, 70 and 90% TSS capture respectively for the three BMP types. It can be seen from this figure that catchment area drained has relatively little effect upon the unit cost effective index CE_i of these devices, with larger surface area forms being almost as costly as much smaller devices.

The basin-type BMP data of Figure 5b is based on essentially earthen structures and assumes runoff volumes equivalent to 2.5 times the mean storm event, a drawdown time of 24 hours for the infiltration and extended detention basins and TSS removal rates greater than 75%. The influence of catchment area on basin-type BMPs is much clearer for these devices (particularly for basins draining up to 8 hectares), with constructed wetlands and wet retention basins having roughly similar costs but being 2 to 3 times more expensive than infiltration or extended detention basins. Whilst this may be the case, and to some extent this is confirmed by the information provided in Figure 4 based on the DayWater methodology, one must remember that there is considerable uncertainty in the cost-performance analysis caused by the aggregation of data derived from several discrete studies into the single line BMP plots shown in Figure 5a and 5b. In addition, the plots reflect storm specific data and could be strongly influenced by the intensity of the storm event and therefore the pollutant concentrations and loadings delivered to and captured by the BMP device.

However, the declining form of the basin-type BMP plots may well be generic and confirm the general validity of single cost function approaches. Irrespective of this, the data in Figures 4 and 5 are complementary and reinforcing indicating quite clearly that the order of relative BMP costing follows:

filter strip < swale < porous paving < infiltration basin/trench < wet ponds.

The cheaper costing of the first three BMPs is at least partially related to their catchment and construction size restrictions in comparison to the larger basin-scale devices. The economies of scale operating for basin-type BMPs resulting from potential storage volumes are also apparent, with relatively small differences (apart from infiltration basins) shown in the costing differential for basins draining catchments greater than 9 hectares.

Lee *et al* (2005) have presented a life cycle BMP optimisation procedure which is also based on cost-effectiveness using the storage-treatment relationship for evaluating performance in respect of long term rainfall time series data. Their conclusions are broadly similar in emphasising the advantageous economies of scale offered by large volume storage basins. It is probably more efficient to use undersized versions of BMP infiltration trenches and porous paving for water quality control in combination with less expensive structures such as dry detention basins, to attenuate and reduce peak runoff. Wossink and Hunt (2003) have also approached whole life costing from a cost-effectiveness basis using unit-cost curves for treatability and catchment area drained. Some of their raw data has been plotted in Figure 6 for retention basins, wetlands and bioretention (combined swale and infiltration trench) BMPs. The results confirm to some extent the economies of scale as catchment area drained increases, even when no statistical relationship was detected between pollutant removal efficiency and catchment size.

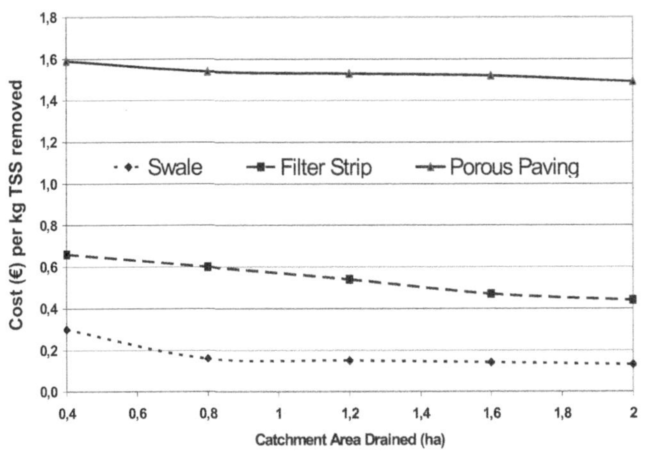

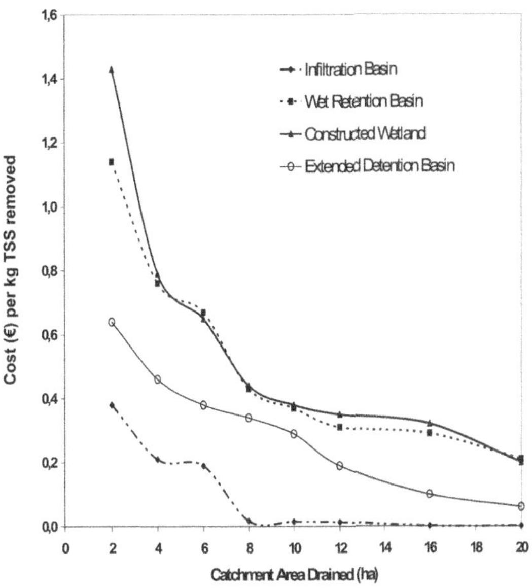

Figure 5a: BMP unit cost-performance for basin-type structures, when considering the TSS removal (€ m^{-2}.TSS kg^{-1}) within different catchment area.

Figure 5b: BMP unit cost performance index for swales, filter strips and porous paving, when considering the TSS removal (€ m^{-2}.TSS kg^{-1}) within different catchment area.

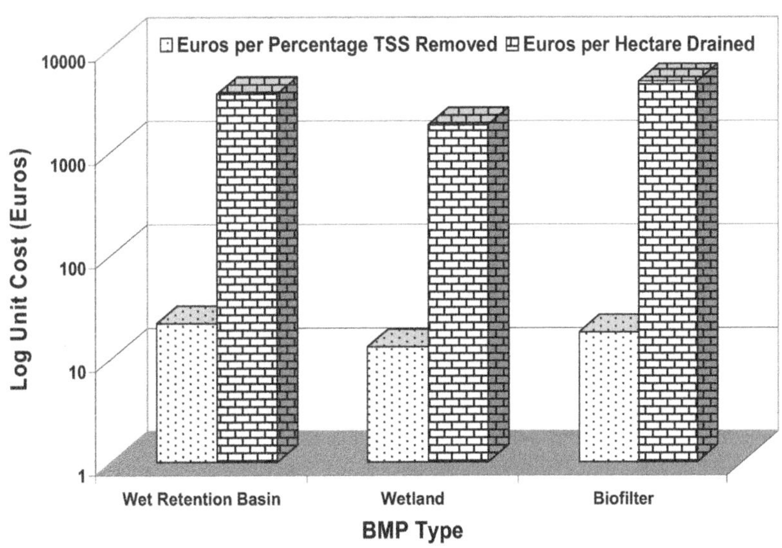

Figure 6: Unit cost comparison for BMPs either per percentage TSS removed or per hectare drained: wet detention basin, wetland and biofilter (combined swale and infiltration trench) (based on data in Wossink and Hunt, 2003).

CONCLUSION

Neither the DayWater Unit Cost nor the Cost-Effectiveness approach take into account the additional secondary benefits that BMPs can give to the lifetime costs which include amenity and environmental benefit values, cleaner receiving watercourses, avoidance of algal blooms and delay in building of water supply infrastructure that could result from stormwater recycling (Chapter 6). If these benefit values were costed into the methodologies, BMPs would undoubtedly provide at least as much overall benefit as traditional approaches across their entire lifetimes and quite probably at all discounting rates. Any small apparent increases over conventional drainage systems in terms of lifecycle costs associated with the source control BMPs resulting from the simple methodologies would therefore appear to be fully justified.

Standardised cost analysis across the entire lifespan of stormwater drainage infrastructure is essential for a viable comparison of the relative sustainability of alternative source control BMPs with conventional piped systems. Costs can frequently be the prime influencing factor for developers in deciding which system to adopt. There is therefore a need to demonstrate a relative costing procedure in a consistent, robust and understandable manner. It may be that a cost-effectiveness approach will be less influenced by variations in national cost regimes than a unit cost approach. In addition, being based on BMP performance, it can provide a clearer picture of the relative long term cost-benefits accruing to specific decisions in terms of environmental outcomes.

REFERENCES

Barret, M.E. (1999) *Complying with the Edwards Aquifer Rules: Technical Guidance on BMPs*. Texas Natural Resource Conservation Commission. Austin, Texas, USA.

Bray, R. (2001) Maintenance of sustainable drainage. In *Proc 1st National Conference on Sustainable Drainage*, Pratt, C.J., Davies, J.W and Perry, J.L (eds):. June 2001. pp. 93–104 Coventry University, Coventry. UK.

Brown, W and Schueler, T. R. (1997). *The Economics of Stormwater BMPs in the Mid-Atlantic Region*. Centre for Watershed Protection, Ellicott City, Maryland. USA.

Ellis, J.B. (2005) *UK BMP CAPEX and OPEX Cost Data*. DayWater WP5, Hydropolis, BMP Catalogue. EU 6th Framework Programme. (www.daywater.org) (verified on July 2007).

Garden, M., Woods Ballard, B and Taylor, L. (2005) Whole life costing of SUDS in Practice. In *Proc. 3rd Nat. Conf. on Sustainable Drainage* Newman, A.P., Pratt, C.J., Davies, J.W and Blakeman, J.M (eds), June 2005. pp. 164 – 173 Coventry University, Coventry. UK.

Heaney, J.P., Sample, D and Wright, L. (1999) *Costs of Urban Stormwater Systems*. Report EPA/R-02/021. Nat. Risk Managt. Res. Lab., US Environmental Protection Agency, Cincinnati, Ohio, USA.

Harlow, C., Landphair, P.D., McFalls, J.A and Thompson, D. (2000) *Design Methods, Selection and Cost Effectiveness of Stormwater Quality Structures*. Report FHWA/TX-01/1837-1, Texas Dept of Transportation, Austin, Texas, USA.

Lee, J.G., Heaney, J.P., Rapp, D.N and Pack, C.A. (2005) Life cycle optimisation for highway BMPs. In *Proc.10th Int. Conf. on Urban Drainage*, 21 – 26 August 2005, Copenhagen, Denmark.

Sample, D.J., Heaney, J.P., Wright, L.T., Fan, S., Lai, F and Field, R. (2003) Cost of best management practices and associated land for urban stormwater control. Amer. Soc. Civil Eng., *J. Water Resources Planning & Managt.*, **129**(1), 59 – 68.

Schueler, T. (1997) *Costs and Benefits of Stormwater BMPs*. Final Report, Centre for Watershed Protection, Ellicott City, Maryland, USA.

Tarnaras, I., Papavasiliou, C., Papatzani, S and Aftias, E. (2004) B*est Management Practices Cost Estimates*. DayWater Deliverable 6.2. EU 6th Framework Programme. (www.daywater.org) (verified July 2007).

Tucker, J., Rozis, N and Rahman, A. (2000) A simple method for life cycle cost assessment of water sensitive urban design. In *Global Solutions for Urban Drainage*. Proc 9th Int. Conf. on Urban Drainage, Strecker, E.W and Huber, W.C (eds) September 2002, Portland, Oregon, USA.

US EPA. (1995) *Economic Benefits of Runoff Control*. Report EPA 841S-95-002. United States Environmental Protection Agency, Washington DC, USA.

Weigand, C., Schueler, T., Chittenden, W and Jellick, D. (1986) Cost of urban runoff quality controls. In *Urban Runoff Quality; Impact and Quality Enhancement Technology*. Urbanos, B and Roesner, L.A (Edits) pp. 366 – 380 American Society Civil Engineers, New York. USA.

Woods Ballard, B., Malcolm, M and Kellagher, R. (2003) *Whole Life Costing for Sustainable Drainage Schemes*. HR Wallingford, Oxon. [See also pp. 181 – 192 In *Proc 2nd Nat. Conf. on Sustainable Drainage*, Pratt, C.J, Davies, J. W., Newman, A. P and Perry, J. L. (eds), June 2003, Coventry University, Coventry, UK.

Wossink. A and Hunt, B. (2003) *The Economics of Structural Stormwater BMPs in North Carolina*. Res. Report 344, Water Resources Research Inst., North Carolina State University, Raleigh, N Carolina. USA.

Young, G.K., Stein, S., Cole, P., Kammer, T., Graziano, F and Bank, F. (1996) *Evaluation and Management of Highway Runoff Water Quality*. Office of Environment & Planning, Federal Highway Administration, Washington DC, US.

Chapter 11

Selection of priority pollutants in the DayWater project

Eva Eriksson, Anders Baun, Peter Steen Mikkelsen and Anna Ledin

INTRODUCTION AND OBJECTIVES

Urban stormwater contains a vast range of different pollutants such as organic matter, nutrients, suspended solids, metals, inorganic trace elements and xenobiotic organic compounds (XOCs). For the evaluation of alternatives applicable for stormwater handling, it is necessary to assess potential hazards and problems related to the water constituents. The types of hazards and problems that should be taken into account are related to the selected strategy for stormwater handling. Hazards regarding exposure of humans, livestock, aquatic and terrestrial organisms, crops and plants have to be considered, as well as technical and aesthetical problems. Hazards that are needed to be taken into account are both acute and long-term effects to living organisms, e.g. toxicity, bioaccumulation, carcinogenicity, mutagenicity, reproduction hazards and endocrine disrupting effects, as well as promoting allergic reactions in humans and the persistence of the compounds in the environment. Precipitation of salts/minerals and corrosion of different types of installations and tubing are examples of relevant technical problems. Bad odour, frothing and discolouration are examples of aesthetical problems to be considered.

The objectives for this part of the DayWater project were:

- To develop a tool that can support the user in the selection of priority pollutants i.e. it is intended to be used as a road map that can be adapted to the special requirements of the individual user,
- To develop a tool that supports the user to identify which XOCs present highest potential hazards,
- To establish a list of selected stormwater priority pollutants (SSPP) suitable for the DayWater project, to be used in other sub projects such as evaluation of best management practices (BMPs) (Chapter 8) and development of the Source & Flux model (SEWSYS) (Chapter 14) as well as a potential stormwater monitoring programme.

The Chemical Hazard Identification and Assessment Tool (CHIAT) developed in the DayWater project, is limited to chemical, technical and aesthetical risks, as the risk related to flooding has been covered in another sub project in DayWater, Flood Vulnerability Assessment Tool (Chapter 13). Microbial risks have been assessed in the Swedish Urban Water Management Programme (UWMP) (Malmqvist *et al.*, 2006) during a science exchange within the CityNet cooperation and the UWMP. Furthermore, an example of the application of CHIAT in a specific case study is presented in Chapter 19.

 DayWater: an Adaptive Decision Support System for Urban Stormwater Management. Edited by Daniel R. Thévenot. ISBN: 9781843391609. Published by IWA Publishing, London, UK.

METHODOLOGY

The tool developed, CHIAT, was inspired by the first two steps in the Technical Guidance Document for approving new chemicals to be introduced in the market (European Commission, 2003), i.e. the hazard identification and the hazard assessment. CHIAT consists of five steps: (1) Source characterisation, (2) Recipient, exposure object and criteria identification, (3) Hazard and problem identification, (4) Hazard assessment and (5) Expert judgement (Figure 1) (Eriksson *et al.*, 2005a).

Furthermore, over 10500 compounds, primarily XOCs, are used or produced in amounts over 10 tonnes per annum in Europe (European Chemicals Bureau, 2006). The XOCs group is heterogeneous and it is therefore difficult to generalise the behaviour of the individual XOCs. Step 3, 'Hazard and problem identification', consist of a tool 'Ranking and Identification of Chemical Hazards' (RICH) which has been developed to rank the XOCs based on their inherent properties in order to identify the most hazardous ones (Figure 1). Thanks to the tiered approach in RICH, it copes with the lack of knowledge of some of the inherent properties in a way that it allows more pollutants to be ranked than other ranking approaches (Baun *et al.*, 2006).

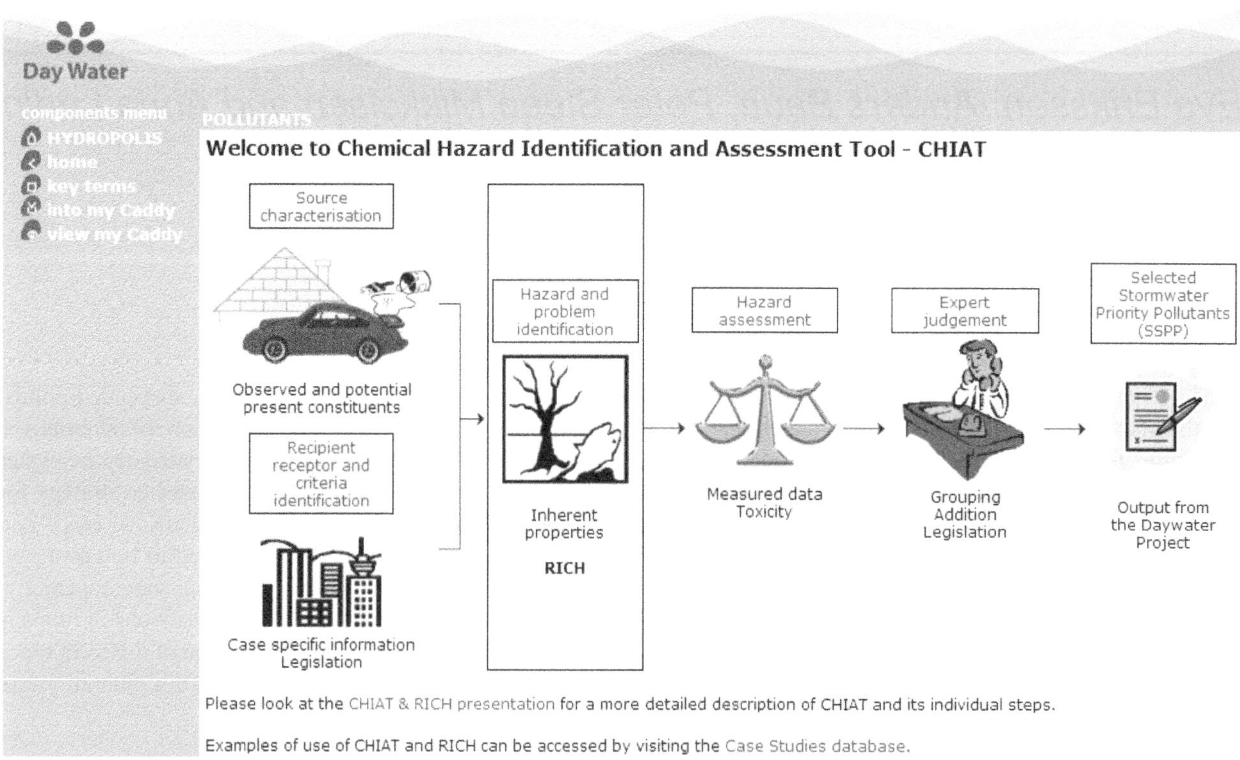

Figure 1: Front page of the Chemical Hazard Identification and Assessment Tool (CHIAT).

Step 1. Source characterisation

Initially, information regarding potential pollutants should be collected. This can be performed by two somewhat different approaches: (i) Searching in the open international literature regarding observations/measurements of pollutants (general water quality parameters, metals, inorganic trace elements and XOCs) in stormwater: this is generating a list of observed constituents (Eriksson *et al.*, 2005a and b) and (ii) Searching in the literature regarding pollutants that potentially could be present in stormwater due to the use of chemical products such as de-icing agents and pesticides in the gardens, releases from materials such as building and road materials, brakes and tires on vehicles, and atmospheric deposition (Figure 2). The aim of this second collection procedure is to identify those pollutants that have not yet been included in monitoring programmes. The output from both operations is a list with all observed and potentially present pollutants: the **potential pollutants** (Figure 1).

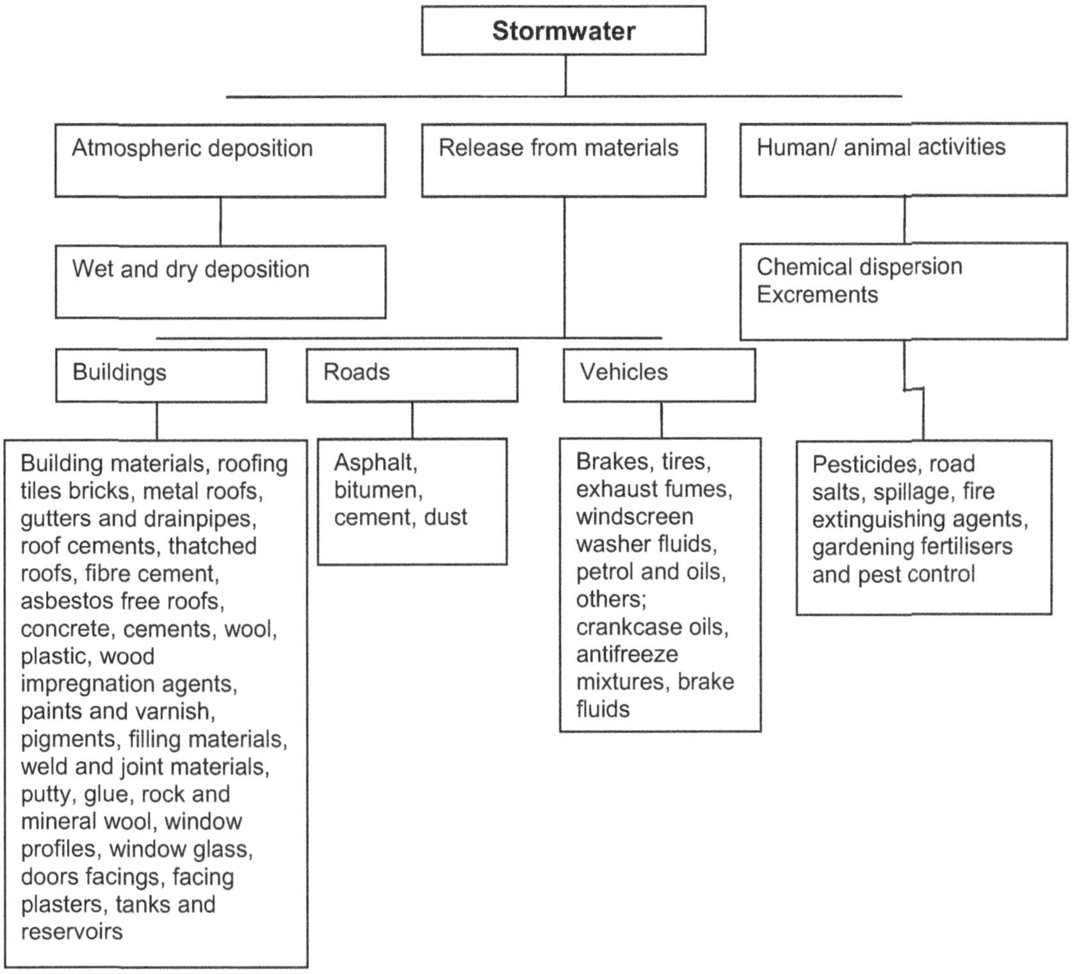

Figure 2: Pollutant sources in urban stormwater.

Step 2. Recipient, exposure object and criteria identification

In this step, the different strategies for handling of the stormwater are evaluated with respect to potential human health hazards, technical and aesthetical problems as well as environmental hazards that could occur due to the presence of the chemical pollutants. Use, treatment or discharge scenarios will be investigated in order to identify the potential exposure routes and what or who are exposed (the exposure objects).

Table 1: Recipients and exposure objects included in DayWater (X; recipient and exposure objects relevant for the strategy described in DayWater).

	Recipients				Exposure objects			
Strategy considered in DayWater	Surface water	Ground water	Soil	Technical installations	Aquatic organisms	Terrestrial organisms[1]	Terrestrial organisms[2]	Humans
1. Stormwater discharge to surface water	X				X			X
2. Stormwater infiltration		X	X				X	X
3. Stormwater use for toilet flushing, laundry wash etc after on-site treatment				X				X
4. Stormwater use for garden watering after on-site treatment		X	X	X		X	X	X

(1) Larger mammals e.g. dogs, horses or other recreational/companion animals; (2) soil dwelling organisms.

Legislation should be reviewed, for each scenario, to elucidate the quality and emission standards to be used in step 4 and 5. Examples of recipients of relevance are water, soil and technical installations such as stormwater utilisation facilities for toilet flushing. Humans as well as aquatic and terrestrial living organisms are examples of exposure objects (Table 1). The output from this step is a list with **recipients, exposure objects and criteria for hazard assessment** (Figure 1), e.g. policy documents, regulation, etc.

Step 3. Hazard and problem identification

All constituents identified as potential pollutants in the first step of the CHIAT procedure (Figure 1) are evaluated in the third step: the hazard and problem identification. The criteria for evaluation are based on environmental fate (sorption, volatilisation, persistence to degradation and bioaccumulation), short-term aquatic toxicity and long-term effects to living organisms (carcinogenicity, mutagenicity, reproduction damage and endocrine disrupting effects as well as promoting allergic reactions in humans).

The hazard and problem identification is performed according to the RICH tool, which up to now has been developed and tested for XOCs (Baun *et al.*, 2006). The tool consists of a decision tree in which hazardous and problematic compounds are identified. In order to visualize the ranking of XOCs, the decision tree can be described as a funnel fitted with several filters (Figure 3). The filters have been set according to specified criteria show in Table 2. The pollutants are first divided up, based on their sorption and volatility and hence they are listed as primary existing in the water phase and/or primary existing in solid phase (sediments, suspended solids etc) or evaporating from the system. RICH presupposes an open system where stormwater is in contact with air during the initial part of the runoff process. The criteria used for the subsequent ranking are based on persistence to biodegradation, potential for bioaccumulation and aquatic toxicity. There are also one on/off filter for technical/aesthetical problems and one filter for long-term chronic effects-filter considering cancer, mutagenic and reproduction hazards, endocrine disruption effects and allergenic effects. The cut-off values used for the shown criteria in the current version of RICH are presented in Table 2. These values can however be changed and thus adapted to a specific scenario. The output is a classification of the chemical compounds in three categories (green, yellow and red) depending on their priority as possible pollutants (Figure 3).

Green compounds are considered as non-priority pollutants: this means that these compounds will be excluded from further evaluation. Yellow compounds are passed on to the next filter. These compounds may or may not be priority pollutants depending on the outcome of the following filtration. Red compounds are considered as priority pollutants. However, green compounds from the phase distribution filter e.g. in the water phase, does not mean that they are non-hazardous but merely that they will not exist in water phase and thus, that they will be tested in the solid phase funnel and thereafter classified as green, yellow or red depending on their fate and effects in a solid phase environment will be carried out.

The inherent properties of the potential pollutants, needed to carry out this step, have to be compiled from the literature (databases and handbooks) (Baun *et al.*, 2006).

Metals and inorganic trace element were not subjected to the RICH tool as the persistence criteria do not apply to these pollutants. Instead, saturation indexes were calculated by geochemical modelling (Parkhurst and Appelo, 2001). Metals and inorganic trace elements receiving a saturation index above 1 (i.e. the concentration of the mineral in the solution is higher than what is theoretically possible during the given conditions) are oversaturated and thus are disposed to precipitation of the corresponding mineral, which could constitute a technical problem. This approach is used for the metals and inorganic trace element instead of the RICH tool used for the XOCs. The compounds that are identified as hazardous or problematic in this step are listed as **potential stormwater priority pollutants** (PSPP).

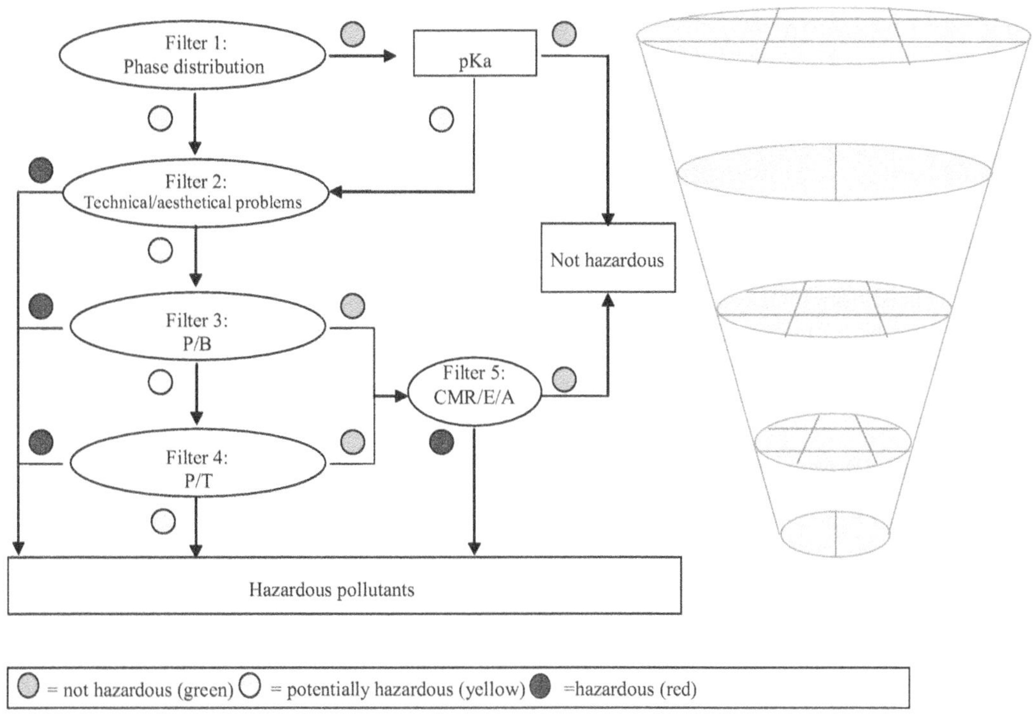

Figure 3: RICH decision tree used for identification of hazardous pollutants: Filter 1 (phase distribution), Filter 2 (technical/aesthetical problems), Filter 3 (persistent and bioaccumulating), Filter 4 (persistent and toxic) and Filter 5 (carcinogenic, mutagenic, reproduction hazardous or endocrine disrupting or allergenic).

Table 2: Criteria and corresponding cut-off levels used the different filters in step 3.

		Ranking filters			On-Off filters	
		Low	Medium	High	On	Off
Phase distribution	K_H	$K_H < 10^{-7}$	10^{-7} $K_H \le 10^{-3}$	$K_H > 10^{-3}$		
	K_D	$K_D < 100$	$100 \le K_D < 1000$	$K_D \ge 1000$		
	Koc	Koc < 500	500 ≤ Koc <5000	Koc ≥ 5000		
	pKa				pKa < 9.5	pKa > 9.5
Technical and aesthetical problems	Precipitation				Sw or SI > 1	Sw or SI < 1
	Discolouration				Colour	No colour
	Odour				Odour threshold	No known odour
Environ-mental fate	Persistence	1. Readily biodegradable* 2. T½ < 40 d 3. Biowin3 weeks & Biowin5 ≥ 0.5	1. Inherently biodegradable* 2. 40 ≤ T½ < 180 3. Biowin3 weeks or Biowin5 ≥ 0.5	1. Persistent* 2. T½ ≥ 180 d 3. Biowin3 months or Biowin5 < 0.5		
	Bioaccumulatio n	BCF<100 Log Kow<3.0	100≤BCF≤5000 3.0≤ Log Kow<4.5	BCF ≥ 5000 Log Kow ≥ 4.5		
Toxic effects	Acute toxicity	$EC/LC_{50} > 100$	$1 \le EC/LC_{50} < 100$	$EC/LC_{50} < 1$		
	Long-term chronic effects**				Known human or animal effects	No known effects on humans or animals

K_H: Henry's law constant i.e. partition coefficient between air and water; K_D: Sorption/distribution coefficient; Koc: Partition coefficient between organic carbon, e.g., organic matter and water (L/kg); pKa: Acid dissociation constant; Sw: Water solubility in mg/L (here used for XOCs); SI: Saturation index; T½: 1st order degradation half-life (days); Biowin: Biodegradability models no. 3 and 5 included in EPI Suite 3.12 (US-EPA, 2005); BCF: Bioconcentration factor in L/kg or g/kg; Kow: Partition coefficient between n-octanol and water; LC_{50}: Lethal concentration (acute test, 50% mortality); EC_{50}: Effect concentration (acute test, 50% effect).

* Classification according to standard OECD/ISO tests for ready and inherent biodegradability.

**Cancer, mutagenic effects, reproductive effects and/or endocrine disrupting effects as well as promoting allergy in humans.

Step 4. Hazard assessment

The hazard assessment is, in general, a comparison between the effect and the exposure levels of a specific pollutant, in order to evaluate if there is an actual hazard under the actual project context. The selected criteria for a hazard assessment are related to the exposure object e.g. for surface waters, toxicity data for fresh water organisms are used whereas for infiltration scenarios, toxicity data for soil dwelling organisms are used. The exposure can be represented by predicted environmental concentrations (PEC), where the values can be based on measured data or model simulations e.g. by SEWSYS (Chapter 14) developed in DayWater. Evaluation of estimated concentrations for which unacceptable effects are not likely to occur can be done by estimation of predicted no effect concentrations (PNEC). PNEC can be calculated from toxicity data using the most sensitive species and application of a safety factor to account for uncertainty (European Commission, 2003). Comparison between the PEC and PNEC values are made in order to determine if the compound should be considered as hazardous for organisms in the environment. The pollutants for which PEC/PNEC ratios are above 1 are classified as **priority stormwater pollutants (PSP)**.

Step 5. Expert judgement

Finally, the expert judgement is performed on the PSP list. The 'expert' is not necessarily a single person but may be a group of decision-makers with different backgrounds (e.g. engineers, politicians, environmental chemists, local stakeholders etc). The idea is that the experts select the PSP for which further action needs to be taken. The expert evaluation may, for example, aim at reducing the number of compounds due to financial limitations in the specific project. Compounds may also be removed based on use patterns in the catchments or grouped based on similarity in structure and fate: in the latter case, an indicator compound may be chosen to represent the whole group. Compounds that are banned from use can also be removed unless certain reasons exist e.g. if continuous emissions still exist (as for example for PCBs). Legislation concerning limit values such as drinking water standards as well as environmental quality standards and emission limit values for watercourses, lakes or the sea, reviewed in step 2 should be used in order to identify compounds that may need to be added to the list. Compounds may also be added if they are priority pollutants present on national/international lists or 'special case' compounds. The output from this step is a list containing those chemical compounds and other parameters that constitute a hazard after evaluation by the expert: the **selected stormwater priority pollutants (SSPP)**.

THE DAYWATER SSPP-LIST

The CHIAT procedure and the RICH tool were developed and tested during the DayWater project. However, RICH was only developed for XOCs and step 4 in the CHIAT procedure was excluded due to the scope of the third aim, to select a general European list of stormwater priority pollutants thus not requiring site-specific data.

Step 1. Source characterization

The literature survey regarding observations and measurements of pollutants in urban stormwater showed that a large number of pollutants have been identified and quantified in stormwater originating from roads, roofs, recreational areas, court yards and airports as well as other urban areas. In all, over 500 general water quality parameters, metals, inorganic trace elements and XOCs have been reported (Makepeace *et al.* 1995; Eriksson et al., 2006a). In order to illustrate the diversity within the XOCs group, a division in subgroups is shown in Table 3. The study of the potentially present pollutants was limited; however, 428 compounds were identified from this part of the study. It is likely that more potentially present pollutants could be identified if the search was extended. It was found that a relatively small number of pollutants belonged to both groups, i.e. compounds that have been identified in stormwater and pointed out as potentially present (Eriksson *et al.*, 2005a). Several PAHs have been found to be present in stormwater (58), and over fifty has also be identified as potentially present pollutants (51), and the overlap between the two groups, i.e. the number of pollutants both potentially present and actually confirmed to be present were 27 pollutants. Similarly, for the pesticides, 115 different pesticides have been detected in urban stormwater whereas only 64 have been found to be potentially present pollutants (Table 3). Hence the overlap between the groups are only 26 pesticides which indicate that there are non urban sources contributing to the load of pesticides in urban stormwater as well as there are pesticides used in the urban environment which it has not yet been measured for in stormwater.

Table 3: Observed and potentially present pollutants in urban stormwater (Eriksson *et al.*, 2005a and Eriksson *et al.*, 2006a).

	Compounds identified in stormwater	Potentially present pollutants	Number of constituents in both categories
Metals and inorganic trace elements	34	17	15
Alkanes	19	17	15
Benzenes	19	37	8
Dioxins/furans	31	9	4
Ethers	8	7	1
Halogenated aliphates	27	25	8
Organotin compounds	0	9	0
Organolead compounds	9	0	0
PAHs	58	51	27
PCBs	14	0	0
Pesticides	115	64	26
Phenols	32	36	20
Phthalates	8	7	5
Others	26	149	7
Total number of pollutants	**400**	**428**	**136**

Step 2. Recipient, exposure object and criteria identification

Surface waters were identified as the major recipients based on the types and number of structural and non-structural BMPs identified in DayWater (Chapter 8) and accordingly aquatic living organisms were the exposure objects of primary interest. Other strategies such as stormwater infiltration, and use for irrigations or ornamental purposes as well as use in households for toilet flushing and laundry wash were considered and the corresponding exposure objects can be found in Table 1. Accumulation of pollutants in stormwater sediments, percolation of pollutants to groundwater and technical/aesthetical issues such as precipitation and discolouration of toilet bowl were considered in second-hand. Relevant legislations were the common European directives, national statutory orders for the countries included in the project and recommendations from pan-European organisations such as OSPAR and HELCOM as well as guidelines from international organisations (e.g. the United Nations and the World Health Organisation). Furthermore, quality criteria, emission standards and environmental objectives as well as information regarding monitoring programmes were collected; detailed information is presented elsewhere (Eriksson *et al.*, 2006b).

Step 3. Hazard and Problem Identification

Inherent data were searched for 300 XOCs, selected among the 650 XOCs identified in step 1. The selection was made by the DayWater project scientific partners and end-users in the first expert judgment meeting and included pesticides (herbicides) and PAHs. Data were found for 233 of these 300 XOCs (Baun *et al.*, 2006). A total of 151 XOCs were identified as potentially hazardous of which 121 were hazardous with regard to the solid phase (sediment, soil and suspended solid) and 55 with regard to the aqueous phase (Table 4). Thirty-three XOCs were found to be hazardous with respect to both phases.

Table 4: Potential stormwater priority pollutants (from Baun *et al.*, 2006).

	Data found	Water phase	Sediment phase
Alkanes	3	0	3
Dioxins and furans	18	0	18
PAHs	36	5	36
PCBs	6	0	6
Pesticides	105	33	37
Phenols	18	5	7
Phthalates	5	2	4
Others	41	10	10
Total number of pollutants	**233**	**55**	**121**

A ranked hazard profile for all the 233 XOCs can be accessed via the DayWater ADSS or directly via the CHIAT homepage (including the PSPP database) (CHIAT, 2006). The pollutants group, name and CAS number are listed jointly with the classification (red, yellow and green) in the different filters (Filter 1: Phase distribution; Filter 2: Problem; Filter 3: P/B (persistent and bioaccumulating); Filter 4: P/T (persistent and toxic); and Filter 5: CMR/E/A (carcinogenic, mutagenic, reproduction hazardous or endocrine disrupting or allergenic), as shown in Figure 4. The profile facilitates a visual way to show in which filter the pollutants have been classified as hazards or acquitted from suspicion of adverse effects. By those means, the process is transparent to the readers.

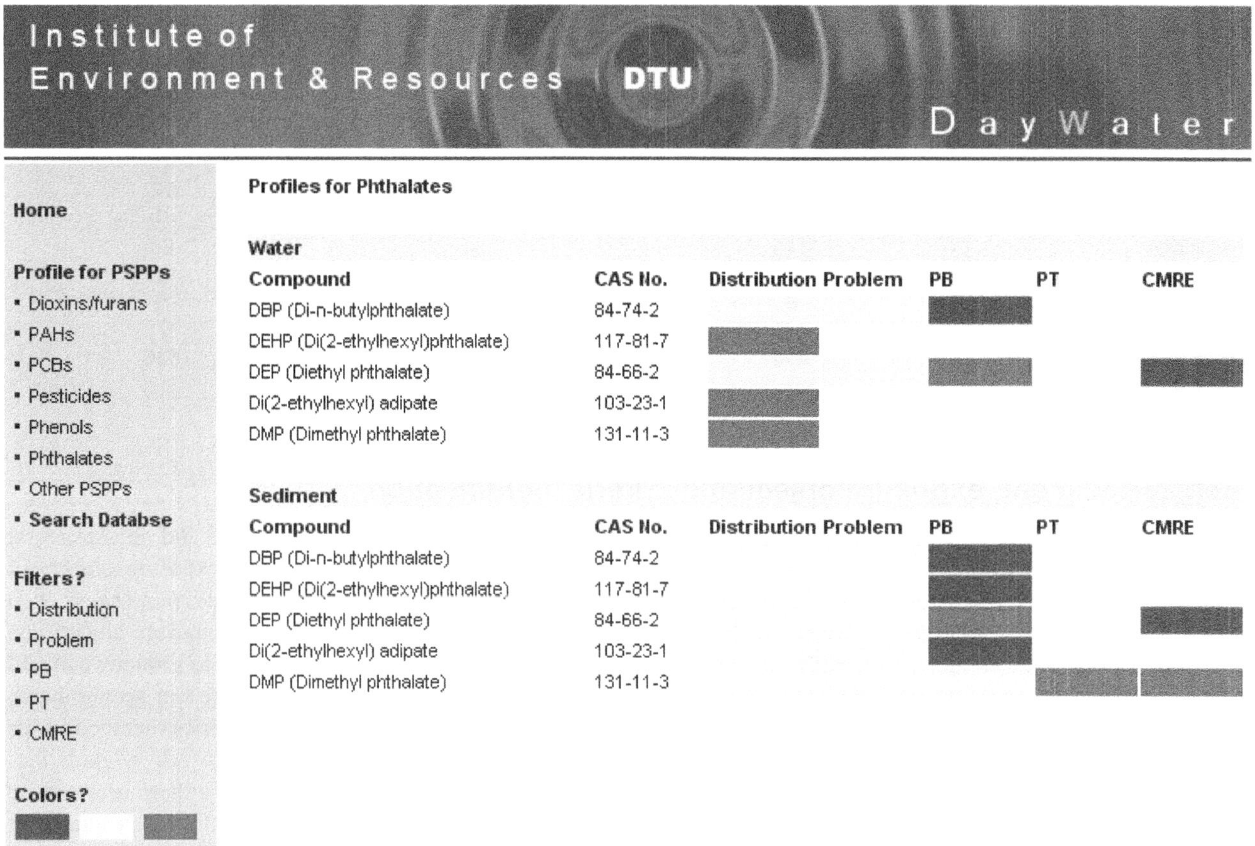

Figure 4: Screen capture from the PSPP database, containing the results from the ranking of 233 XOCs: profile for phtalates (CHIAT, 2006).

Saturation indexes were calculated for metals and inorganic trace element based on concentrations measured in urban stormwater (Eriksson *et al.*, 2006a) by geochemical modelling (PHREEQC, Parkhurst and Appelo, 2001). Three scenarios were evaluated: toilet flushing at 20°C and laundry wash at 40 and 60°C. Minerals receiving a saturation index above 1 were oversaturated and are thus disposed to precipitation. Speciation calculations confirmed the ionic charge of the relevant metal species in the pH range in stormwater.

Step 4. Hazard assessment

The hazard assessment step was excluded in the development of the SSPP list. However, examples of site specific hazard assessments can be found in the DayWater case study database (described in Chapter 5) and the European Commission has guidelines for environmental exposure assessment and effects assessments for aquatic, terrestrial, air and marine compartments as well as in sediments and sewage treatment plants (European Commission, 2003).

Concentration ranges of pollutants in urban stormwater are compiled in Eriksson *et al.* (2006a) and are available as background in the BMP catalogue (Chapter 8). Furthermore, the STORM-SEWSYS model (Chapter 14) can generate estimated concentrations whereas toxicity data can be compiled from the literature (databases and handbooks).

Step 5. Expert judgment

A total of three DayWater expert judgment meetings (Step 5) concentrated on agreement of a SSPP-list. The first of the three meetings included partners and end-users from eight different European countries and aimed at discussing the need of various pollutants in different tools being developed within the project and cases studied, as well as consent to which basic water quality parameters that should be included and which XOCs to focus on in the hazard identification. The second meeting was held with scientists specifically working with chemical risk assessment in order to make a review and selection of the output from the hazard and problem identification, step 3, for the organic compounds. The final decision of the SSPP was made during the third expert meeting in which interested partners and end-users participated. A comprehensive analysis was made of the information originating from the previous two meetings and focus was kept on the present key pollutants. The discussion addressed different aspects: (i) pollutants should be harmonised with relevant national and European legislation, (ii) banned or phased-out compounds should not be included unless continuous emission still prevailed, (iii) commercial analyses should be available and (iv) politically "hot" pollutants should be considered (Eriksson *et al.*, 2007).

Table 5: List of Selected Stormwater Priority Pollutants (indicator parameters).

Type	Name	Abbreviation	CAS number	Present on EU's WFD list of priority substances (EC, 2001)
Basic water quality parameters	Biological oxygen demand	BOD	-	
	Chemical oxygen demand	COD		
	Suspended solids	SS	-	
	Nitrogen	N	-	
	pH			
	Phosphorus	P	-	
Metals	Cadmium	Cd	7440-43-9	X
	Chromium as Chromate	Cr(VI)	11104-59-9	
	Copper	Cu	7440-50-8	
	Nickel	Ni	7440-02-0	X
	Lead	Pb	7439-92-1	X
	Platinum	Pt	7440-06-4	
	Zinc	Zn	7440-66-6	
PAHs	Naphthalene		91-20-3	X
	Pyrene		129-00-0	X[1]
	Benzo[a]pyrene	BaP	50-32-8	X
Herbicides	Pendimethalin		40487-42-1	
	Phenmedipham		13684-63-4	
	Glyphosate		1071-83-6	
	Terbutylazine		5915-41-3	
Miscellaneous	Nonylphenol ethoxylates and degradation products e.g. nonylphenol and carboxylates	NPEO	25154-52-3, 104-40-5, 27986-36-3 1916-45-9 etc.	X[2]
	Pentachlorophenol	PCP	87-86-5	X
	Di(2-ethylhexyl) phthalate	DEHP	117-81-7	X
	Polychlorinated biphenyl 28	PCB 28	7012-37-5	
	Methyl *tert*-butyl ether	MTBE	1634-04-4	

([1]) Included as a part of a summary parameter (PAH).
([2]) Nonyl phenols are included on the EU's list of priority substances.

Grouping and selection of PAH indicators was performed for PAHs based on their chemical properties i.e. ring structure, phase distribution and long-term chronic toxicity (Table 5). Selection of herbicides was made based on European consumption statistics in addition to the inherent properties, while selection of heavy metals, car catalyst metals and inorganic trace elements were based on metal speciation. Some non-pesticide and non-PAH XOCs originating from the output from step 3. These were pollutant from stormwater relevant sources and they have been repeatedly observed in stormwater as well as included in national and international "hot topic" discussions. Some basic parameters (BOD, COD, SS, N, P and pH) were, during the first expert meeting, also added based on the need for this kind of general knowledge when evaluating the quality of stormwater, treatment efficiency in structural BMPs and long-term trends. Information and justification for each individual SSPP were collected in datasheets as shown in Figure 5 (DTU, 2005; Eriksson *et al.*, 2006c).

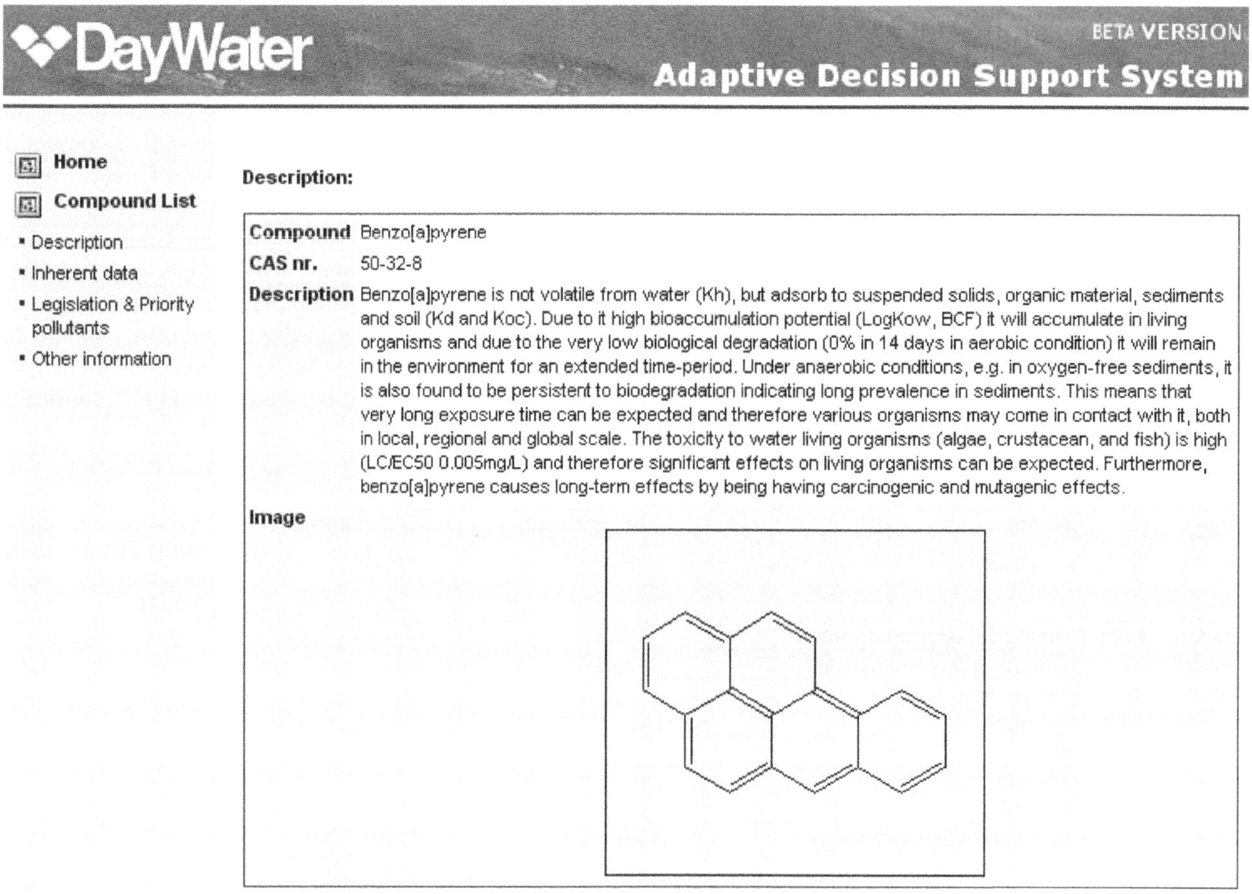

Figure 5: Screen capture from the DayWater SSPP database on Benzo[a]pyrene: similar datasheets are available for all the parameters/substances defined in Table 5 (DTU, 2005).

Discussion

The presence of many of the proposed SSPP (Cd, Ni, Pb, PAHs, DEHP, nonylphenols and pentachlorophenol) in the EU's list Water Framework Directive 'Priority substances in the field of water policy' (European Commission, 2001) is a clear evidence of their importance and of the conformity of the proposed methodology to EU's suggestions (Eriksson *et al.*, 2006b). However, the EU does not have a corresponding list for sediments or soil, hence the DayWater list complements by including compounds sorbing to the solid phase (Table 5). In comparison with several national regulations, international guidelines and non-governmental initiatives for pollutant control as well as two large European monitoring programmes it was clear that several new pollutants have been included in the DayWater SSPP-list (Eriksson *et al.*, 2006b). These are platinum, herbicides (pendimethalin, phenmedipham and terbutylazine) and MTBE, which indicates that existing monitoring programmes may required revision.

The CHIAT methodology and the RICH tool has been applied in two real life cases in Sweden i.e. the Stockholm New toxins – New tools project (Chapter 19) and the Malmö Municipality Stormwater Action Programme (Ledin *et al.*, 2007). Furthermore, the RICH tool jointly with the hazard assessment step has been used to evaluate the pollutants included in an existing monitoring programme applied in Harrestrup Å (Denmark) in cooperation with Copenhagen Energy. The most important priority pollutants were pointed out for Harrestrup Å thus facilitating a revision of the monitoring programme as described in the DayWater ADSS Case Study Database.

CHIAT covers chemical, technical and aesthetical risks, which is reflected in the SSPP list (Table 5). In order to fully assess the hazards and risk connected with stormwater management other tools and approaches are required. For example, the microbial risk related to human contact with faecal contaminated water. This issue is addressed in (Malmqvist *et al.*, 2006) based on a science exchange of microbial and chemical risks between DayWater and UWMP. Additionally, the risk related to flooding is of growing importance due to the recent years massive flooding in Europe and the impact of climate change (Grum *et al.*, 2005). In DayWater, this issue is addressed in the Flood Vulnerability Assessment Tool (Chapter 13) and in the risk-screening tool FLEXT (Chapter 15).

CONCLUSIONS

In DayWater, a methodology for identification and assessment of chemical hazards (CHIAT) has successfully been developed (Eriksson *et al.*, 2005a). It can be used for design of a new or revision of an old monitoring programme, to evaluate and compare stormwater treatment options and to identify pollutant for other source control options such as phase-out and substitution. The tool for identification and ranking of organic chemicals (RICH) has proven to be able to point out the potentially most hazardous pollutants (Baun *et al.*, 2006).

Experiences with application of CHIAT have shown that, in a transparent way, it can help the user to prioritize pollutants within a large numbers of organic compounds. In the end, it can yield a limited list of priority pollutant chosen for each specific scenario at a local, regional or pan-European scale (Chapter 19 and the ADSS Case Study Database described in Chapter 5).

A list of 25 SSPP consisting of both general water quality parameters, metals and XOCs have been developed. In the process of selecting such SSPP, CHIAT was used as a road map. Step 4 was omitted and Step 5 was applied in three iterations that included different experts depending on the scope of the expert meeting. The list of SSPP has served as input to several other tools and initiatives in DayWater (Eriksson *et al.*, 2007). An in-depth review of pollutants in stormwater and compounds potentially present based on their sources revealed that stormwater could be heavily contaminated (Eriksson *et al.*, 2005a and Eriksson *et al.*, 2006a). A comparison of the DayWater SSPP with existing regulation, guidelines and monitoring programmes showed that some new, not previously regulated or monitored pollutant have been included among the SSPP (Eriksson et al., 2006b).

The information and chemicals properties of the SSPP can be accessed via the DayWater ADSS or directly at http://www.er.dtu.dk/daywater. The ranked hazard profiles for the 233 pollutants evaluated can be accessed via the DayWater ADSS or directly at http://chiat.er.dtu.dk.

REFERENCES

Baun A., Eriksson E., Ledin A. and Mikkelsen P.S. (2006) A Methodology for Ranking and Hazard Identification of Xenobiotic Organic Compounds in Urban Stormwater. *Science of the Total Environment*, **370**, 29-38.

CHIAT (2006) Chemical Hazard Identification and Assessment Tool. Available at http://chiat.er.dtu.dk (verified July 2007).

DTU (2005) DayWater Compound list and database. Available at http://www2.er.dtu.dk/daywater/complist.php (verified July 2007).

Eriksson, E., Baun, A., Mikkelsen, P.S. and Ledin, A. (2005a) Chemical hazard identification and assessment tool for evaluation of stormwater priority pollutants. *Water Science and Technology*, **51**(2), 47-55.

Eriksson E., Baun A., Mikkelsen P.S. and Ledin A. (2005b) A methodology for selecting and evaluating priority pollutants. In: 10th International Conference on Urban Drainage, Abstract Collection, Copenhagen, Denmark, August 21-26, 2005. Institute of Environment & Resources, Technical University of Denmark. ISBN 87-89220-80-3.

Eriksson, E., Mikkelsen, P.S. and Ledin, A. (2006a). Characteristics and quality of urban runoff – a review, Institute of Environment & Resources, Technical University of Denmark, in preparation.

Eriksson; E., Baun, A., Mikkelsen, P.S., and Ledin, A. (2006b) New stormwater priority pollutants - Comparison with international classifications, quality criteria and monitoring programmes. Institute of Environment & Resources, Technical University of Denmark, in preparation.

Eriksson, E., Baun, A., Scholes, L., Ledin, A., Ahlman, S., Revitt, M., Noutsopoulos, K. and Mikkelsen, P.S. (2007) Selected stormwater priority pollutants — a European perspective. *Science of the Total Environment*, **383**(1-3), 41–51.

Eriksson, E., Baun, A., Mikkelsen, P.S. and Ledin, A. (2006c) Selected stormwater priority pollutants (SSPP) – Introduction and database. DayWater deliverables D4.3 & D4.4. Available at http://www.daywater.org (verified July 2007).

European Chemicals Bureau (2006) European chemical Substances Information System (ESIS). Available at http://ecb.jrc.it/esis/ (verified July 2007).

European Commission (2001) Decision No 2455/2001/EC of the European Parliament and of the Council of 20 November 2001 establishing the list of priority substances in the field of water policy and amending Directive 2000/60/EC.

European Commission (2003) Technical Guidance Document in support of Commission Directive 93/67/EEC on Risk Assessment for new notified substances, Commission Regulation (EC) No 1488/94 on Risk Assessment for existing substances and Directive 98/8/EC of the European Parliament and of the Council concerning the placing of biocidal products on the market (2nd edition).

Grum, M., Jørgensen, A.T., Johansen, R.M. & Linde, J.J. (2005) The effect of climate change on urban drainage: An evaluation based on regional climate model simulations. In: 10th International Conference on Urban Drainage, 21-26 August 2005, Copenhagen/Denmark, 10ICUD, Draft Papers. CD-ROM, Institute of Environment & Resources, Technical University of Denmark & Section of Environmental Engineering, Aalborg University, Kgs. Lyngby.

Ledin, A. Eriksson, E., Flygare, S., Runesson, L., Baun, A., and Mikkelsen, P.S. (2007) Selection of stormwater priority pollutants for the City of Malmö, Sweden. In: Proceedings of NOVATECH 2007, 6th international conference on sustainable techniques and strategies in urban water management, 25-28 June 2007, Lyon/France, 1475-1482.

Makepeace, D.K., Smith, D.W. and Stanley, S.J. (1995). Urban Stormwater quality: Summary of contamination data. *Critical reviews in Environmental Science and Technology*, **25**(2), 93–139.

Malmqvist, P-A., Heinicke, G., Kärrman, E., Stenström, T.A. and Svensson, G. (2006) *Strategic Planning of Sustainable Urban Water Systems*, IWA Publishing, London, UK, ISBN: 1843391058.

Parkhurst P.L. and Appelo C.A.J. (2001) PHREEQC - A Computer Program for Speciation, Batch-Reaction, One-Dimensional Transport, and Inverse Geochemical Calculations, Available at http://wwwbrr.cr.usgs.gov/projects/GWC_coupled/phreeqc/index.html (verified July 2007).

US. EPA (2005) EPI Suite v3.12. Available on http://www.epa.gov/oppt/exposure/pubs/episuite.htm (verified July 2007).

Chapter 12

Risk and vulnerability

Peter Steen Mikkelsen and Jean-Marie Mouchel

INTRODUCTION

Risk assessment can be a time consuming and costly task expanding to an extent that exceeds the available resources, because many risk problems include a large range of hazards and effects (Chapter 24). Thus for preliminary analysis of potential risks it may be wise to start with simpler methods that focus on well-chosen risk indicators. Hazard analysis has been the initial step in risk analysis for decades. It is however important to realize that the same hazard will not produce the same effect on all systems. Therefore vulnerability is introduced as to express how vulnerable a risk object is towards a potential risk. It is the combination of the occurrence of a hazard and the vulnerability of the system that results in the effect.

The drawing shown in Figure 1 illustrates the action oriented risk terminology used in the DayWater project that includes vulnerability while being still in agreement with the most common terminology in the risk assessment field (Christensen *et al.,* 2003; Hauger *et al*, 2006b). In this context hazard and vulnerability identifications and assessments are initial screening steps (see the top and the bottom of the Figure 1), which may be followed by more comprehensive risk characterisation if judged necessary.

RISK INDICATORS

The traditional definition of risk in the engineering community is that risk is equal to the probability of occurrence of an event multiplied with the consequence, or effect, of an event:

Risk = P (occurrence of harmful event) X Effect

The traditional risk definition is easy to understand but in practice difficult to use as well as communicate to lay people. Therefore, we have adopted a more symmetrical risk definition in the DayWater project, where risk is a function of the hazard related to the risk source and the vulnerability related to the risk object:

Risk = f (hazard, vulnerability)

This way of interpreting risk is incorporated in e.g. French legislation (Chabert, 2003), and it is currently becoming more and more common in many different engineering and environmental risks assessment fields. Distinguishing between hazard and vulnerability (and between risk sources and objects) is advantageous because these concepts relate directly to possible actions to mitigate risks. Another advantage is that the vulnerability concept relates well to how stakeholders perceive risks. This primary interest of most laypersons is not: “what hazard am I exposed to?” but “will I

© 2008 IWA Publishing. *DayWater: an Adaptive Decision Support System for Urban Stormwater Management.* Edited by Daniel R. Thévenot. ISBN: 9781843391609. Published by IWA Publishing, London, UK.

be injured or will I suffer from economic loss?" In some cases it may be possible to mitigate the risk based on carefully studying the vulnerability, without estimating the risk formally. Thus, it is clear that getting basic information about a wide range of risk indicators relevant for management decisions is of primary importance. Hauger *et al.* (2005, 2006a) discuss this in further detail.

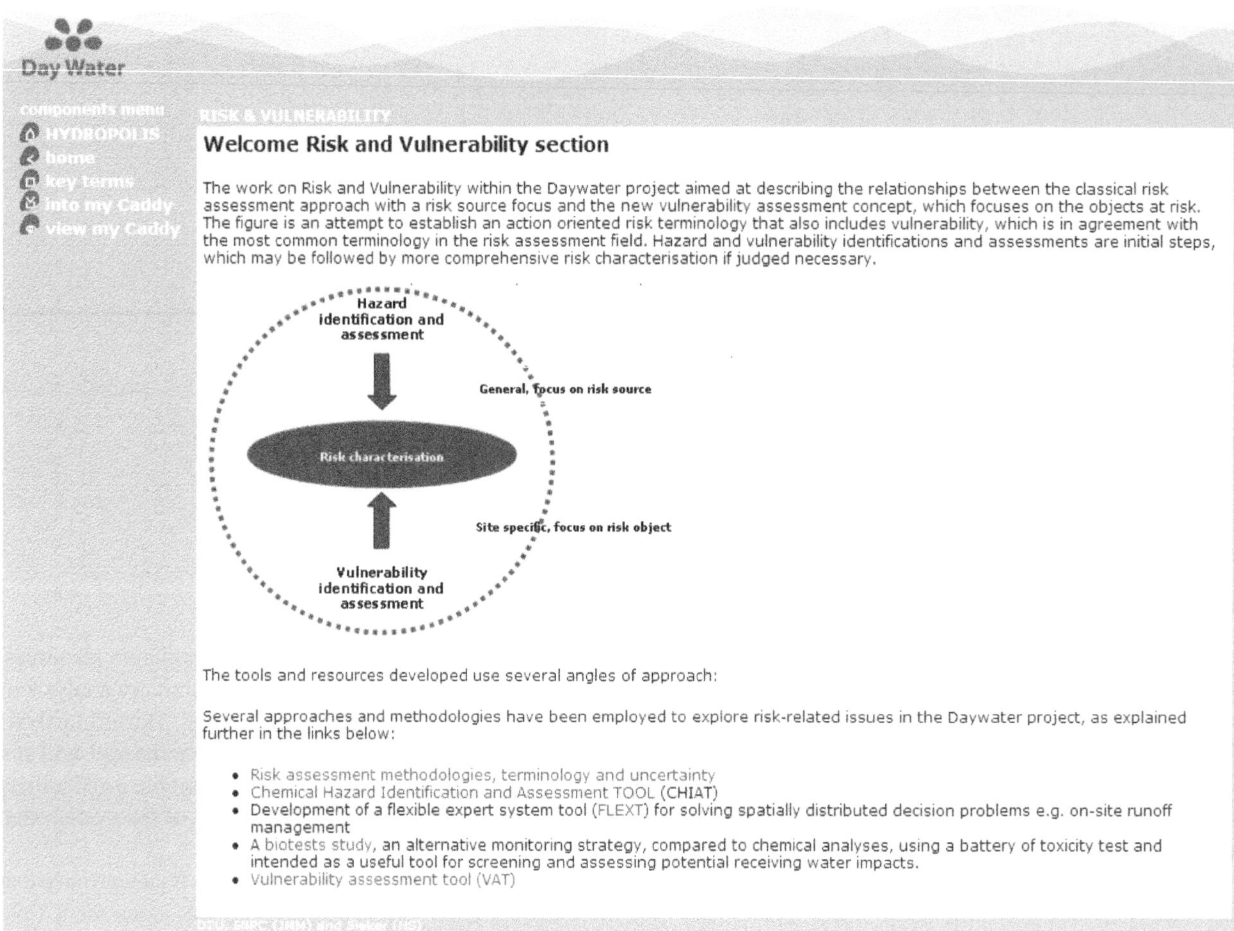

Figure 1. Relationship between hazard identification and assessment, vulnerability identification and assessment, and risk characterisation, as addressed in the DayWater project (Chapters 1 and 3).

HAZARD AND VULNERABILITY SCREENING

Hazard identification and assessment is normally of very general character, focusing on the risk source, and are not site specific. Typical hazard assessments focus on e.g. frequency of earthquakes in an area or the probability of a river overflooding its banks. In the case of urban runoff the hazard assessment mainly concerns the structure of rain events and their frequency. Although some important questions remain open, such as the influence of climate change on rain events or the return period of very rare events and the real meaning of such a concept, very significant pieces of work by hydrologists and meteorologists enable probabilistic evaluations of the rainfall levels at various time scales from hours to decades. In the DayWater project hazard identification and assessment focuses on the environmental risks imposed by hazardous chemical constituents in stormwater (Eriksson *et al.,* 2005). Further information about this and the corresponding developed DayWater tools can be found in Chapter 11 describing the 'Selection of the priority pollutants' tile of the Hydropolis front page (Chapter 3).

Vulnerability identification and assessment is used for answering questions focused on the risk object which is much more site specific than the hazard assessment, e.g. "how much will my house be damaged by a flooding?", "will my family be affected by the smoke from the nearby industrial smokestack?" or "will my crop be damaged by a storm?". In the DayWater project vulnerability identification and assessment focuses on two different application areas: ecological effects from discharges of chemical constituents in stormwater (Baun *et al.,* 2005) and flooding in urban areas. Figure 2 illustrates the Vulnerability Assessment Tool (VAT) developed to allow users to reflect on what indicators of flooding that may be important in the local context concerned. Further information about this tool can be found in Barroca *et al.* (2006) and in Chapter 13.

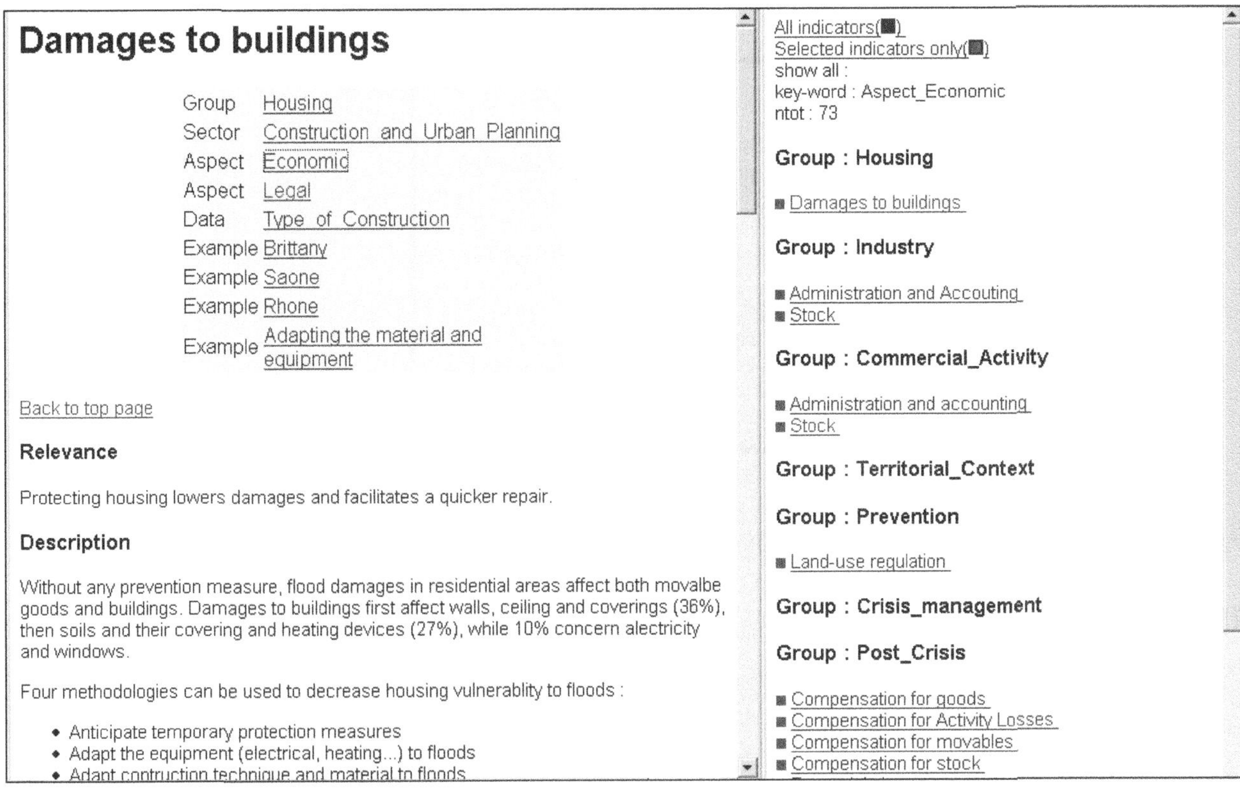

Figure 2. Screen capture from the VAT tool giving information about vulnerability indicators for flooding damage to buildings. Available at www.h2o.net/daywater/ (verified September 2007).

IMPLICATIONS OF UNCERTAINTY FOR MANAGEMENT

Hazards, vulnerabilities and risks are all connected with substantial amounts of uncertainty, and a concept for coping with uncertainty in stormwater management has therefore been developed. The basic principle is illustrated in Figure 3: the "level of uncertainty" concept developed by Walker *et al.* (2003) is combined with the "matrix of decision making" originally proposed by Christensen (1984) into a tool that can be used when reflecting on appropriate decisions and management means under different levels of uncertainty.

Three levels of uncertainty are defined: (i) statistical uncertainty that can be described adequately in statistical terms, (ii) scenario uncertainty that represents a level beyond statistical uncertainty where only assumptions can be made or scenarios drawn up, and (iii) recognised ignorance where the basis for making rational or optimal decisions is weak or absent. The axes of the matrix basically distinguish situations where there is agreement, alternatively disagreement about the goals of a decision and correspondingly where the knowledge required to achieving the goal is available or not. The four quadrants in the matrix thus describe very different situations where "wise decisions" are made very differently using different decision making tools.

Quadrant 1 (agreed goals, known means, statistical uncertainty) is the typical "home ground" of engineers: the important functional relationships are understood and rational decisions and predictive engineering designs are made using deterministic models. In quadrant 2 (disagreement about the goal, known means, scenario uncertainty) the task is to find agreement about the goals or to reach a compromise that satisfies different stakeholders with different scenarios in mind. Stakeholder participation is therefore important, and 'process managers' become important in addition to the engineers who can deliver concrete information. Quadrant 3 (agreed goals, unknown means, assumption uncertainty) is where researchers typically operate: decisions can be made based on different assumptions (empirical-iterative design), or investments can be directed towards creating more knowledge. Quadrant 4 (disagreement about goals, unknown means, recognised ignorance) is particularly interesting in the context of the DayWater project, because many real-life projects involving stormwater source control involve a lot of chaos. There are typically two possible strategies to pursue: trying to move the problem into another quadrant or trying to act in this chaotic environment by making "gentle decisions" or "no regret decisions", i.e. there should be much more emphasis on learning and interaction with stakeholders than on designing master plans that will probably never be implemented in practice. Further information about the developed uncertainty concept can be found in Hauger *et al.* (2006b).

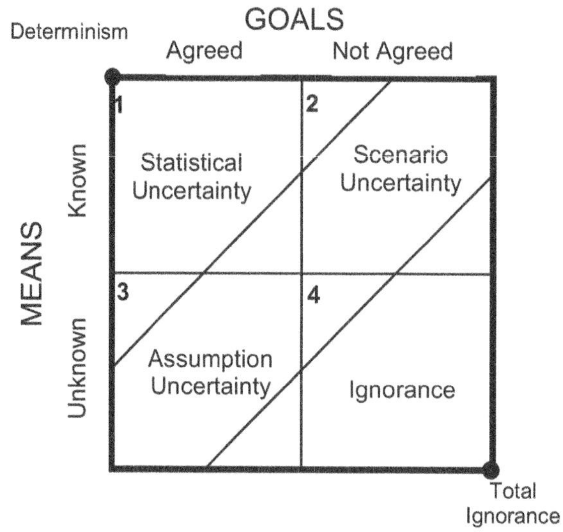

Figure 3. Matrix of decision making, combining "levels of uncertainty" with "knowledge about goals and means".

REFERENCES

Barroca, B., Hubert, G. & Mouchel, J.-M. (2006) Flood vulnerability tool. DayWater project, deliverable D4.6. www.daywater.org (verified July 2007).

Baun, A., Qualmann, S., Eriksson, E., Scholes, L., Revitt, M., Seidl, M. and Mouchel, J.-M. (2005) Application of a battery of biotests for toxicity characterization of stormwater from four European catchments. DayWater project, deliverable D4.5. www.daywater.org (verified July 2007).

Chabert, L. (2003) Report inventory and characterization of risks. DayWater project, contribution to task 4.1. www.daywater.org (verified July 2007).

Christensen, K.S. (1984) Coping with uncertainty in planning. *Journal of the American Planning Association*. **51**(1) 63-73.

Christensen, F.M., Andersen, O., Duijm, N.J. and Harremoës, P. (2003) Risk terminology – A platform for common understanding and better communication. *Journal of Hazardous Materials*, **A103**, 181-203.

Eriksson, E., Baun, A., Mikkelsen, P.S. and Ledin, A. (2005) Chemical hazard identification and assessment tool for evaluation of stormwater priority pollutants. *Water Science and Technology*, **51**(2), 47-55.

Hauger, M.B., Ledin, A. and Mikkelsen, P.S. (2005) Risks and problems in urban stormwater management from a pan European end user perspective. In: Eriksson, E., Genc-Fuhrman, H., Vollertsen, J., Ledin, A., Hvitved-Jacobsen, T. & Mikkelsen, P. S. (eds.) 10th International Conference on Urban Drainage, 21-26 August 2005, Copenhagen/Denmark, Institute of Environment & Resources, Technical University of Denmark.

Hauger, M.B., Mouchel, J.-M. & Mikkelsen, P.S. (2006a): Indicators of hazard, vulnerability and risk in urban drainage. *Water Science & Technology*, **54**(6-7), 441-450.

Hauger, M.B., de Roo, C., Geldof, G.D. and Mikkelsen, P.S. (2006b) Whitepaper on risk perception, risk assessment and risk management in the DayWater context. DayWater project, deliverable D4.1. www.daywater.org (verified July 2007).

Walker, W.E., Harremoës, P., Rotmas, J., van der Sluijs, J.P., van der Asselt, M.B.A., Janssen, P. and Krauss, M.P. (2003) Defining uncertainty – a conceptual basis for uncertainty management in model-based decision support. *Integrated Assessment*, **4**(1), 5-17.

Chapter 13

Flood Vulnerability Assessment Tool (FVAT)

Bruno Barroca, Jean-Marie Mouchel, Thomas Bonierbale and Gilles Hubert

INTRODUCTION

It has been generally agreed (CGCT, 1996) in the recent years that efficient solutions against flooding in urban areas should be based on a wide set of alternatives including source control of runoff waters. A large number of techniques are currently proposed promoting retention, infiltration and local use of rain water for specific purposes (garden watering, toilet flushing etc…); they include infiltration trenches, storage on roofs, porous pavements, wetlands etc... (Chapter 8) These techniques aim at reducing runoff by water flow rerouting, utilisation and reduction of non-permeable surfaces: consequently they reduce flood hazard in urban areas. Furthermore, an additional objective is to reduce the need for huge and costly infrastructures required to transport water flows during heavy rain events. When the infrastructure already exists, source control should prevent its saturation and downstream floods or negative impacts on ecosystems (particles flushing, heavy pollutant inputs). When the infrastructure does not exist, e.g. in new estates, smaller infrastructures will be promoted, and the flooding risk may be displaced upstream. The zero hazards do not exist. The complex and better-distributed solutions, which are promoted nowadays for urban runoff management, will reduce hazard and, in some situations, distribute this hazard through the whole territory. In addition to hazard reduction, the urban manager must be aware of the need to tackle the problem and to seek at reducing vulnerability.

Risk is a combination of 'hazard' and 'vulnerability' (Chapter 12). We can define 'hazard' like "a threatening event, or the probability of occurrence of a potentially damaging phenomenon within a given time period and area" (WIN, 2007). When a hazardous event ('hazard') occurs, resulting damages will depend on the element at risk. The 'element at risk' is commonly considered as the population, buildings and civil engineering works, economic activities, public services and infrastructure, etc… exposed to hazards.

'Vulnerability' is a recent concept, which is more and more widely used in risk analysis. It is understood as the complement to hazard, required to define risk. 'Vulnerability' ideally defines what would be the resulting damages for any possible occurrence of hazard. Vulnerability is the "susceptibility to degradation or damage from adverse factors or influences" (US-EPA, 1997).

In spite of various regulations, institutional standards are not always applied. This is perceived as a lack of willingness on the decision level (Shrubsole *et al.* 1997; Pottier, 2000). Integration of flood risk in territorial management needs a better knowledge of the vulnerability. Vulnerability becomes a statutory value if it is based on public participation and not only a subject of discussion after flooding events (Morris-Oswald and Sinclair, 2005).

According to Comfort *et al.* (1999), studies should focus on the local level: "Investment in risk reduction is likely to be most efficient and effective when directed toward improving local capacity to act in coordinated ways to achieve this community-wide goal; the link between policy and practice in disaster mitigation needs to be established at the local

 DayWater: an Adaptive Decision Support System for Urban Stormwater Management. Edited by Daniel R. Thévenot. ISBN: 9781843391609. Published by IWA Publishing, London, UK.

level". But urban stakeholders which should take into account flood vulnerability at this local scale do rarely have flood vulnerability assessment tools (FVATs) or methods. The potential users (engineers, risk managers, real estate developers, etc.) of such a FVAT tool are managing floods caused by exceptional precipitation or overflows. A wide range of private people or institutions is seeking for an extremely flexible tool for flood management. The objective of this FVAT would be to provide a methodological frame to allow users to choose relevant indicators.

This chapter is structured in five sections with an introduction in section 1. Section 2 shows the scientific and operational context of the vulnerability analysis and a brief review of the currently used methodology. Section 3 presents a new FVAT tool and methodological elements for the identification of vulnerability indicators. The objectives of the tool are then:

- to give information about the various aspects of vulnerability,
- to show the relations between indicators (enhanced or decreased vulnerability),
- to illustrate the development of indicators,
- to help the end-user in the development of its own set of indicators, the idea being that all aspects of vulnerability should be covered (or not covered for some specified reason).

It follows a description of the FVAT use, the indicator classification in trees and the graphical representation. The section 4 discusses the contribution of the FVAT tool (www.h2o.net/daywater/) within the decision making process. Section 5 is the concluding part of the paper.

DEALING WITH VULNERABILITY

The three complexity aspects of vulnerability

The word 'vulnerability' has been used by the geographs of the Chicago University since the first half of XXth century, referring to the Mississippi valley (AFPCN, 2005). In practice, there are many different definitions of vulnerability. Such diversity result from the specific aspects of the vulnerability which depend on the type of study, on the results required (damage evaluation or urban planning project), on the kind of flood hazard (flash flood or slow-flood), on the spatial and temporal scale of study, on the specificity of the study area and on the temporality (prevention, crisis, post crisis). Some definitions focus on the protection of individuals, others on the maintenance of economic activities or the protection of the environment. Finally, vulnerability definition is also linked to the particular system functions analysed (Parent *et al.,* 1999).

In this paper, we cope with the complexity of vulnerability definition, at least at three levels:

(1) The first complexity of 'vulnerability' concerns the event itself: hazard is complex and can take various forms. It cannot be described in a simple way, as the various scales, which can be used, do not refer to the same damages, and hence do not have an impact of same intensity. In the case of flooding, water level, immersion velocity, flow, water quality and frequency of events are example criteria used to evaluate the 'hazard'. This is called "external complexity".

(2) The second complexity of 'vulnerability' concerns the various types of functions of the element at risk, which can be affected by the 'hazard'. Effects on shelter constructions, transportation, industrial activity, health services do not have the same impact in terms of vulnerability. The time scales of the effects may be very different. Short-term effects may cause human casualties or direct costs to economic activities, whereas on longer time scales costs for maintenance will be considered more important. This is called the "first inherent complexity of vulnerability".

(3) The third complexity aspect of 'vulnerability' is connected to the intelligence of the element at risk and relations between the object at risk. Being aware of the risk, in all its components, individuals or societies are able to react. They may organise rescue systems, modify urban development plans, or change building techniques. This is called the "second inherent complexity", which is even more complex to assess.

A brief literature review of vulnerability evaluation

In the project reported in this chapter, a set of risk factors was identified, starting from the existent knowledge, mainly in the domain of risk analysis but also in the natural hazard literature. Historically, the evaluation of vulnerability was conceived to support technical and financial choices for protection against floods (Parker *et al.,* 1987; Penning-Roswell *et al.*, 1992; Blong, 2003; D'Ercole, 1994; Leone *et al.,* 1996). Therefore, the first vulnerability analysis linked to flood hazard were done in the USA and Great Britain during the seventies (Foster, 1976; Flax *et al.*, 2002; Penning-Roswell and Chatterton, 1977).

Today, risk management is an object of social debate, which must lead to explicit collective choices (Strategis, 2002; Lefevre, 1995). The evaluation of the socio-economic impacts of floods should not continue to be locked up in cost-benefit logic (DEFRA, 2002; Gendreau, 1999; Dauphine, 2001). These methods must include a quantitative and qualitative analysis (Flax *et al.,* 2002e; ODPM, 2001). One of them is the toolbox 'Vulnerability and Capacity Assessment' (VCA) (IFRC, 1996), identifying potential threats, vulnerabilities, assessing capacities and strengths. The purpose of this type of evaluation is to estimate as well as possible the space distribution of element at risk, their importance to networks (Felts, 2002) and to strategic crisis-element, as well as their vulnerability (Godin *et al.,* 2004). A risk calculation can show the possibilities for vulnerability reduction (local territorial strategies, vulnerability assessment of buildings, temporary relocation and reception capacity) (Ledoux, 1995; IFRC, 1996).

Some documents identify the different types of building vulnerability: volcano (Pomonis *et al.,* 1999), tephra fall (Spence *et al.,* 2005), flood (Kelman and Spencer, 2003; Valencia *et al.,* 2004) and seismic vulnerability (Mebarki and Valencia, 2004). Some identify elements likely to affect human vulnerability and classify the building stock (Spence *et al.,* 2004). Others determine respective levels of vulnerability of people and buildings: for landslide the results are given in Glade (2003), for risk due to tunnelling and excavation in Zihri (2004). Others present a conceptual framework for addressing temporal variation in natural risk (Wisner *et al.,* 2004; Hufschmidt *et al.,* 2005). Various recent works suggest the use of a unique and aggregated value or 'index' to quantify the damages: this is actually the case for the "Damage Index 1999" and the "Central Damage Values" which uses an average rate of damage (NHRC, 1999). This method has been widely used in Australia by the Insurance Foundation. The Environmental Vulnerability Index (EVI) belongs to a new generation of tools: it supports the assessment and gives advice to decision-makers. Furthermore, it intends to combine measures of economic and social vulnerability. Again, some studies described the most important factors and proposed methods for an index of economic vulnerabilities for different countries (Briguglio, 1995). Another method (not specific to natural hazard) presents the social and economic adaptive capacity: this is a point of departure for the construction of vulnerability indices (Yohe and Tol, 2002).

This brief review shows that a set of tools and procedures exists for the quantification of the vulnerability and for the identification of areas at risk. At the same time, no method exists to select vulnerability factors specifically for a particular case study. Broadest knowledge on vulnerability and its contributing factors are to be privileged in order to develop the sum of independent actions and, at the same time, a unit of inter-connected actions (prevention, protection, management of crisis, repair, reparation, etc).

Vulnerability includes several dimensions (social, cultural, economic, ecological etc), and then indicators may potentially be numerous. Moreover, data, which sustain the indicators, are often dispersed within various services or administrations database (Pornon, 1998). The aim of a new vulnerability tool should be to capitalize those data and to translate them into knowledge.

A NEW SUPPORT TOOL FOR FLOOD VULNERABILITY ANALYSIS

Description of the support tool

As it mentioned above, the objective of the FVAT tool is to provide a methodological frame to allow users to choose relevant indicators. It does not provide indicators but keys to design them (Barroca, 2006). The final FVAT tool proposes indicator sheets with such keys. Several groups of indicators are referring to widely shared functions of urban systems (housing, commerce and industry) and to risk specific aspects (territorial context and the various times of crisis management). Those functions are simplified, as much as possible, while demonstrating their relevance in case studies. A comprehensive list of vulnerability indicators would be irrelevant for a given situation (e.g. vulnerability indicators referring to non-existing types of land-use). The methodology to evaluate them is strongly context-dependent (e.g. characteristics of housing used to evaluate vulnerability strongly depend on local architectural traditions).

A possibilistic approach could be chosen: it would require an extensive review of the hazard completed by an extensive analysis of vulnerabilities inside the urban territory, whatever the existing or already revealed damages. Attractive to some extent, we consider this approach as too much theoretical, uncertain and lengthy. In this work, a phenomenological or empirical approach was adopted, starting from examples of damages and classifying them into indicators. The FVAT tool organizes information regarding vulnerability and proposes a browsing trip between indicator sheets and related keywords. It was conceived to be easily improved through interactions with end-users and the inclusion of additional examples. Such an empirical approach should be more practical and satisfactory for the end-user (engineers, risk managers, real estate developers, etc.).

The proposed FVAT software has been developed using PHP/HTML languages, attempting to reach these goals:

- give information about the various aspects of vulnerability,
- demonstrate the relationships between indicators (enhancing or decreasing vulnerability),
- illustrate the development of indicators,
- help the end-user in the development of its personal set of indicators, where all aspects of vulnerability should be covered (or not covered for some specified purpose).

Each indicator is described in one HTML page. Classified groups of keywords cross-referenciate the set of indicators: main indicator group (Figure 1), economic sector, 'water aspects' (Chapter 6), referred data, relevant case studies...(Chapter 5). The keywords can be used to browse into the set of pages and to learn about the application of indicators to assess flood vulnerability. In terms of software, the FVAT tool consists of a set of HTML/PHP files which can be viewed on any Web browser. The set of files is generated from a single XML document.

Methodological aspects for indicator identification

The method for identification of indicators results out of a literature review, which was based on complementary levels: the local work bibliography, the scientific bibliography, and the research project bibliography. We seek here to synthesise and identify the expertise coming from:

- literature available on local level: it contains expert reports, notes, studies, diagnostic etc... related to a specific context. (Ayral, 2002),
- scientific literature: it includes internationally and nationally reviewed articles in order to introduce methodological elements,
- research project reports: they include contributions from vulnerability analysis related to other areas and other programs not being the subject of scientific publication (European framework, National and Regional research programs),
- case studies conducted within the framework of this work in order to improve the literature database (e.g. Lefort, 2003; Lefort, 2004; Pottier *et al.*, 2004).

Such indicators need to be measurable on a local scale, according to final vulnerability characterisation objectives, i.e. in agreement with National regulations. Thus, users must consider logistic and technical limits related to the study of a catchment scale. The associated data sources are identified and updating the database is considered to be needed at regular intervals. For the indicators that require specific data, the methodologies needed to collect these data must be known.

Organisation and use of the indicators

The indicators are structured in a tree configuration. Five main branches include secondary fields, and more subdivision can be added. Each of the five modules (Figure 1) contains a set of indicators related to the studied field. The user selects the content from various modules. Some examples are provided to help organising indicators adapted to specific needs and to a specific case. Thus, the user decides which indicator needs to be considered in his specific case.

Organisation of the tree and representation of interactions

On Figure 1, the tree structure shows the heritage concept (branch / sub-branch / leaf) but not the complex interactions, which can exist between the indicators. Thus, those interactions between the leaves must be enlightened, in order to allow more flexible browsing. For this reason, each leaf also belongs to several horizontal groups and thus it is possible to extract all the leaves belonging to a specific group. Inside any group, the leaves are connected. For example, a user will be able to extract the vulnerability indicators related to land management, in the same way another user might be interested in his specific case study indicators (which can be the vulnerability to a specific flood). There are several types of groups, based on a type of fundamental argument.

- case studies: a series of indicators is presented with their interconnections within detailed application to case studies (Chapter 5),
- sectors, i.e. planning management, construction, environment, safety, and communication,
- 'water aspects', i.e. legal, economic, aesthetic, moral (Chapter 6),
- data, i.e. which indicators can be evaluated when a certain data type is available (e.g. socio-economic, insurance),
- examples: case studies were the indicator is relevant.

Each leaf gives information on the indicator and the group it belongs to. From a design point of view, the Graphic User Interface (GUI) consists of two windows, a right menu with the tree arranged according to the main branches or vertical groups (automatic system referring to single leaves), and a mainframe with the details of the selected indicator (Figure 2).

Clicking on one of the horizontal groups or branch inside the 'leaf' highlights the corresponding leaves inside the tree, making it possible for the user to browse the tree. Clicking on a 'leaf' opens the window presenting the leaves. This FVAT is available both an English and French version from the DayWater ADSS (http://www.daywater.cz) or directly from the software host website http://www.h2o.net/daywater/.

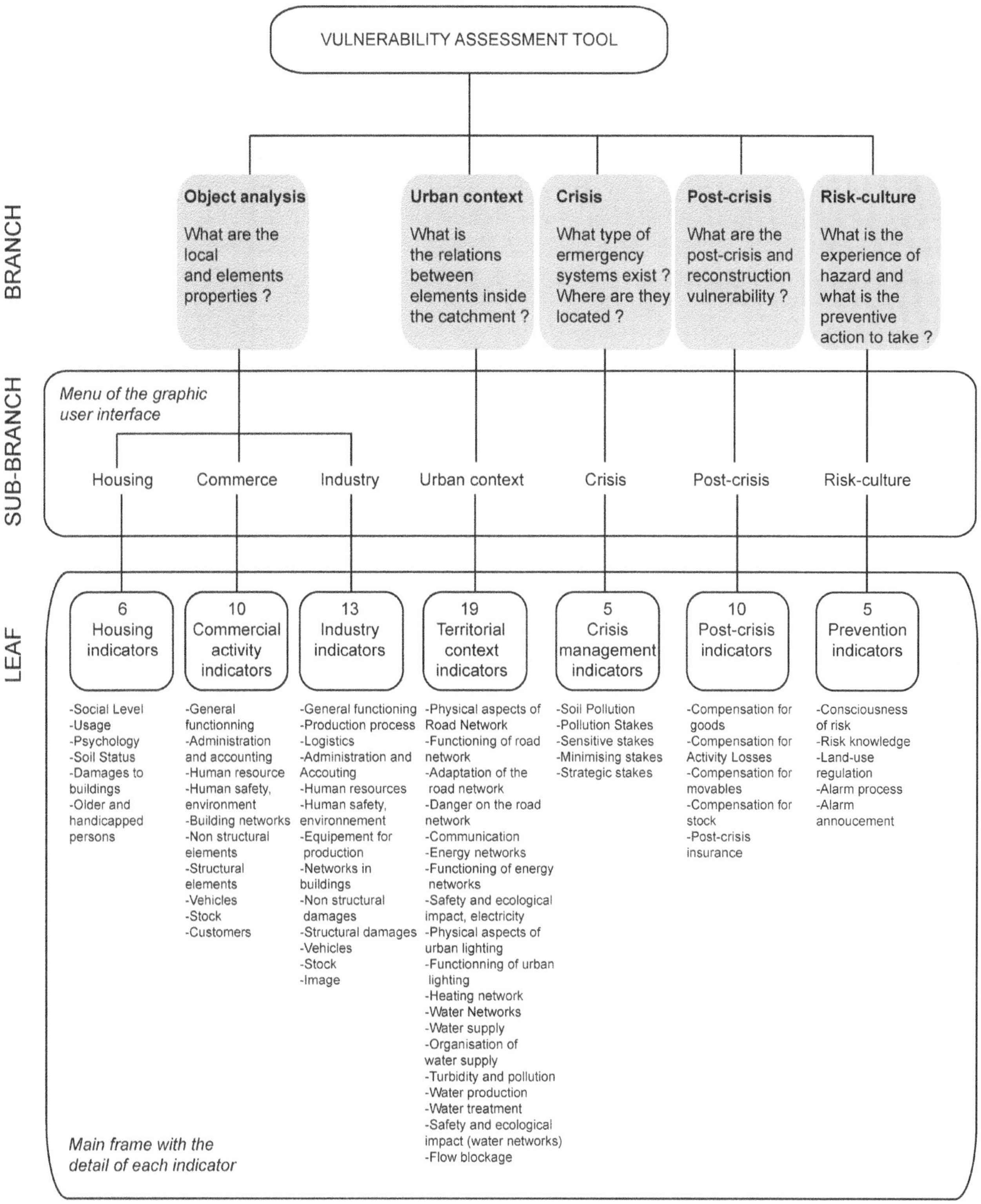

Figure 1: Proposed set of indicators for the flood vulnerability assessment tool (FVAT.

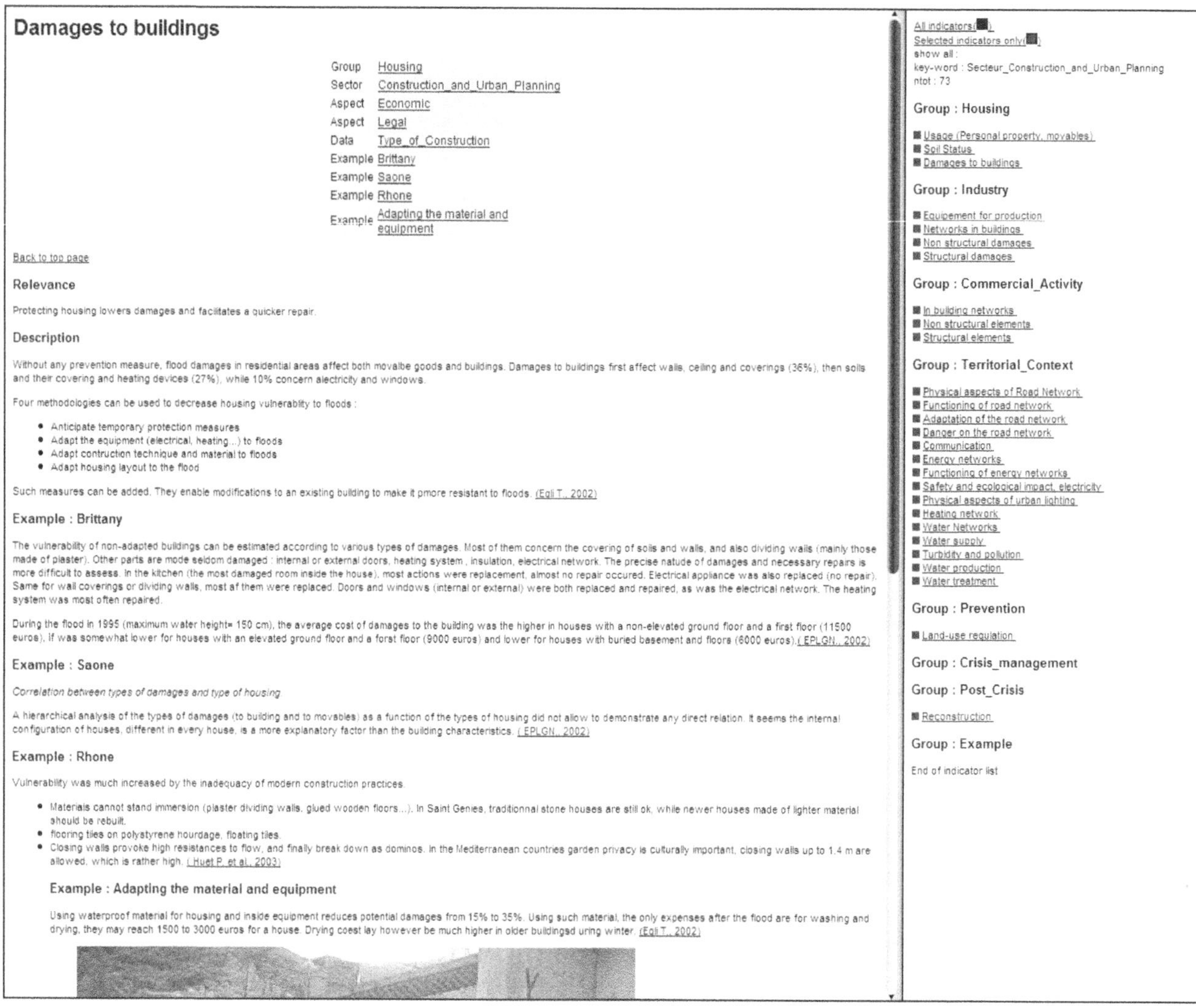

Figure 2: Screen capture of the FVAT tool: damages to buildings.

DISCUSSION

The interesting point about vulnerability is that it can be examined at different levels and scales for different issues. It can be used to look at a single issue such as a building, or to assess a complex entity such as a town. Here is presented an illustration based on a virtual case. In this case, local authorities need to manage a stormwater hazard, for which *a priori* knowledge does not reveal human vulnerability. Two projects are considered, both of them deal with a commercial area and are faced to the same hazard: the first one is located on a new urban area and the second is part of an urban renewal project.

In the first case, dealing with a new project offer the possibility to fit the constructed elements (structural or non structural elements, buildings networks, physical aspects of roads network, etc.) and the operation of the activities (general functioning, human safety environment, stocks, etc.) to the considered hazard.

In the case of urban renewal, the possibility to fit the constructed elements is circumscribed by the presence of buildings or networks: the vulnerability parameters which are likely to be relevant will relate to prevention and crisis management indicators referring to the vulnerability assessment tool.

Local decision makers need tools to manage urban systems under their responsibilities. Flood vulnerability assessment tools (FVAT) often deal with specific aspects of vulnerability (building, economic, social etc). Within those aspects, indicators and processing rules are developed. Processing rules are the result of experts' works. A drawback is that users of these assessment tools may express concerns about an apparent complexity for first-time application. Then it is usually assumed that users must be involved, as soon as possible, in the development of FVAT. The proposed FVAT involves the user from the definition of variables; according to indicators construction method of Lajzerowicz *et al.*

(1998), who defines a five step method to construct indicators: (i) identify objectives, (ii) define variables, (iii) choose relevant parameters, (iv) construct indicator and (v) validate indicator.

The proposed FVAT provides assistance in the further development and use of indicators within various decision-making levels. Each vulnerability assessment tool deals frequently with a defined set of indicators. Usually, the set refers to a specific decision level. In fact, three levels of decision (operational, functional and strategic) appear to be relevant in urban engineering (FCM & NRC, 2003):

- indicators for the 'operational' level are sustained with raw data on infrastructure assets; short-term decisions are made by using information at this level;
- 'functional' indicators concern combined data from different kinds and/or origins: inhabitants living in flooding area; these indicators could also be an integration of several operational indicators; a functional indicator provides managerial-level municipal decision makers (e.g. city engineer, public works manager) with an overview of an infrastructure asset condition, state or value;
- 'strategic' indicators provide global assessment of abstract objectives (e.g. protect population against flooding) or more practical ones (less than 24 hours to restore power network); they are set and reviewed by the highest level of municipal decision makers.

The description of each indicator within each group assists FVAT users to design their own set of indicators. The user may direct the design either to available data categories (type of construction, employment etc...) or either to one or more specific sectors (construction and urban planning, assistant etc...). According to the decision level, the spatial scale might be different. However, the authors aim to works at urban scales and not elements scales. Thereof, indicators chosen will be of functional or strategic level. As the FVAT tool is a checklist, the final user is in charge of the clear definition of each indicator at the relevant level.

Decision makers cannot deal with too many indicators. Moreover, dealing with a single indicator could lead to an incongruous simplification of a complex problem. The aim of an indicator is to convey information to a target. An indicator performs two main functions: an integration of many parameters and a decomplexification of the assessment's results (Bonierbale, 2004). In order to achieve these two functions, the number of indicators should be optimized (OECD, 1993; Weber & Lavoux, 1994; ISO, 1999). The result might be confusing and the overall perception required obscured when too many indicators are selected. On the other side, aggregating all the parameters into a single indicator or index does not represent a situation in its totality and leads to an information loss (Chatagnon, 1999).

Dealing with indicators leads to limits (Bouni, 1996). Indicators are not neutral neither objective, inasmuch as they are corresponding to a specific reading of reality which is based on a theoretical model or a specific knowledge. Each of them is not able to take into account the whole complexity of the vulnerability. Then, they convey a partial vision of reality. At last, as the data are incomplete a strong pragmatism is linked to the indicators.

Acquisition, management and updating of geographical information remain difficult, but are essential for the evaluation of indicators. In future development, public participation and Geographical Information System (GIS) should be coupled into a 'Public Participation Geographic Information Systems' (PPGIS) (Jankowski and Nyerges, 2001). These devices use geographic information or geographic technologies, developed partly by and for the public, with an aim of supporting the participation of the public in the local processes of territorial management (data, cartography, space analysis, territorial decision etc) (Edwards and Ligoza, 2004). PPGIS could be seen as the most popular and current solution.

CONCLUSION

Knowledge of the process of 'vulnerability' identification represents a very important contribution to decrease and control the land damage caused by natural hazard. The purpose of the flood vulnerability study is to limit weakness and integrate flood risk into urban development. The proposed flood vulnerability analysis tool (FVAT-www.h2o.net/daywater/) creates an integrated vision of the risk for its users: the project aims at gathering various visions of 'vulnerability' and shows as well that the analysis of risk must not be disconnected from possible interactions with territorial identity.

The DayWater decision making systems proposes, in a integrated manner, a series of tools to assess the potential of source control measures and evaluate their feasibility as well as the possible route for their implementation into the local social system. The primary objective of source control measures is to prevent runoff, although they can deserve additional purposes based on the use of harvested water or groundwater recharge (Chapter 6). The decision process must be obviously fed with information regarding the techniques to be implemented (Chapter 8), but it should not ignore the expected benefits and possible additional means of reaching them. Describing 'vulnerability' provides a qualitative answer to both issues, the damages are thus better identified and, once identified, means to reduce the 'vulnerability' of the object at risk can be proposed and compared. Such 'alternative' methods with reference to source control methods were not the central issue of the DayWater project, but they needed to be considered. This is the reason why the Flood Vulnerability Assessment Tool (FVAT) was designed in a very versatile manner, with a major objective of organising information related to the 'vulnerability' story rather than providing a well established methodology to quantify 'vulnerability'. It is an open tool which cans easily incorporate additional 'stories' describing the vulnerability of urban areas and available means and techniques which were implemented to reduce it.

REFERENCES

AFPCN Association Française pour la Prévention des Catastrophes Naturelles (2005) Risques et vulnérabilité [Risks and vulnerability]. Conference-discussion, Ecole nationale des Arts et Métiers, Paris.

Ayral, P.A. (2002) Terminologie en science du risque : Recueil de definitions [Terminology in risk science : definitions], GEM risques report, 87 p.

Barroca, B. (2006) Risque et vulnérabilités territoriales: les inondations en milieu urbain [Risk and territories vulnerabilities: Urban areas flooding.] PhD Thesis, University of Marne la Vallée, 370 p.

Blong, R. (2003) A review of damage intensity scale, *Natural hazards* **29**, 57-76.

Bonierbale, T. (2004) Eléments pour l'évaluation de la qualité environnementale des systèmes d'assainissement urbains [Elements for assessing the environmental performance of urban sanitation systems]. PhD Thesis, University of Marne la Vallée, 265 p.

Bouni, C. (1996) Développement durable et macrosystème d'information: des comptes d'environnement à l'aide multicritère à la decision [Sustainable development and macro information system : environmental accounting using multicriteria decision making]. PhD Thesis, Université Paris I Panthéon Sorbonne, 540 p.

Briguglio, L. (1995) Small island developing states and their economic vulnerabilities, *World Development*, **23**(9), 1615-1632.

Chatagnon, N. (1999) Développement d'une méthode d'évaluation de la qualité environnementale des bâtiments au stade de la conception [Development of an environmental quality assessment of buildings at their planning stage]. PhD Thesis, University of Savoie, 344 p.

CGCT Code Général des Collectivités Territoriales (1996), Article L.2224-10. Direction des journaux officiels, Paris, 684 p.

Comfort, L., Wisner, B., Cutter, S., Pulwarty, R., Hewitt, K., Oliver-Smith, A., Wiener, J., Fordham, M., Peacock, W. and Krimgold, F. (1999) Reframing disaster policy: the global evolution of vulnerable communities *Global Environmental Change* Part B: Environmental Hazards, **1**(1), 39-44.

D'ercole R.(1994) Les vulnérabilités des sociétés et des espaces urbanisés: concepts, typologie, modes d'analyse, *Revue de Géographie Alpine*, **LXXXII** (4), 87-96.

Dauphine, A. (2001) *Risques et catastrophes: Observer, Spatialiser, Comprendre, Gérer* [Risks and disasters: observe, localise, understand, manage]. Armand Colin, Paris, 288 p.

Edwards, G. and Ligoza, G. (2004) A formal model for structuring local perceptions of environmental space, *Cognitive processing*, **5**, 3-9.

EVI, Environmental Vulnerability Index http://www.vulnerabilityindex.net/ (verified September 2007).

FCM & NRC (2003) Developing indicators and benchmarks: a best practice by the national guide to sustainable municipal infrastructure. InfraGuide Best Practice Reports, 22 p..

Felts, D. (2002) *Vulnérabilité des réseaux urbains et gestion de crise* [Vulnerability od urban networks and crisis management], Centre d'Études sur les réseaux, les transports, l'urbanisme et les constructions publiques, 78 p.

Flax, L.K., Jackson, R.W. and Stein D.N. (2002) Community Vulnerability Assessment Tool Methodology, *Natural Hazards Review* **3**(4), 163-176.

Foster, H.D. (1976) Assessing disaster magnitude: a social science approach, *Professional Geographer* **28**(3), 241-247.

Gendreau, N. (1999) La méthode inondabilité. In *Le coût du risque... L'évaluation des impacts socio-économiques des inondations*, Hubert G., Ledoux B. (eds), Presses des Ponts, Paris, 123-127.

Glade, T.(2003) Vulnerability assessment in landslide risk analysis, *Die Erde*, **134**, 121-138.

Godin, C., Halbecq, X., Le Bris, A., Mallet, C. and Martin, A. (2004) Identification des enjeux exposés au risque d'inondation sur la Loire Moyenne [Identification of the elements at risk to flood hazard on the middle catchment of the river Loire]. SIG/Carto report, Cycle des Ingénieurs diplômés de l'Ecole Nationale des Sciences Géographiques, 2 volumes, 63 and 36 p.

Hufschmidt, G., Crozier, M. and Glade, T. (2005) Evolution of natural risk: research framework and perspectives. *Natural Hazards and Earth System Sciences*, **5**, 375-387.

IFRC International Federation of Red Cross and Red Crescent Societies (1996) *Vulnerability and Capacity Assessment*, October 1996, http://www.ifrc.org/ (verified September 2007).

ISO (1999) *Environmental management, Environmental performance evaluation, Guidelines*. International Standard, ISO 14031:1999, AFNOR, Paris, 35 p.

Jankowski, P. and Nyerges, T. (2001) *Geographic Information System for group decision making*, Taylor and Francis, London, 273 p.

Kelman, I. and Spencer, R. (2003) A limit Analysis of Unreinforced Masonry Failing Under Flood Water Pressures, *Masonry International*, **16**(2), 51-61.

Lajzerowicz, N., Gattino, B. and Cerutti, O. (1998) *Construction d'indicateurs et tableaux de bord synthétiques* [Construction of aggregative indicators and scoring board]. AFNOR, Paris, 29 p.

Ledoux, B. (1995) *Bilan de la mise en œuvre des méthodologies d'élaboration des études de vulnérabilité dans le cadre des PER entre 1984 et 1994*, Paris, Ministère de l'Environnement, DPPR, SDPRM, 160 p.

Lefevre, C. (1995) La recomposition territoriale en question: position d'acteurs, Lyon, *Revue de Géographie de Lyon*, **70**(2) 159-168.

Lefort, E. (2003) Dynamique territoriale et vulnérabilités aux inondations: Exemple du Val de Saône-Mâconnais. Bacchelor's report, Université de Versailles Saint-Quentin-en-Yvelines, 177 p.

Lefort, E. (2004) Evaluation des vulnérabilités aux inondations dans le cadre de la gestion opérationnelle des risques et du développement local (application au bassin versant de l'Orge aval). Master's report in Environmental Science and Technology, ENPC-ENGREF-University Paris 12, 43 p.

Leone, F., Aste, J.P. and Leroi E. (1996) L'évaluation de la vulnérabilité aux mouvements de terrain, *Revue de géographie alpine*, Grenoble, **84**(1), 35-46.

Mebarki, A, Valencia, N. (2004) Informal Masonry Structures: Seismic Vulnerability and GIS maps. *Masonry International Journal*, British Masonry Society, UK, **17**(1). 18-25.

Ministry of Agriculture, Fisheries and Food (1999). *Flood and Coastal Defence Project Appraisal Guidance, Part 3: Economic Appraisal.* Guide of the Flood and Coastal Defence with Emergencies Division, 80 p + annex.

Moris-Oswald, T. and Sinclair, A.J. (2005) Values and floodplain management: Case studies from the Red River Basin, Canada, *Environmental Hazards,* **6**, 9-22.

NHRC Natural Hazards Research Center (1999) Australia; NHQ *Natural Hazards Quarterly*, **5**(4), http://www.riskfrontiers.com/nhq/nhq5-4tables.htmh

ODPM Office of the Deputy Prime Minister (2001) *Multi-criteria analysis: a manual.* Department of the Environment, Transport and the Regions, London, 158 p.

OECD Organisation for Economic Co-operation and Development (1993) *Core set of indicators for environmental performance reviews.* Paris, 41 p.

Parent, C., Spaccapietra, S. and Zimanyi, E. (1999) Spatio-Temporal Conceptual Models: Data Structures + Space + Time. *Proceedings of the 7th ACM Symposium on Advances in GIS*, Kansas City, Kansas, 26-33.

Parker, D., Green, C. & Thompson, P. (1987) *Urban Flood Protection Benefits, a project appraisal guide (Red Manual),* Gower Technical Press, Aldershot, 284 p..

Penning-Roswell, E. C., Chatterton, J. B. and Green C. (1992) *The Economics of Coastal Management, a manual of Benefits Assessment Techniques (Yellow Manual).* Belhaven Press, London, 380 p.

Penning-Roswell, E. C. and Chatterton, J. B. (1977) *The benefits of flood alleviation – A manual of assessment techniques(Blue Manuals),* Gower Technical Press, Aldershot, 297 p.

Pomonis, A., Spence, R. and Baxter, P. J. (1999) Risk assessment of residential buildings for an eruption of Furnas Volcano, São Miguel, the Azores, *Journal of Volcanology and Geothermal Research*, **92**(1-2), 107-131.

Pornon, H. (1998) Le marché des SIG. Evolutions et perspectives. *Revue XYZ*, **75**, 68-70.

Pottier, N. (2000) Risques d'inondation, réglementations et territoires. *Hommes et Terres du Nord,* **2**, 93-101.

Pottier, N., Lefort, E., Vine,t F. and Barroca, B. (2004) L'évaluation des vulnérabilités territoriales pour l'aide à la gestion des inondations par les collectivités locales, in *Contraintes environnementales et gouvernance des territoires*, Scarwell H.J., Franchomme M. (ed.), éditions de l'Aube, p. 44-53.

Shrubsole, D., Hammond, V.J., Kreutzwiser, R. and Woodley, I. (1997) Assessing Flooplain Regulation in Glen Williams, Ontario, Canada, *Journal of Environmental Management,* **50**, 301-320.

Spence, R.J.S., Baxter, P.J. and Zuccaro, G.(2004) Building vulnerability and human casualty estimation for a pyroclastic flow: a model and its application to Vesuvius, *Journal of Volcanology and Geothermal Research*, **133**(1-4), 321-343.

Spence, R.J.S., Kelman, I., Baxter, P.J., Zuccaro, G and Petrazzuoli, S. (2005) Residential building and occupant vulnerability to tephra fall, *Natural Hazards and Earth System Sciences*, **5**, 477-494.

Strategis (2002) *Etude de faisabilité d'un observatoire des enjeux et de leur vulnérabilité aux inondations* [Feasibility study of an observatory of stakes and their flood vulnerability], Phase 2 report: literature survey, Paris, Ministère de l'environnement, DPPR, SDPRM, August 2002, 46 p. + annexes.

US-EPA United States Environmental Protection Agency. (1997) *Terms of environment: Glossary, Abbreviations and Acronyms*, revised December 1997, 65 p.

Valencia, N, Mebarki, A. and Salagnac, JL. (2004) Risques naturels et SIG Application à la vulnérabilité des ouvrages en maçonnerie. XXII *Rencontres Universitaires de Génie Civil AUGC*, Université Marne la Vallée, Champs sur Marne, France, June 3-4, 2004 [CD-Rom].

Weber, J.-L. and Lavoux, T. (1994) *Réflexion sur les critères de définition et de choix des indicateurs d'environnement* [Criteria for definition and selection of environmental assessment indicators]. Orléans : Institut Français de l'Environnement, 54 p.

WIN Wild Information Network for Risk Management (2007) Multi Lingual Risk Management GLOSSARY - English, French, German, Spanish, Romanian. Report: IST Integrated Project No FP6-511 481, Deliverable D2204, 396 p.

Wisner, B., P. Blaikie, Cannon, T. and Davis I. (2004) *At Risk: Natural Hazards, People's Vulnerability and Disasters*, 2nd ed. Routledge, London, UK, 496 p..

Yohe, G. and Tol, R.S.J. (2002) Indicators for social and economic coping capacity: moving toward a working definition of adaptive capacity, *Global Environmental Change*, **12**(1), 25-40.

Zihri, G. (2004) Risques liés aux ouvrages souterrains: constitution d'une échelle de dommages [Risk related to underground cavities: the constitution of a damage scale], INPL PhD thesis, Nancy, 226 p.

Chapter 14

Source and flux modelling in urban stormwater management (STORM/SEWSYS): application examples in Germany and Sweden

Stefan Ahlman, Colette de Roo, Gilbert Svensson and Heiko Sieker

SOURCE AND FLUX MODELLING CONCEPT

A large number of numerical models are available for simulating quantity and quality aspects in stormwater systems (See *e.g.*: Achleitner and Rauch, 2005; Calabro, 2001; Ruan, 1999; Wong et al., 2002). The models can be used as design tools, decision support tools or even both, and they vary both in terms of complexity and input data requirements. In terms of mechanisms, computer models can be classified as deterministic, stochastic or conceptual. Hydraulic models like InfoWorks (www.wallingfordsoftware.com), MOUSE (www.dhigroup.com) and SWMM (www.epa.gov/ednnrmrl/models/swmm/) rely on deterministic physically based formulations of the modelled processes, whereas hydrological models describe the processes in a more conceptual way. Conceptual models are normally based on simple yet sound physical concepts and relatively limited data resources. Compared to deterministic models they are more efficient, *i.e.* timesaving and less data intensive, but it should be stated that hydraulic and hydrological models have different purposes. Simple and well-established hydrological principles together with systems analysis usually form the description of a conceptual model. As such, the hydrological model STORM and the substance flow model SEWSYS have been chosen as suitable building parts of the proposed sources and flux modelling (SFM) tool. The STORM model acts as the platform for the SFM tool. To enhance the stormwater quality part of STORM, the substance flow concept of SEWSYS has been integrated into STORM. The main feature of the new SFM tool is the ability to simulate scenarios of different stormwater source control practises, including both quantity and quality aspects.

The model STORM

The STORM software is a hydrological model for simulating stormwater runoff and pollution load (IPS, 2003). STORM is a modern 32-bit software developed in Microsoft® Visual C++, an object-oriented programming language that facilitates further development and allows for rapid simulations. The main window of STORM is shown in Figure 1. To the left is an "explorer window" where parameters and drainage elements are listed. The drainage structure of the studied catchment(s) is built up in a graphical way in the window to the right using nodes and links. STORM is available in both English and German versions. A demo version of STORM can be downloaded from the website http://www.sieker.de.

STORM is predominately a planning tool for urban water management but also contains features for designing single best management practices (BMPs), *e.g.* using design storms to compute necessary storage volumes. Rainfall-runoff transformation is conceptualised using either linear cascade storages or time-surface functions. The model contains conventional drainage elements, both decentralized and centralized, to treat and handle runoff in conventional sewer systems and alternative stormwater management system. The drainage elements include retention tanks, trough-flow tank, retention sewer and overflows as well as infiltration facilities, green roofs or cisterns.

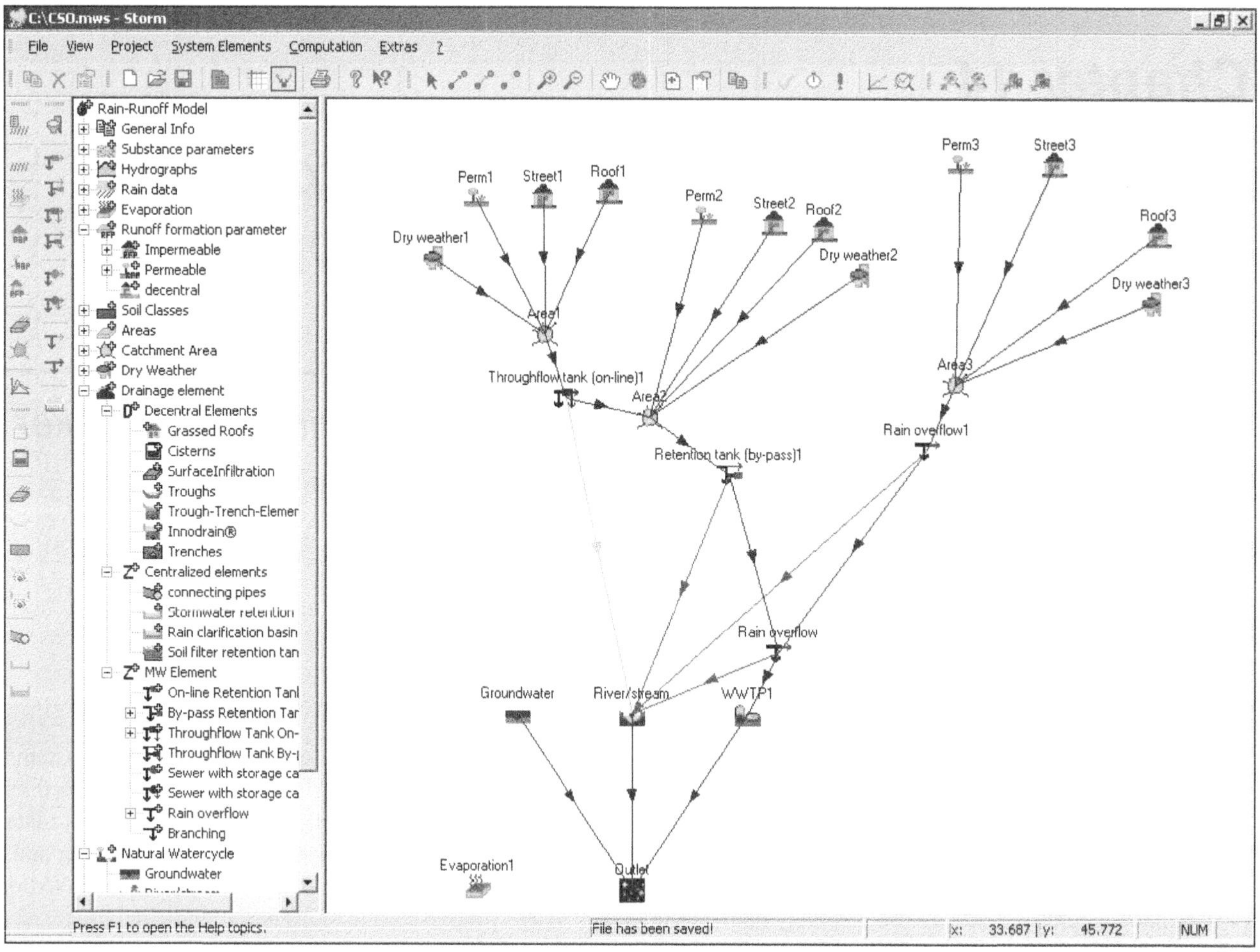

Figure 1: STORM user interface.

The results from STORM focus on overall water balances and BMP performance with respect to quantity and pollution transport. User-defined standard concentrations of pollutants are used as input in the model to calculate pollution loads. The runoff processes in the catchments are simulated dynamically using a time-series of rainfall data. The time resolution in a rain file also determines the time-step in the model. Usually a time step of 5 to 15 minutes is used. A prerequisite for the long-term simulation is that the input data consist of actual measured rain gauge readings in their natural chronological progression. STORM also includes a snow-module with accumulation and melting routines based on the work in DayWater by Camilla Westerlund and Maria Viklander (chapter 22).

The model SEWSYS

The pollutants in urban stormwater contribute significantly to the total pollutant mass flow in the urban sewer system. To address this issue the new sources and flux model has incorporated a stormwater pollutant generator module from the SEWSYS model.

SEWSYS is a model developed in MATLAB/Simulink® for simulating substance flows in urban sewer systems (Ahlman, 2006). The present version of SEWSYS handles both domestic wastewater and stormwater. In the stormwater quality module, the pollutants from sources like dry deposition, traffic and construction materials are generated and accumulated during dry weather until they are washed away during rainfall.

The model STORM/SEWSYS – an SFM tool

The relative importance of different sources for stormwater pollution can vary greatly between catchments. Depending on local conditions, pollution sources can be quite different when looking at catchments of different sizes and different regions. This variability is included in the SFM tool. The different stormwater pollutant sources and pathways are summarised in Figure 2. This descriptive concept has formed a framework in the development of the pollutant generator in STORM/SEWSYS.

The user of the model has the option to employ either standard concentrations or the pollutant generator in SEWSYS to calculate pollution loads. If the user chooses the SEWSYS option, it will be possible to analyse what kinds of substances are present in the catchment, their sources, transport routes and the sinks in which the substances finally reside. At present STORM/SEWSYS handles sources and fluxes of 22 substances and indicators that have been identified in the DayWater project as relevant in stormwater quality management (Eriksson *et al.*, 2005). These selected stormwater priority pollutants (SSPP) include nutrients, heavy metals, PAHs, pesticides and various other ubiquitous organic compounds (see also Chapter 11).

To calculate pollution loads using the SEWSYS pollutant generator, the model requires catchment-specific parameters including traffic load and the percentage of heavy vehicles. The distribution of impervious area between road, roof and other areas is also required. The model also needs information about the different roofing materials used in the catchment, such as zinc (galvanised) and copper surfaces. The pollutant generator module in STORM/SEWSYS obtains this data from specific values set by the user for each sub-catchment, preferably generated by a GIS. An interface with common GIS-Software ESRI® (www.esri.com) and Intergraph® (www.intergraph.com) is available.

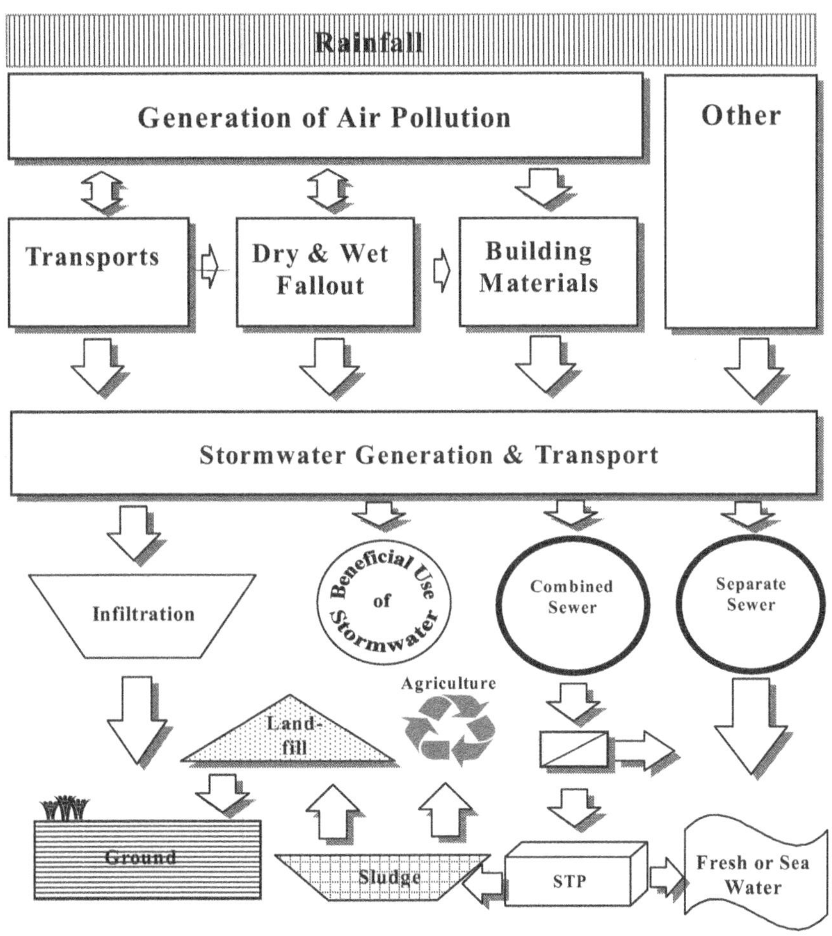

Figure 2: Stormwater pollutant sources and pathways.

The model input database

Apart from previously described catchment specific data, the pollutant generator module also requires background pollutant data. This is imported into STORM/SEWSYS from a Microsoft® Access database and contains general pollution figures, *e.g.* corrosion rates and emissions for different kinds of material (Figure 3).

Each substance is linked to different source parameters in the database. Data for the pollutant sources have been obtained from literature studies and other existing Swedish models (Ahlman and Svensson, 2002). The user chooses an appropriate data set from the database to be used in the simulations. The imported data can be modified within the STORM/SEWSYS environment. At present pollution data for catchments in Sweden, Denmark and Germany have been entered into the database. More data have been collected within the DayWater project and afterwards as the model is used in several studies over Europe.

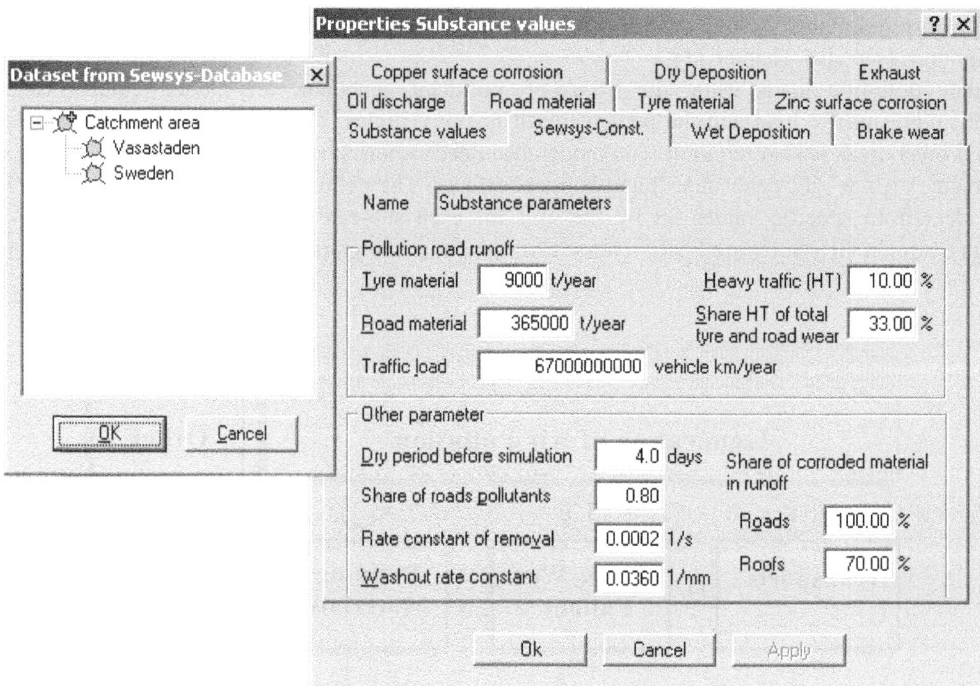

Figure 3: Input data for the pollutant generator module.

MODEL OUTPUTS AND NEW POSSIBILITIES

The new STORM/SEWSYS model provides simulation techniques to evaluate hydrological cycle changes and hydrological risks in urban catchments. These changes can be caused by the non-existence or application of stormwater source control measures. The model matches the spatial scale and integration level that govern the implementation of source control. From a stormwater quantity aspect, the focus is on the functional behaviour of structural BMPs, interactions with changing groundwater levels, increased runoff via building drainage and implications for existing downstream wastewater systems.

Traditionally in stormwater quality models, including the present version of STORM, the water quality part in the model incorporates standard concentrations as input for user-defined pollutants. Using standard pollutant concentrations for different land uses is useful because it makes the model easier to work with and requires less input data. However, in terms of pollutant load from stormwater runoff, the appropriate long-term strategy would be to avoid, replace or reduce the sources of stormwater pollution. To be able to achieve quality source control, *i.e.* to reduce the pollutants at the source, it is necessary to have a more detailed pollutant generation and transport model. The sources for pollutants from different activities in the urban area have to be separated in their respective origin, *i.e.* material corrosion, brake wear, tyre wear etc. The pollutant generator module in STORM/SEWSYS makes this possible. A selection of model outputs have been summarised in Table 1. In addition, results from long-term STORM/SEWSYS simulations enable statistical analysis and the calculation of hydrological data such as hydrographs and runoff volumes.

Output from the model can also be exported and imported to other models as input data. The user can for example store output data like discharge and pollution concentration from STORM/SEWSYS in CSV-files by using a *Hydrograph*-object. These can be imported to hydrological models, river quality models or even hydrodynamic models.

Table 1: Selected model outputs from STORM/SEWSYS.

Hydrology		Runoff hydrograph	time; m^3/s
		Runoff volume	m^3/year
Pollution Loads	Wet deposition	A number of substances	g/year
	Roads	A number of substances	g/year
	Roofs	A number of substances	g/year
	Other	A number of substances	g/year
Pollution Sources	Roads	A number of sources	%
	Roofs	A number of sources	%
	Other	A number of sources	%

INTERACTION WITH THE DAYWATER ADSS

This section describes the role of STORM/SEWSYS within the DayWater ADSS. The ADSS is described in Chapter 3 and 4 of this book. Information about the STORM/SEWSYS tool is found in the tile 'Modelling Tools' in the HYDROPOLIS page of the ADSS. A short description of the tool in terms of necessary input data, produced output data and benchmarks that can be calculated is available in this section. There is also a tutorial/description of the model, as well as examples from real case studies. The model can also be downloaded via a link to the IPS web server (http://www.sieker.de).

An external ADSS component and a stand-alone computer software, STORM/SEWSYS can communicate with the ADSS via XML-protocol. It is possible to import a vector with the names and units of priority pollutants from the ADSS. After modelling existing situations or planned scenarios, the simulation results *e.g.* water balances or pollutant loads can be exported to an XML-file again. Within the ADSS this information can be used to fill out the decision matrix and serve as a base for a Multi-Criteria Analysis. The XML-protocol was developed within DayWater by Heiko Sieker (IPS) and Tomas Metelka (DHI).

Other interactions with the ADSS are given by the BMP catalogue (Chapter 8) which can be used to develop scenarios for the simulation and the risk screening tool FLEXT. How FLEXT has been used to develop scenarios for the Wupper catchment is shown in Chapter 15 of this book. The potential maps for on-site BMPs produced with FLEXT can be imported with the GIS interface included in STORM.

APPLICATIONS

As one of the external tools of the DayWater ADSS (Chapter 3), the STORM/SEWSYS model has undergone further testing during the year 2005. The DayWater end-users Wupperverband in Germany and Stockholm Water Co. in Sweden have been involved in the testing phase. The Wupper case from Germany describes a sources and flux modelling study in a large catchment with both combined and separate sewer systems. The Hammarby Sjöstad case describes the use of the SEWSYS approach to calculate pollution load and sources for a smaller separate catchment in the city of Stockholm.

The Wupper case

The water board of the river Wupper (Wupperverband) was chosen as one of the end-users of the DayWater Project. As each end-user tested a segment of the ADSS, the Wupperverband was chosen to test one of the modelling tools. In this specific case, the STORM/SEWSYS model was applied to a major part of the urban areas in the river basin of the Wupper. The results were an extensive picture of the sources of water volumes and pollutants and their fluxes through the urban areas towards the river Wupper. Thus providing a powerful visual tool to understand and positively influence decision making in urban storm water management. A process that in this case included quantified results, which supported the application of, decentral measures for stormwater management in the area (See Figure 4 for an example).

Figure 4: Example of a decentral measure for urban stormwater management (INNODRAIN Element, same principle as a trough-trench system).

Wupperverband

The Wupperverband is one of the largest water boards in the federal state North Rhine Westphalia in Western Germany (Figure 5) and has been assigned, as a public corporation, the responsibilities for managing all water quality and quantity issues in the river basin of the Wupper. Members of the water board are municipalities, counties, water suppliers and wastewater companies, as well as industries situated in the river basin. The Wupperverband is very engaged in ensuring an integral and sustainable environmental development in the entire river basin, which covers an area of 813 km^2.

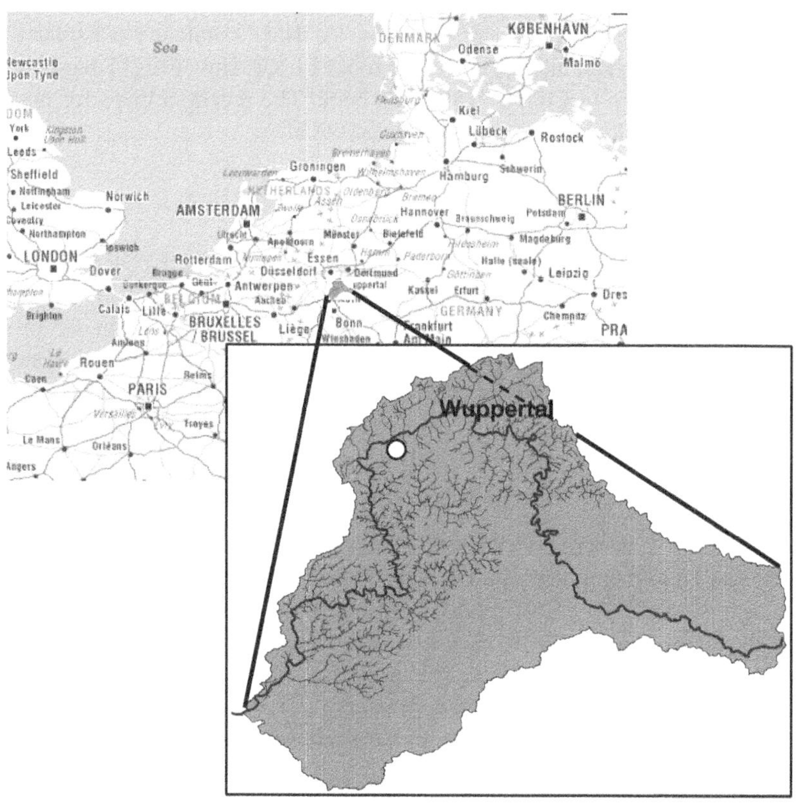

Figure 5: The river basin of the Wupper situated in North Western Europe.

Project area

The exact project area chosen for the STORM/SEWSYS models is the middle and lower Wupper, which covers roughly half of the total river basin of the Wupper (Figures 5 and 6). In this area, six urban catchment areas including their wastewater treatment plants (WWTPs) are situated along the Wupper River. Among these urban areas are the cities Leverkusen, Remscheid and Wuppertal, all of which have between 100,000 and 200,000 inhabitants. Smaller models were made for the remaining urban catchments Burg, Schwelm and Radevormwald.

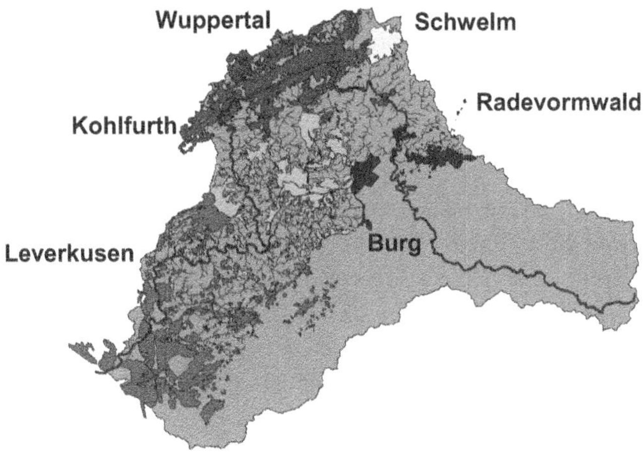

Figure 6: Graphic representation of the six submodels for the STORM/SEWSYS testing procedure with the end-user Wupperverband.

Objectives

The area was found very suitable to use for testing the STORM/SEWSYS model. First of all generally good input data for a source-flux simulation was available. More importantly, however, because the Government of North Rhine Westphalia has adopted strict regulations regarding pollutant loads towards water bodies. Since the STORM/SEWSYS model allows decision makers to increase their insight in the sources of specific compounds, it becomes possible to make aimed modifications to the system, including the introduction of decentral stormwater measures, and thus disconnect very specific surfaces that contribute to the pollutant load of a certain hazardous compound.

Another good reason for the preparation of the STORM/SEWSYS models in the river basin of the Wupper is the fact that many stakeholders operate in the area, with a variety of agendas and accordingly many conflicts of interest. The liabilities for the different aspects of stormwater management for instance are divided between the Wupperverband and the municipalities, but obviously cannot be dealt with separately. This implies constant negotiation and finding compromises, for which an easily adaptable model with a strong visual side can serve as a good means of communication.

Since the STORM/SEWSYS model can be used on different spatial levels and with different components, it is always advisable to define precisely which objectives, available inputs and expected outputs comply with the specific catchment area and wishes of the user. For the Wupperverband and its catchment areas, the questions and answers were the following:

- Which objectives do you have? Creating a solid basis for negotiation and discussion between stakeholders in the river basin of the Wupper.
- Which output do you expect? Quantified results on impacts of different possible scenarios for storm water management.
- Will you use your STORM/SEWSYS results for other simulations? As input for a hydrological model for the receiving waters.
- Which input data do you have? Varied as they are being maintained by different stakeholders, but generally good availability.
- Which SFM elements do you need? Long term simulation of rainfall-runoff processes, including scenarios with non-structural BMPs, pollution loads/fluxes included in the model, exporting tools.

The six Wupper models

To create a set of models for the six urban catchment areas in the river basin of the middle and lower Wupper, the following input sources were available: flow charts, digital maps, GIS data, rainfall time series, other simulation results and calibration data. Especially useful for the setting up of the models was the availability of extensive GIS data on the whole river basin of the Wupper through a special website managed by the Wupperverband, the so-called FluGGS tool (http://fluggs.wupperverband.de/).

To show that different approaches are possible with the STORM/SEWSYS model, for making a useful simulation, the input data, as well as the calibration data for three of the catchment areas are mentioned below. Those three were chosen, based on their distinctly different input and calibration data sources.

Simulation model Leverkusen

- Flow charts, databases with data on BMPs, NASIM (hydrological) model transferred to STORM/SEWSYS for part of the model structure,
- GIS data for visual background and rain gauges locations,
- Database with WWTP performance figures for first calibration,
- MOMENT (urban hydrology) model output as reference simulation.

Simulation model Wuppertal

- Extensive detailed GIS data for allocating types of surfaces to the system and BMPs (geographically), databases for BMP characteristics,
- GIS data for visual background and rain gauges locations,
- WWTP and BMP performance data as calibration data.

Simulation model Burg

- NASIM model transferred to STORM/SEWSYS for model structure,
- GIS data for visual background and rain gauges locations,
- NASIM (hydrological) model output as calibration data.

An example: the STORM/SEWSYS Model for the catchment area of WWTP Leverkusen

As an example the model for the catchment area connected to the WWTP of Leverkusen can be presented by several input, simulation and output windows, pictures and graphics (Figure 7). This catchment area is the most complex system of all six studied catchment areas and is made up by the urban areas of six different municipalities. As mentioned above, for Leverkusen different input data sources for different 'corners' of the catchment area were available. For this purpose, a compromise had to be found between the levels of detail and the time that was available to investigate the present gaps.

Results

With the STORM/SEWSYS simulation models the current status of the systems in the Wupper River Basin including decentral measures have been set up. Their performance was simulated and calibrated and, in the case of Leverkusen, compared to an alternative simulation with the urban hydrological model MOMENT. The outcome of this comparison was very satisfying, because both models represent the performance of the urban water system to the same extent. However a great advantage of using the STORM/SEWSYS models is, that scenarios including decentral measures and determination of reduction of priority pollutants can be tested. This assessment was not possible with other available hydrological models, like MOMENT. It was also determined which influence certain measures of disconnection would have on the pollution loads towards the Wupper River.

The information generated in this end-user test has proved to be very useful when dealing with the requirements of the EU Water Framework Directive (WFD) and other local legislation or guidelines, like the above-mentioned specific regulation on pollution load in North Rhine Westphalia.

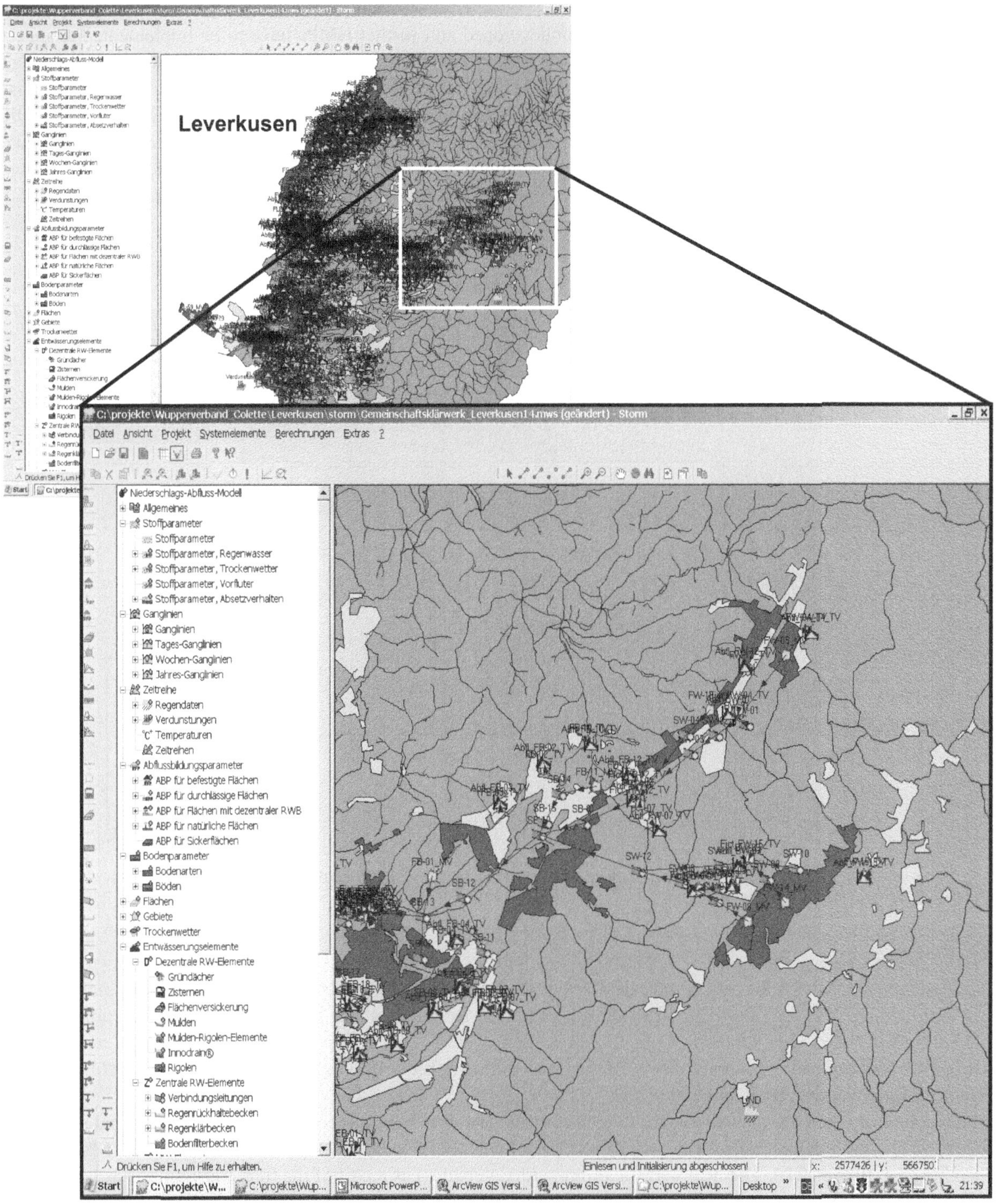

Figure 7: Graphic representation of the STORM/SEWSYS model for the catchment area connected to the WWTP of Leverkusen as it appears in the graphic mode of STORM/SEWSYS (with detail).

The Hammarby Sjöstad case

This is an ongoing project for housing in the city of Stockholm, as 30,000 people will be living in this area by year 2015 (Figure 8). The area covers about 130 ha. The area has got its own local environmental program concerning, among other issues, also stormwater. Originally, the environmental program stated that all stormwater should be handled locally with attenuation and also infiltration if possible. All stormwater from traffic areas should be treated and a number of BMPs have been constructed so far and further will be added.

In Hammarby Sjöstad, there is now a need to evaluate the environmental program and the existing BMPs. Around two thirds of the constructions in the area remains undeveloped and further BMPs have to be implemented to meet the local environmental program. The Stockholm Water Co. has some specific questions about stormwater management in the area: What is the effect of the built BMPs? Could the effect have been achieved in another way? How polluted is the stormwater? What are the most important sources for stormwater pollution in the area? To help answering these questions a STORM/SEWSYS model has been set-up for the whole catchment area. To compare the stormwater quality of the mixed runoff with only road runoff a small sub-catchment called Mårtensdal was also included in the analysis. In this sub-catchment runoff from a 1-hectare road area is conveyed to a treatment system consisting of a sedimentation tank and infiltration surface.

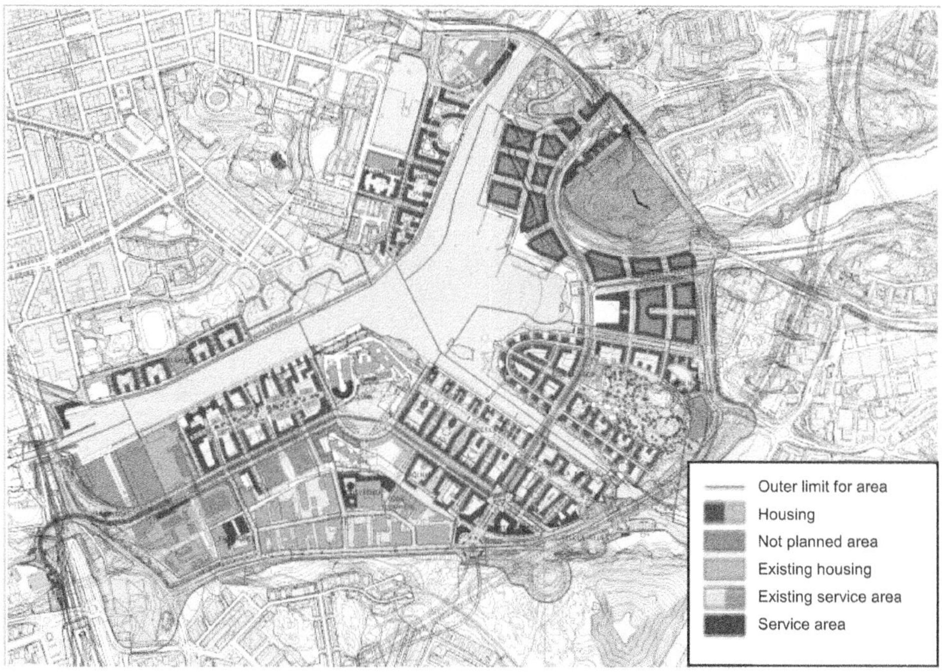

Figure 8: The area Hammarby Sjöstad in Stockholm, Sweden.

GIS analysis

As the first step an analysis in a geographical information system (GIS) was performed to find the necessary input data for the STORM/SEWSYS model. The GIS data was extracted from a database run by the different city administrations (Stockholm Water Co., City Planning Office, Traffic Office etc.). The traffic load in the catchments was calculated utilising GIS data on traffic volumes. The GIS data on building structures was supplemented with a field study of the different roofing materials used in the area.

The results from the GIS analysis for the two catchments are shown in Table 2 and their respective land use is shown in Figure 9. The roofing materials for the whole catchment was found to be 75 % galvanised or painted, 6 % copper and 19 % other (*e.g.* tiles or felt). The heavy traffic (commercial vehicles) as a part of the total traffic load has been assumed to be 5 percent for both catchments.

Table 2: STORM/SEWSYS input data from GIS analysis of Hammarby Sjöstad and Mårtensdal catchments.

	Hammarby Sjöstad	Mårtensdal	Unit
Total area	130	2.07	ha
Pervious area	43	1.29	ha
Impervious area	87	0.78	ha
Roads	15	0.56	ha
Roofs	22	0	ha
Other impervious	50	0.22	ha
Traffic load	127,000	6,600	vehicle_km day^{-1}

Figure 9: Land use from GIS analysis (whole catchment to the left, sub-catchment Mårtensdal to the right).

STORM/SEWSYS model

The physical structure of the catchments has been implemented and built in STORM/SEWSYS based on the results from the GIS analysis (Figure 10). Stormwater quality is calculated in the model using the database on emission factors and the dataset for Hammarby Sjöstad, which has been added previously to the database.

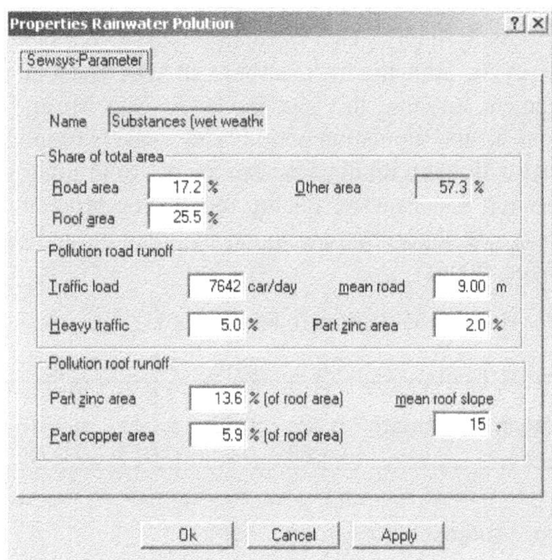

Figure 10: The properties for the impervious area of Hammarby Sjöstad as it is implemented in the model.

The rain data used in the simulations is from a rain gauge operated by Stockholm Water Co., located about 4 kilometres from the Hammarby area. The rain series contains rain data for 10 years (1995-2004) and has an average annual rainfall of 469 mm. Evaporation in the Stockholm area is set to 444 mm per year, *i.e.* 1.2 mm per day (SMHI, 1980). The time-step used in the simulations is 5 minutes.

Results

The STORM/SEWSYS has the capability to calculate and display a lot of results regarding both stormwater quantity and quality. In the presentation of results from this study, the main focus is on the pollution load and sources. However, the first result presented in Figure 11 is an example of quantitative output data and shows a statistical evaluation of the rain series used in the simulations. An inter-event time of 240 minutes has been used as dividing criterion and the events are sorted after rain depth (column "precipitation height" PH in Figure 11). This output is especially valuable when structural BMPs have been modelled because it shows the return period for the rain events.

No.	Date	Time	Duration [min]	max. intensity [l/(s ha)]	mean Int. [l/(s ha)]	PH [mm]	n-tr exist [1/a]	tr exist [min]
1	16.06.1997	07:15:00	2130	12.77	3.65	46.69	0.06	16.80
2	01.05.2003	01:15:00	1095	18.83	6.33	41.57	0.16	6.30
3	06.07.1998	09:00:00	600	87.57	10.24	36.86	0.26	3.88
4	27.04.1995	18:15:00	1785	10.57	3.35	35.92	0.36	2.80
5	25.07.2000	08:30:00	2250	10.37	2.64	35.60	0.46	2.19
6	22.07.2003	14:45:00	285	139.83	20.65	35.31	0.56	1.80
7	17.07.1995	08:45:00	1125	20.57	5.13	34.61	0.65	1.53
8	07.06.2004	14:45:00	840	29.10	6.86	34.57	0.75	1.33
9	17.04.1999	10:45:00	2355	9.43	2.40	33.90	0.85	1.17
10	29.09.1995	10:45:00	1080	17.77	5.20	33.71	0.95	1.05
11	11.06.1995	19:15:00	720	16.67	7.73	33.39	1.05	0.95
12	05.05.1996	04:15:00	2220	20.40	2.39	31.87	1.15	0.87
13	10.01.2002	01:15:00	975	15.57	5.03	29.41	1.25	0.80
14	03.12.1999	21:30:00	750	12.77	6.20	27.91	1.35	0.74
15	11.12.1999	20:30:00	1230	11.10	3.56	26.30	1.45	0.69
16	28.09.1995	05:30:00	1185	16.67	3.69	26.20	1.55	0.65
17	11.06.2004	17:15:00	465	93.40	9.06	25.28	1.65	0.61
18	22.09.1999	05:45:00	1290	20.33	3.21	24.88	1.75	0.57
19	27.08.2001	03:15:00	600	36.50	6.78	24.43	1.85	0.54
20	13.06.2002	20:30:00	960	22.17	3.94	22.68	1.94	0.51
21	09.05.1997	10:15:00	1215	18.67	2.94	21.44	2.04	0.49
22	13.06.1997	22:30:00	1080	31.93	3.26	21.14	2.14	0.47
23	24.01.2002	11:45:00	855	11.67	3.94	20.19	2.24	0.45
24	07.11.2000	22:30:00	960	7.23	3.49	20.11	2.34	0.43
25	12.07.1998	02:00:00	525	26.00	6.37	20.07	2.44	0.41

Figure 11: Rainfall statistics for the 25 largest events in the Stockholm series (1995-2004).

The stormwater quality data calculated with the model is shown in Table 3 for the whole catchment and in Table 4 for the sub-catchment with only road runoff. The substances displayed here are copper, zinc, lead, cadmium, phosphorus, nitrogen and total polycyclic aromatic hydrocarbons (PAH). The total pollution loads have been divided by the impervious area of the catchment to show the specific loads. Site Mean Concentration (SMC) is the volume weighted event mean concentrations for the simulation period. This value is usually a good quality check of the model's output because it can be compared with concentration measurements from other studies in similar catchments. It was unfortunately not possible to carry out a stormwater quality monitoring programme in the area. However, the SMCs from the model are within the range of concentrations found in literature and databases (Brezonik and Stadelmann, 2002; Smullen *et. al*, 1999; StormTac, 2006).

Table 3: Stormwater quality data for Hammarby Sjöstad (mixed runoff).

	Cu	Zn	Pb	Cd	P	N	PAH
Load [kg/year]	25	59	3.6	0.14	31	616	0.24
Load per unit area [kg/imp. ha/year]	0.28	0.68	0.041	0.0016	0.36	7.1	0.0027
SMC [mg/L]	0.088	0.21	0.013	0.0005	0.11	2.2	0.0008

Table 4: Stormwater quality data for Mårtensdal (only road runoff).

	Cu	Zn	Pb	Cd	P	N	PAH
Load [kg/year]	0.57	1.05	0.033	0.0014	0.85	5.6	0.006
Load per unit area [kg/imp. ha/year]	0.73	1.34	0.041	0.0017	1.08	7.0	0.0081
SMC [mg/L]	0.23	0.42	0.013	0.0005	0.34	2.2	0.0025

Figure 12 shows a comparison of the SMCs from STORM/SEWSYS model assessment for the catchments in Hammarby Sjöstad. The figure also displays the modelled SMC for a catchment called Vasastaden located in the city centre of Göteborg, Sweden, where the SEWSYS model has been tested previously. The comparison reveals that road runoff (Mårtensdal) is the most polluted stormwater, this is especially the case for copper, zinc, phosphorus and total PAH. Stormwater from Hammarby Sjöstad and Vasastaden has about the same quality; the differences for copper and zinc are explained by the fact that there are more copper and zinc roofs in Vasastaden. Göteborg has a higher average annual rainfall (about 700 mm) that dilutes the concentrations for cadmium and nitrogen because these substances mainly depend on dry deposition of atmospheric fallout.

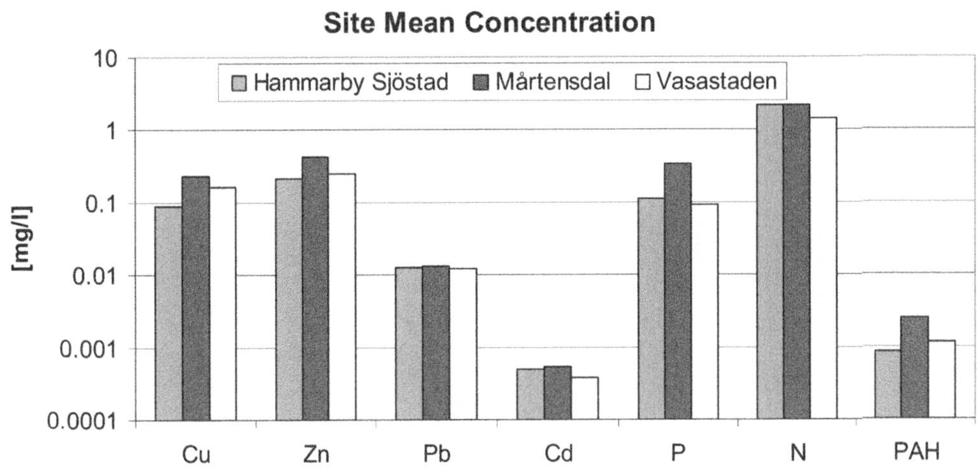

Figure 12: Comparison of Site Mean Concentrations (SMC) for the catchments in Hammarby Sjöstad and Vasastaden.

Another comparison between the catchments is made in Figure 13 where the average annual specific loads are shown for each type of pollutant. It can be seen that stormwater from roads has a higher specific load than stormwater from mixed land use in Hammarby Sjöstad. But the differences are about the same as for the comparison of SMC in Figure 12. For the substances copper, lead and cadmium stormwater from Vasastaden is more polluted per area than any of the catchments in Hammarby Sjöstad. At least for copper this can be explained by the larger amount of copper roofs in Vasastaden: about 19 % of the total roof area has copper as roofing material.

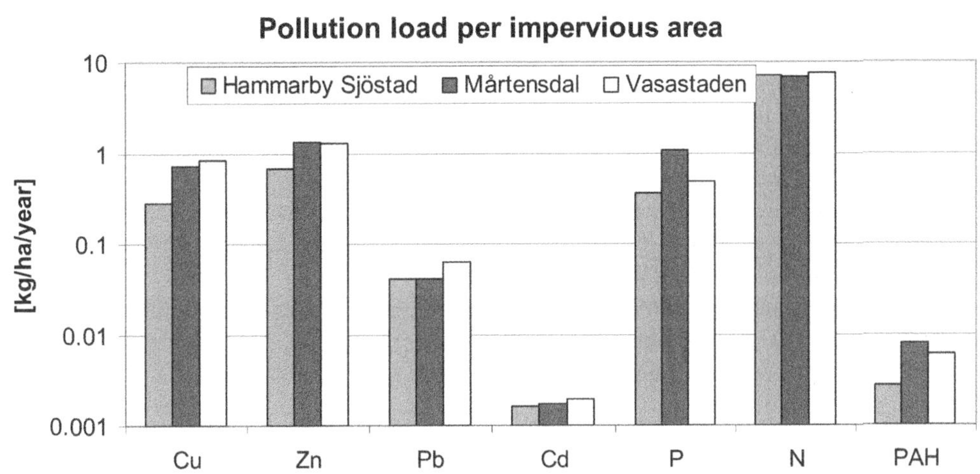

Figure 13: Comparison of the average annual specific load for the catchments in Hammarby Sjöstad and Vasastaden.

In relation to the simulated results for pollution load and concentrations, it is important to know where the pollution actually comes from. The framework of SEWSYS is based on emission factors for different sources of stormwater pollution and the total load of a substance is therefore built up by contributions for each source related to that specific substance. The distribution of pollution sources for the whole catchment Hammarby Sjöstad is shown in Figure 14. For lead, cadmium and nitrogen the dominant source of pollution is atmospheric deposition (wet and dry fallout). This presents some difficult challenges if there is a wish by the city to reduce these substances by qualitative source control. Atmospheric deposition is a particularly diffuse source coming from activities on local, regional, national and even global scale and a reduction implies the co-operation from many stakeholders. The same difficulties apply for the pollution source brake wear, which is a major source for copper in stormwater. But regarding PAH in tyres there has been a successful campaign in Sweden and nowadays there are many tyres for sale that have a lower content of carcinogenic PAHs (Wik and Dave, 2005). In terms of pollution reduction in the Hammarby Sjöstad catchment it would be most efficient to implement measures concerning building materials. Such measures could be reduction of the zinc content in roofing paint and ensuring that the asphalt used in the paved roads has a minimum content of PAHs.

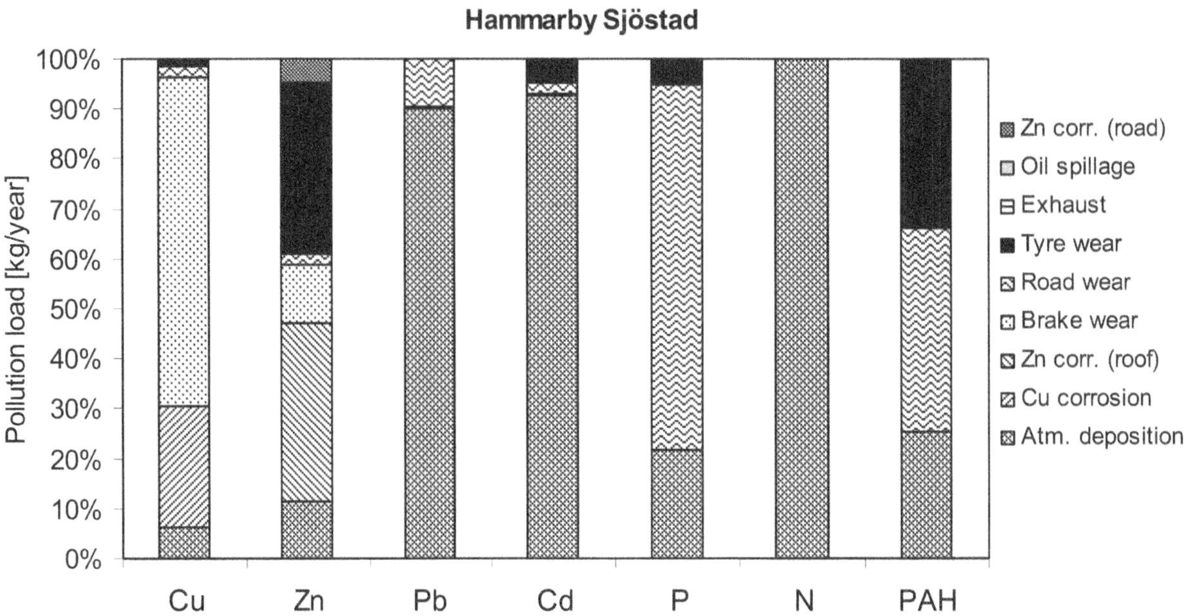

Figure 14: Pollution source distribution for an average year in Hammarby Sjöstad.

CONCLUSIONS AND FINAL REMARKS

The STORM/SEWSYS model combines two modelling approaches into a stand-alone tool containing simple and well-established hydrological and hydraulic principles. The integrated model provides simulation techniques to evaluate changes to the hydrological cycle and hydrological risks in urban catchments.

Based on the level of detail in the model's output for stormwater pollution, it is clear that the model represents a useful tool for simulating and evaluating quality source control measures. With the additional incorporation of structural BMPs such as swales, soakaways, wetlands and ponds, the modelling framework in STORM/SEWSYS will provide a powerful simulation tool with a wide range of applications within sustainable stormwater management.

The sources and flux model STORM/SEWSYS:

- gives overall water balances and BMP performance with respect to water and pollution quantity,
- provides catchment-specific pollution data,
- identifies the important sources of pollution in a catchment,
- makes it possible to study the effects of both structural and non-structural measures.

One interesting future development of STORM/SEWSYS would be the ability to model the fate, behaviour and removal potential of xenobiotic organic compounds (XOCs) in stormwater BMPs. A literature survey by Eriksson *et al.*, (2005) revealed that at least 656 XOCs could be present in stormwater. The survey shows that the assessment of chemical hazards will be a key aspect in future urban stormwater management. A tool for chemical hazard identification and assessment, called CHIAT, has been developed within the research project DayWater (Chapter 11). This generic tool can be applied for evaluation of different strategies for sustainable handling of stormwater, focussing on identification and assessment of problematic or hazardous compounds. The aim of the hazard assessment step would then be to generate data for site specific comparison of different stormwater handling scenarios, preferably using the PEC/PNEC ratio. PECs can be calculated using either measured data (if available) or by modelling. STORM/SEWSYS has the potential for feeding the CHIAT hazard assessment with data.

Future work on STORM/SEWSYS will also focus on the interaction with natural catchments. A component for soil water modelling is already included, so that the hydrological influence of stormwater runoff from urban areas on natural rivers can be analyzed. An extension for the modelling of pollutant fluxes from diffusive sources *e.g.* from farmland would make STORM a valuable tool for river basin management as prescribed by the EU-WFD.

The SFM-tool STORM/SEWSYS is further described at the DayWater website (http://www.daywater.cz). The same website also includes a link to download the prototype version of the model.

REFERENCES

Achleitner, S. and Rauch, W. (2005) CITY DRAIN © - A simulation software for integrated modelling of urban drainage systems. In: *Proceedings of the 10th International Conference on Urban Drainage*, E. Eriksson, et al. (Eds.), August 21-26 2005, Copenhagen, Denmark.

Ahlman, S. (2006) *Modelling of substance flows in urban drainage systems*. PhD Thesis, Dept. of Civil and Environmental Engineering, Division of Water Environment Technology, Chalmers University of Technology, Göteborg.

Ahlman, S. and Svensson, G. (2002) Modelling substance flows in urban sewer systems using MATLAB/Simulink. In: *Proceedings of the Ninth International Conference on: Urban Drainage*, September 8-13 2002, Portland, OR, USA.

Brezonik, P. L. and Stadelmann, T. H. (2002) Analysis and predictive models of stormwater runoff volumes, loads, and pollutant concentrations from watersheds in the Twin Cities metropolitan area, Minnesota, USA. *Water Research*, **36**(7), 1743-1757.

Calabro, P. S. (2001). Cosmoss: conceptual simplified model for sewer system simulation; A new model for urban runoff quality. *Urban Water*, **3**(1-2), 33-42.

Eriksson, E., Baun, A., Mikkelsen, P. S. and Ledin, A. (2005) Chemical hazard identification and assessment tool for evaluation of stormwater priority pollutants. *Water Science And Technology*, **51**(2), 47-55.

IPS. (2003). STORM Manual *Ingenieurgesellschaft Prof. Dr. Sieker mbH.* Berlin.

Ruan, M. (1999) Continuous simulation of CSO emissions. *Urban Water*, **1**(3), 201-205.

SMHI. (1980) *Sveriges vattenbalans, Årsmedelvärden (1931-60) av nederbörd, avdunstning och avrinning. [Sweden's Water Balance, Annual Mean Values (1931-60) of precipitation, evaporation and runoff.]* Rapport RMK 18/RHO 21. Author: B. Eriksson, Norrköping. (In Swedish).

Smullen, J. T., Shallcross, A. L. and Cave, K. A. (1999) Updating the U.S. Nationwide Urban runoff quality data base. *Water Science and Technology*, **39**(12), 9-16.

StormTac. (2006) *Standard Concentrations*.http://www.stormtac.com/page2_stormtac.htm (verified August 2007).

Wik, A. and Dave, G. (2005) Environmental labelling of car tires--toxicity to Daphnia magna can be used as a screening method. *Chemosphere*, **58**(5), 645-651.

Wong, T. H. F., Fletcher, T. D., Duncan, H. P., Coleman, J. R. and Jenkins, G. A. (2002) A model for urban stormwater improvement conceptualisation. In: *Proceedings of the Ninth International Conference on: Urban Drainage*, September 8-13 2002, Portland, OR, USA.

Chapter 15

Integrated planning of stormwater infiltration measures: Flext A Tool for Scenario Development and Vulnerability Mapping, including Wupper River Case Study

Zhengyue Jin, Colette de Roo and Heiko Sieker

INTRODUCTION

On-site rainwater management is becoming a significant and prevalent supplement method in urban drainage planning and has many advantages compared to conventional discharge processes. The question of whether a piece of private property or public space can be disconnected from the sewer system and infiltrated on-site, however, depends on the evaluation of multiple parameters. A systematic and automated evaluation with the computer is not possible at present. This would be rather sensible, in particular, when one considers planning on-site stormwater management measures for large urban catchments.

A transparent, knowledge-based and spatial decision support system (DSS) called Flext was conceptualised and developed as an instrument to automate the evaluation of on-site stormwater management measures. This instrument integrates a knowledge-based system or an expert system (ES), because the solutions to the above-mentioned problems require dynamic, experiential and professional knowledge.

The ES developed in this task, however, has a distinctive feature compared to conventional ESs: as it expected to be used as a DSS, the knowledge base must be transparently available to the user, as well as modifiable by him. For this purpose, a GIS-integrated expert system tool was developed that fulfils the requirements mentioned above. The overall solution consists of ES tool + 'GIS platform' + 'Knowledge Base'.

The established knowledge-based spatial DSS for the Wupper catchment was applied to the prepared GIS databases respectively and the results were presented graphically with a GIS. For each evaluation, the fundamental result is in fact a map of the spatial distribution of possible on-site stormwater management measures. In the database, the expert system also indicates information regarding upon which influencing factors an evaluation was made. Likewise, comments about missing data in certain sub-areas are pointed out too.

The developed transparent knowledge-based DSS give the engineer a flexible and general instrument for planning on-site stormwater management measures.

 DayWater: an Adaptive Decision Support System for Urban Stormwater Management. Edited by Daniel R. Thévenot. ISBN: 9781843391609. Published by IWA Publishing, London, UK.

Integrated planning of on-site stormwater management measures

With the development of the economy, the migration of people from rural to industrialised areas, *i.e.* urbanisation, accelerates worldwide. As a result, large amounts of natural areas, such as meadows, forests, agriculture lands, etc. become impervious, due to the construction of apartments, factories, institutions, sports and entertainment buildings or yards, roads, streets etc. For example, nowadays in Germany, statistically about 122 ha of natural area are daily consumed for new apartments and streets, of which around 50% becomes impervious (Sieker *et al.*, 2002). In fact, the urbanisation process is occurring even more rapidly in developing countries, like China, India, etc. Following text cited from "the Key facts about cities, issues for the urban millennium, United Nations Environment Programme" should give a general idea of the world urbanisation process: "In 1950 less than one person in three lived in a town or city. Today nearly half the world's population is urban. By 2030 the proportion will be more than 60 per cent. Virtually all population growth in the next quarter-century will be in urban areas in the less developed countries".

Urbanisation, as a symbol of economical development, brings out many negative effects. One of the most direct problems is the overloading of the city sewer system, which is at present the prevailing urban water drainage system in the world. The overloading of an existing sewer system is due to more and more surface runoff converged from continuously newly connected impervious areas.

For a combined sewer system, that is, rainwater and wastewater drained together in one pipe system, the overloading means on one hand the hydraulic overload in different points inside the pipe system, on the other hand the overloading of the wastewater treatment plant (WWTP) at the end of the pipe system, or more overflow of the combined sewer (CSO) directly into the receiving water. For separate sewer systems, that is, rainwater drained in its own pipe system, the overload on one hand means the hydraulic overload at different points inside the pipe system, and on the other hand more rainwater without any treatment directly into the receiving water. So in both cases, urbanisation will cause hydraulic overloading inside the pipe system and a higher pollution load in the receiving water.

To solve the hydraulic overload problem in the city, the conventional method is rehabilitation of the sewer system by enlarging relevant pipes. However, this actually worsens the second problem, i.e. receiving water pollution, because more and more rainwater or combined sewage will be directly drained into the receiving water. Although the CSO can be reduced through construction of central water retention basins, there is the same problem as for the existing sewer pipe system, *i.e.* the retention basin must be continuously enlarged, or new retention basins continuously constructed, with the continuous increase of inflow from the sewer system. This is furthermore a problem of available space to construct the retention basins and the high cost of construction for basins and pumping stations.

Aside from the water pollution problem, the continuous enlargement of sewer pipes could transfer the flood problem caused by urbanisation from inside the city to nearby streams or rivers. This is especially significant in some small river basins with high urbanisation. Another negative effect of draining rainwater totally through a pipe system is that the water balance in the drained area will be worse because of the reduction of evapotranspiration and groundwater recharge. The reduction of evapotranspiration affects the local climate to some extent, while the reduction of groundwater recharge endangers sustainable social development regarding water resources.

New concept on Urban Stormwater Management

Considering the above-mentioned problems originating from urbanisation and the shortcomings of the conventional solution, a new concept of urban water management, the decentralised rainwater management, which aims to reduce surface runoff to the existing sewer system during rain events, is being gradually put into practice (Chapter 8). For example, Sydney, Australia, has a stormwater system that is completely separate from the sewerage system by using on-site detention (OSD) measures (Beecham *et al.,* 2005). In USA, many states have a stormwater best management practice (BMP) manual which includes among others detailed instructions to the selection, design and construction of the most suitable on-site stormwater management measures. On-site stormwater management becomes popular nowadays also in many other countries. Figure 1 shows two photos of practical examples of on-site stormwater management structures in Auckland, New Zealand (On-site stormwater management manual, Auckland, New Zealand)

Depression Storage Gravel Swale In Car Park

Figure 1: On-site stormwater management measures applied in Auckland, New Zealand.

In Germany, on-site stormwater management, especially on-site infiltration has been extensively researched and widely put into practice. Researchers have developed many new concepts on on-site infiltration and have accumulated a lot of theoretical and practical knowledge in this aspect. For example, Figure 2 demonstrates schematically the application of the combination of on-site infiltration and the traditional underground pipe drainage, which introduces the concept of the incomplete infiltration of on-site infiltration measure, which is otherwise usually considered only as complete infiltration method.

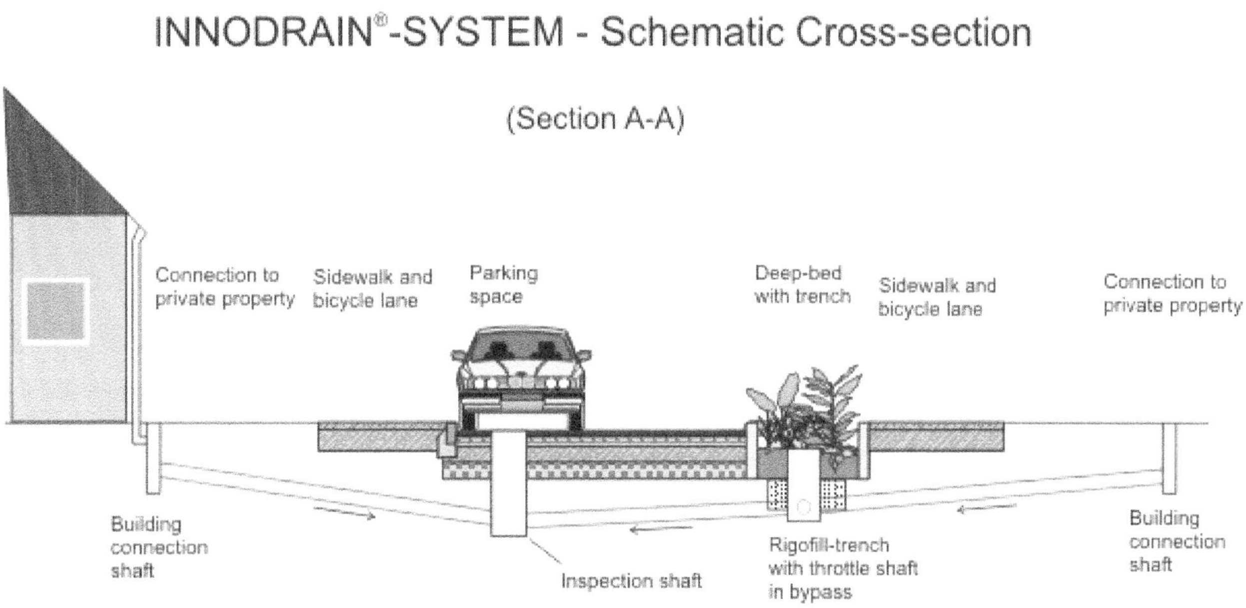

Figure 2: Scheme of the cross section of an INNODRAIN-SYSTEM.

Integrated planning of on-site stormwater management measures in whole urban stormwater drainage system

Among on-site stormwater management measures, rainwater utilisation is a relatively simple concept. Its application is rather limited to household water use or garden irrigation, etc. The boundary conditions on installation of a rainwater utilisation system depend mainly on the integration of the system (for example, a storage tank) with the local environment and the construction and operation cost.

A decentralised rainwater retention basin functions much like a central retention basin, but functions on a reduced scale and inside a drainage catchment. The effects of decentralised retention basins include:

- relieving the hydraulic overloading of the sewer system,
- relaying the concentration of runoff into the receiving water and

- reducing the combined overflow (when the system is the combined sewer system).

Whether a decentralised retention basin should be constructed to relieve the overloading of sewer pipes mainly depends on the available natural area, the management of the retention basin, cost and construction issues.

On-site rainwater infiltration tries to make the best use of the storage capacity of local soil. The rainwater from a local impervious area is collected and drained into an infiltration structure. The collected water will either immediately infiltrate into the surrounding soil or infiltrate after temporary storage. In this way, constructed areas suitable for this type of treatment can be disconnected from the sewer system and drained on-site.

Obviously, on-site rainwater infiltration measures presented on Figures 1 and 2 are more often applicable than rainwater utilisation for their suitability to disconnect more types of built-up areas and their larger flexibly installed than a decentralised retention basin, considering the requirement for natural area and for integration into the surrounding environment. The foremost advantages that make infiltration measures desirable are:

- reducing inflow to the existing sewer system, hence reducing flooding inside the city or contributing to flood control in rivers in or near cities,
- contributing to the cleanliness of the rainwater by filtering rainwater through vegetated shallow ponds and/or active soil layers,
- increasing groundwater recharge and to some extent also evapotranspiration. In this regard, on-site rainwater infiltration manages rainwater in a near-natural way.

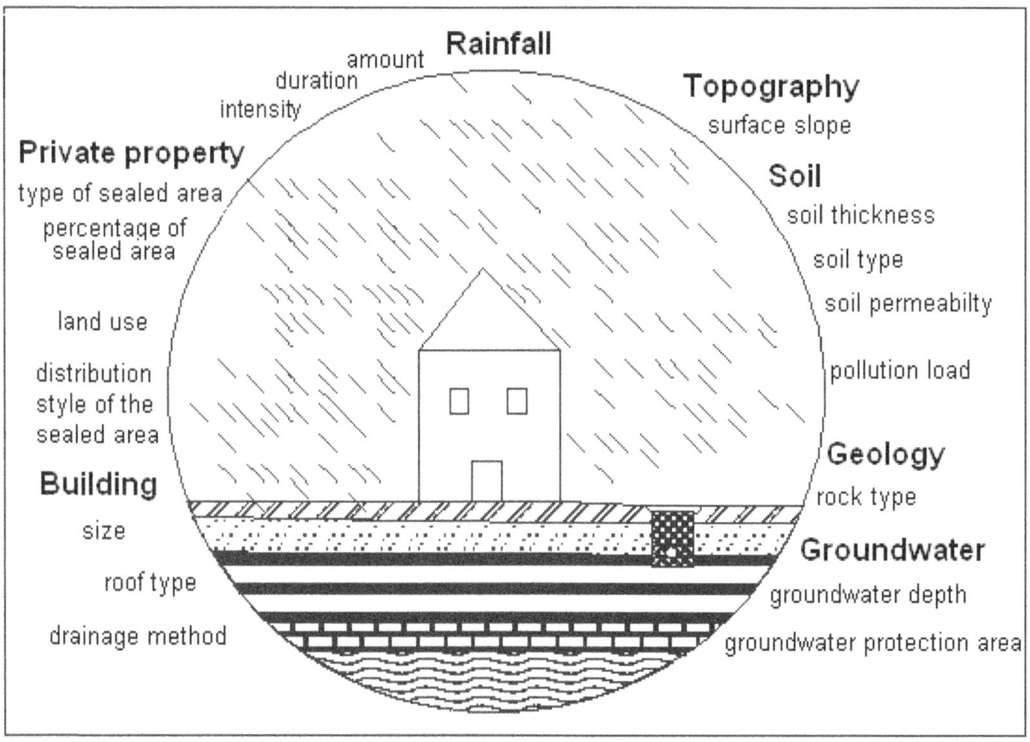

Figure 3: Different influence factors on selection of on-site stormwater infiltration measures.

However, compared to the other two types of decentralised rainwater management, the application of on-site infiltration measures depends not only on factors related to construction and cost, but also influencing factors such as soil type, soil permeability, groundwater depth, stormwater pollution load etc. (Figure 3). The decision on whether on-site stormwater infiltration is suitable for a given area and if yes, which type of infiltration measure is the optimal one for that area involves relatively complicated, multi-disciplinary and rather experienced judgement.

The complexity of this decision or selection will be especially remarkable if one considers that such decision has to be made individually for numerous small areas (such as a set of houses or a single one) during an integrated planning of the stormwater drainage for a whole urban catchment. Moreover, such decisions need usually repeated under the change of the numerous planning constraints, because a planning needs normally the comparison of the hydrologic and hydraulic effects of different scenarios.

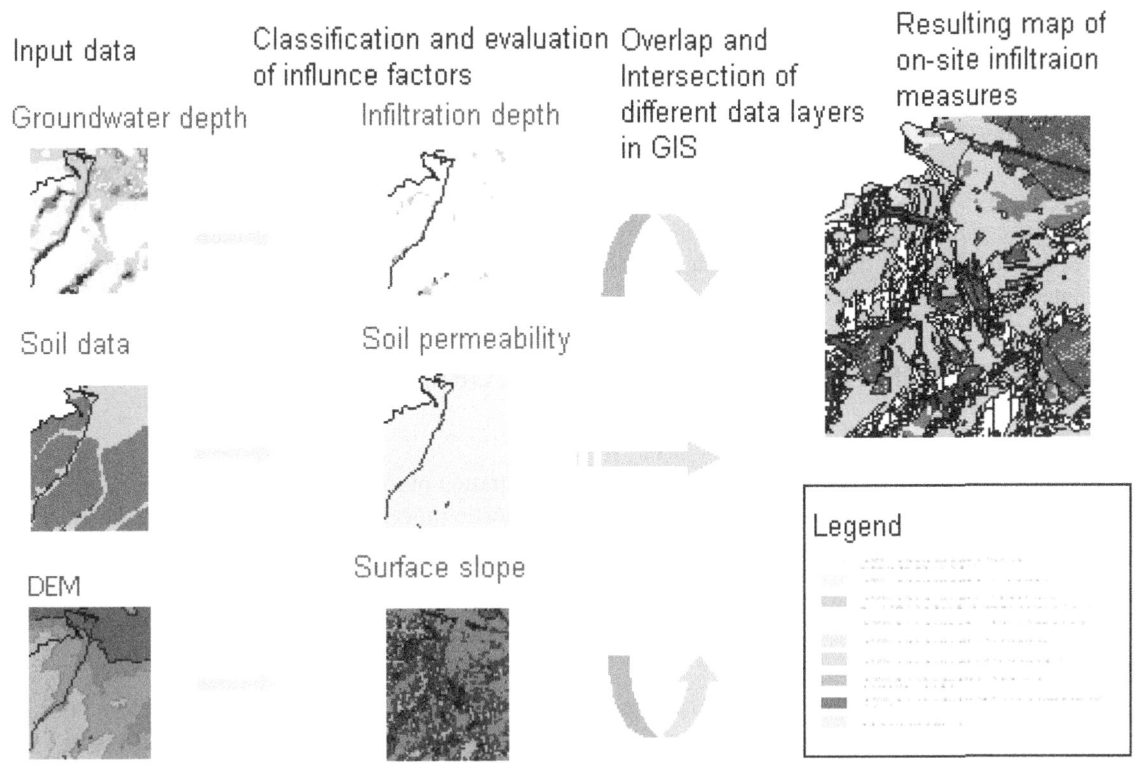

Figure 4: A schematic demonstration of the procedure of planning of on-site infiltration measures.

A schematic presentation (Figure 4) of the general procedure of a planning of on-site infiltration measures for an urban catchment could demonstrate the complexity of the process. First, the different digital input data (GIS data), such as groundwater depth, soil permeability, etc. are first processed and classified according to their individual boundary conditions for each of stormwater infiltration measures. Some factors may not be directly available and hence must be derived from other original data, for example the surface slope data is usually derived from digital elevation model (DEM). The classified spatial data are normally distributed in different style and hence hold in different layers in a GIS project file. Therefore they must then be spatially overlapped and intersected with each other to obtain a new spatial distribution (i.e. a layer), in which data of all factors are enclosed and consistent (that is, values of each factor keep consistent in every object of the layer).

Because of the spatial intersection of many layers, the number of objects in the resulting layer could be hundreds of thousands. This means that there must be hundreds of thousands decisions to be made for one planning scenario. Moreover, each decision process is dynamic, because effects of some factors on the decision of the infiltration measure may depend on what other parameters have currently affected the decision. For example, when the groundwater depth is relatively high, then the on-site infiltration structures with underground storage facilities (such as a trench) are usually ruled out. And hence the further evaluation of permeability of deep soil layer is not necessary. Such dynamic unstructured analysis is difficult to be carried out through macro operations in a database record set or Excel sheets, which could be the way dealing with such numerous decisions. Therefore the selection of a suitable on-site infiltration measure needs then to be analysed manually for one object after another: this is actually not feasible.

Decentralised water management planning aiming to preventive flood control in river watershed

Similarly to on-site stormwater management in urban area, the preventive decentralised measures in a river catchment scale aims to reduce as much as possible on-site surface runoff, in the whole river catchment, during rain events. One of the decentralised management concepts is the increase of natural water infiltration in agricultural lands, forest area, etc. This concept tries to make best use of the water-retention capacity of soil in the whole catchment and concerns every patch of land where potential exists for storing rainwater during rain events. Comparing to conventional river engineering, such as canalisation, this preventive decentralised measures on one hand mitigate potential floods to some extent and on other hand increase base flow as well as soil moisture after flood. The special characteristics of this kind of preventive decentralised measure is that it makes contribution to flood control without just shifting the problem from upper reaches to lower reaches of a river basin, which could be so by conventional river engineering measures, like canalisation (Sieker, 2002).

However, the planning of the decentralised measures is based on appropriate evaluation of soil water-retention potentials of the widely diversified situation in whole catchment. The evaluation involves not only the soil properties, surface slope, etc, as in the evaluation for urban area, but also the land use, such as agricultural lands or forest area, natural protection area, etc. Different land uses require different evaluation regulations. Therefore, one of such planning task involves already different domain knowledge. Similarly, this planning involves also massive evaluations of small areas distributed in a whole river basin.

METHODOLOGY FOR THE FLEXT TOOL DEVELOPMENT

Why a knowledge-based system is applied?

The analysis for the optimal type of decentralised stormwater infiltration measure for a given situation in urban area or the soil water-retention potential in an agricultural land is a diagnostic process, i.e. evaluating a series of factors from different aspects against their critical values on different measures step by step. This process demonstrates the following characteristics:

(1) The considered factors and their critical values can not be determined by pure theoretical considerations, but often depend on researchers' knowledge or experience and even on the current political laws in relevant aspects, public acceptance.
(2) The availability of the data of all influencing factors cannot be guaranteed and varies form project to project.
(3) Some factors may represent different objectives that cannot be unified, therefore, a trade-off between objectives is often needed in an analysis; such a trade-off is inherently subjective and relies on experience or a political viewpoint.

Because of these characteristics, the analysis for decentralised stormwater infiltration does not rely solely on fixed solution logic or algorithms, but relates to variable knowledge. This task can only be fulfilled by relevant experts or trained experienced engineers. However, due to the massive decisions for just one planning scenario, it is not a task that can be done manually. It can neither be automated with the help of the macro functions available in database programs such as MS Access or Excel, due to its unstructured characteristics. In addition, because the knowledge involved in the decision procedure may probably change from one planning scenario to another or from one project to another, it can neither be fixed encoded as a normal program to facilitate automation of the decision processes.

Fortunately, with the development of computer science and the artificial intelligence (AI) research, human knowledge or expertise in a specific domain can be programmed in a so-called ES or knowledge-based system, which can then be used instead of the actual experts to provide expert-level advice or recommendations on the designed problem to users, such as planners or decision-makers. Moreover, an ES is usually so constructed that its knowledge and the solution logic (i.e., the process of reasoning using the specified knowledge) are separated. The solution logic is then fixed coded as a program while the knowledge is rather written in an easy understandable style, as a supporting knowledge base. In this way, the specific knowledge can be easily modified to suit the situation change of a problem domain for which the ES is built, while the coded program of the solution logic remain unchanged. Hence a knowledge-based system or an ES is ideal for solving the specified decision problem of dynamic and unstructured properties.

Development of an GIS integrable Expert System Tool

Although the ES makes the change of the specific knowledge easier than in normal computer programs, the knowledge of an ES is usually not opened to users, needless to say that it can be changed or modified by users. So-called knowledge engineers usually perform modifications of the specific knowledge. However, this does not meet the requirement of the problem defined in this study.

Therefore, to achieve the goal of this study, the first task was to develop an ES tool with a simple but flexible knowledge formulation system. This tool should facilitate easy understanding and editing of knowledge in any of its specific knowledge base, so that decision-makers can direct manipulate the pre-defined knowledge of a built ES. In addition, the tool should also be integrated in a GIS platform to be able to process the spatial oriented problem. Finally for facilitating the preparation of the necessary spatial input data, a GIS extension program providing automation of some spatial operations, such as, intersection or union of many spatial distributions, is also necessary. The final integrated model should be consisted of 'ES tool' + 'GIS extension' + 'GIS platform'.

Development of a knowledge base for selection of on-site rainwater infiltration measures

After the development of the integrated model, i.e. 'ES tool' + 'GIS extension' + 'GIS platform', the empirical knowledge related to the selection of on-site stormwater infiltration measures for a given urban catchment, *i.e.* the Wupper River basin, were analysed and processed towards a DSS. The developed DSS was then used in the practical planning projects and the results were explained in detail. In order to evaluate the sensitivity of the planning results on one of the criteria, *i.e.* the soil permeability K_f, the knowledge base was changed according to a series of designed K_f values, different planning results or scenarios were then obtained and evaluated.

Basic components of an expert system

In an ES, human expertise or knowledge in solving a specific problem should be well represented in one or more special formalisms that computer can understand and apply appropriately in its reasoning process towards a solution to the designated problem. The well-represented knowledge is saved as so-called knowledge base, which is one of two basic components of an ES. The other basic component of an ES, called 'inference engine', helps computer to understand the knowledge in the knowledge base. In other words, the inference engine interprets the knowledge and controls the computer reasoning during the operation of an ES. Apart from these two basic components, a so-called user interface is also important to eventually put a constructed system into successful use and to provide a human-computer dialog on requesting data inputs, displaying the explanations of reasoning as well as results.

Characteristics of the Flext tool

The 'ES tool' + 'GIS extension' + 'GIS platform', called Flext tool, is simple and flexible. Its main properties are listed as follows:

(1) it is adequate and flexible enough for the representation of the desired knowledge for the specific problem in this study,
(2) it has a easy understandable representation formalism (syntax in mathematical expressions, explicit logical rule structure in trees or networks),
(3) it provides friendly user interface and hence support question-specific graphic question interface,
(4) it provides a opportunity to debug the formulated knowledge base,
(5) it has a high-level knowledge acquisition interfaces,
(6) it can be integrated into other applications, for example, GIS platform,
(7) it is able to integrate external applications, for example, a professional simulation model,
(8) it is able to manipulate database operations, such as SQL search,
(9) it is convenient to input from or output to external data files, such as text files, databases.

In summary, the Flext tool has a simple knowledge representation mechanism (conventional syntax, structural rules system, automatic question processing system) and an advanced knowledge acquisition system (i.e., table and graphic edit interface).

INTEGRATION OF FLEXT INTO GIS PLATFORM

Concept of integration of an ES into GIS platform

In the sample application of this study, GeoMedia Professional (a GIS product of Intergraph lnc.) is selected as the GIS platform for integrations. GeoMedia Professional enables users to install their own utilities developed in Visual Basic, etc. as so-called user commands. It also provides dynamic libraries with a large amount of functions on spatial operations or computations and database operations for users to develop their commands. In such commands, relevant procedures can be written to capture events such as a mouse click on an object of a feature, etc. In reaction, the user commands may execute its programmed process and return the relevant results from the executed process. In this way, the developed ES tool Flext can be integrated into GeoMedia Professional; *i.e.* embeds its functions in a user command source code. Hence, any problem-specific ES (i.e. loading Flext with a specific knowledge base) can be integrated into GeoMedia Professional.

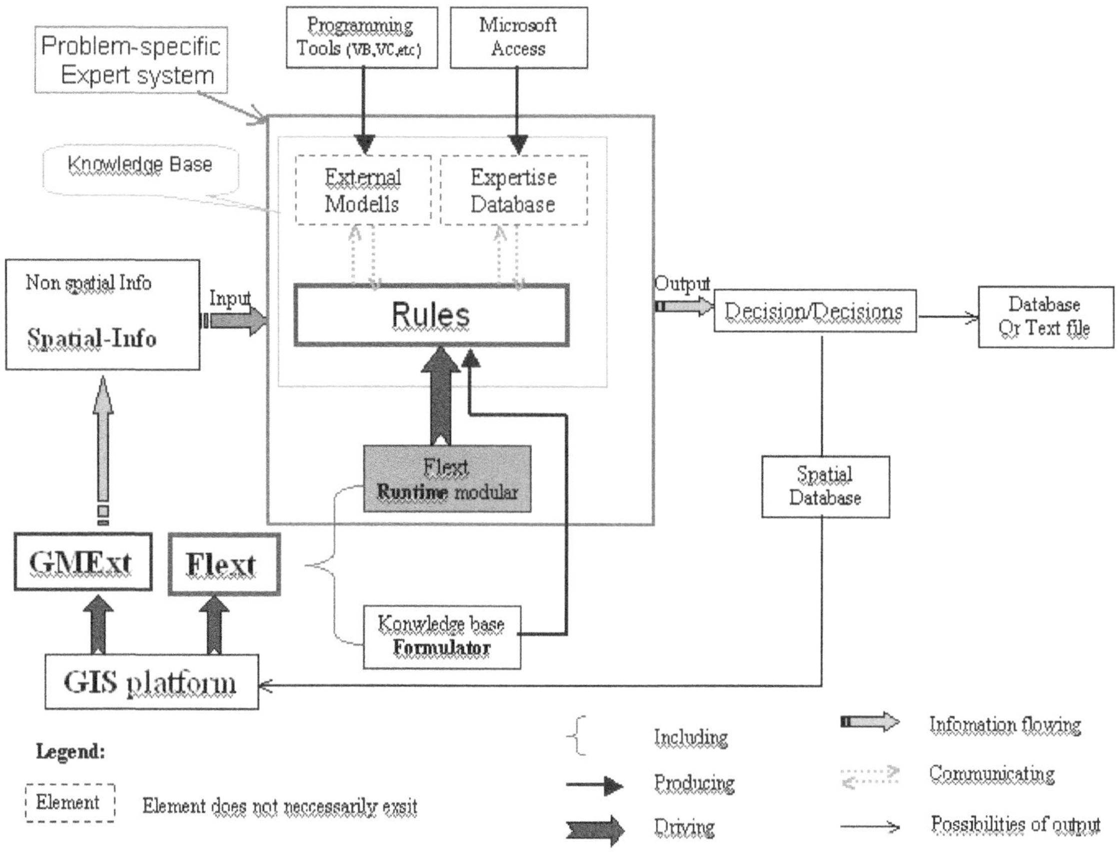

Figure 5: Scheme of integration of Flext into GIS platform.

In this study, except for Flext, a GIS extension programme GMExt, for pre-processing the spatial data for the ES, is also integrated into GeoMedia Professional through a user command. Figure 5 shows the scheme of the integration of Flext and GMExt into a GIS platform.

Running procedure of the Flext integrated model

One can now easily start the knowledge base formulator from GeoMedia Professional to edit one's problem-specific knowledge base, which usually consists of a rules system and, maybe, external professional models and expertise databases referenced in the rule system (Figure 6). Users must prepare their necessary external models either by programming them in conventional programming language like Visual C++ or Visual Basic, or by obtaining them from other sources. The necessary expertise database can be prepared by using MS Access.

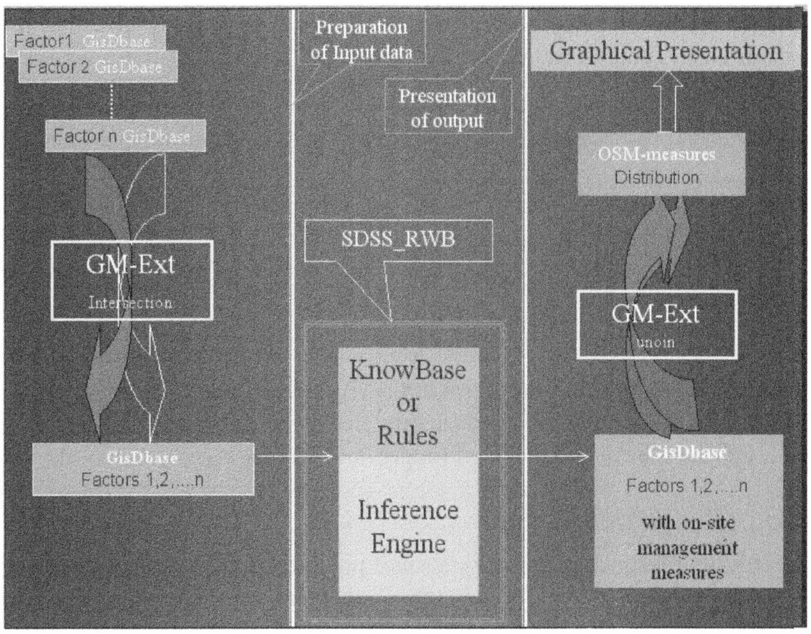

Figure 6: Schematic presentation of the application of a GIS integrated expert system for the planning of on-site stormwater infiltration.

The formulated knowledge base and the Flext Runtime modular (inference engine and interface) form a problem-specific ES or knowledge-based DSS. This system can be activated though menu commands of GeoMedia Professional (user commands) to process for all objects of a selected feature. Clicking on an object of a spatially distributed feature can also trigger the runtime module. For example, in reacting to a mouse click, the integrated knowledge-based system is executed with the necessary inputs read from the pre-matched fields of the record related to the clicked object. The results or decisions for this object are then displayed in a docking window.

DEVELOPMENT OF A TRANSPARENT SPATIAL DECISION SUPPORT SYSTEMS FOR ON-SITE URBAN STORMWATER MANAGEMENT PLANNING: FLEXT

For the development of a transparent spatial DSS on the planning of the on-site stormwater infiltration measures, the necessary case-specific knowledge was first analysed and clearly formulated in actual project contexts. The transformation of the human knowledge into a knowledge base for Flext and the application of the developed system were then tested on on-site stormwater management planning in the basin of the Wupper River.

Within decentralised rainwater management measures, the on-site rainwater infiltration measures, which serve as making good use of the water storage capacity of local soil, are of special interest in this study. Among on-site rainwater infiltration measures, some are designed such that the runoff from the connected catchment (local impervious area) can be totally infiltrated underground and later either percolated further into groundwater bodies or evapotranspired before the next rain event. However, these so-called complete infiltration measures are sometimes not justified in soil with very low permeability. In such situations, measures aimed at partial infiltration are then considered.

Classification of affecting factors of decentralised stormwater management

Although the main constraints of some on-site infiltration structures have already been described above, factors related to overall decision or selection of a stormwater infiltration measure, for a given area, are classified below and explained in detail.

Factors affecting on-site water retention

On-site water retention is the essential pre-condition for decentralised stormwater management. Therefore, related factors are investigated and evaluated first. Water retention means, in this case, immediate on-site rainwater infiltration or infiltration after temporary storage. It is principally influenced by the following factors:

- Permeability (K_f) of the surrounding soil: a parameter that indicates how fast the infiltrated water will be transported underground. Table 1 shows the K_f criteria on each infiltration measure according to German practice.

Table 1: Criteria of soil permeability coefficient K_f (m.s^{-1}) on different on-site stormwater infiltration measures.

Decentralised rainwater infiltration measures	Permeability (m/s)
Surface infiltration	$K_f \cong 2*10^{-5}$
Swale infiltration	$K_f \cong 5*10^{-6}$ Can be smaller when large pervious area available
Swale-trench element without drainage	$K_f \cong 1*10^{-6}$
Swale-trench-system with throttled drainage	No limitation

- Porosity (n) of the soil layer: indicates the extent of the void space of soil.
- Groundwater table: reflects the thickness of the unsaturated layer, and hence the local water retention capacity,.
- Thickness of soil: indicates the distance to hard (impervious) rocks.

Factors concerning pollution dispersion

The infiltration of collected rainwater will inevitably transport pollutants underground at meantime, even though there are vegetated filtering processes in the swale and the active soil layer at the bottom of the swale, so the infiltration measures should be avoided for such environments as:

- Well zones where groundwater is pumped for drinking water, since the demand on water quality is then very high.
- Highly polluted areas such as old industrial areas, industry deposits, roadways with high traffic volumes, etc., because the high pollution loads are expected to be present in the stormwater runoff from such area.

Factors affecting the construction of on-site infiltration measures

- Surface slope concerns both the installation of the infiltration basin and the inlet of the stormwater runoff from connected impervious area to the basin: the steeper the slope, the higher the cost of the construction of the infiltration basin is.
- Thickness of soil layers: described above, it influences also the construction of the infiltration structures, for example, the underground trenches.
- Available natural area: for swale and swale-trench system, an existing natural area (at least 10 percent of the planned connected impervious area) is a pre-condition of installing these measures.

Factors concerning other utilisation of the site or aesthetics

When planning on-site infiltration measures, it should be considered whether the local original utilisation or function is affected by the installation of the stormwater infiltration measures, such as the basement, foundation of buildings, railways or highways, etc. The main indicating factor to evaluate these effects is the local groundwater table (described above). According to ATV (German Water Association, Germany), the critical distance between the bottom of the infiltration structure and the groundwater surface is 1 m.

The effects on aesthetics depend mainly on whether there is enough green space available. Ideally, the design of the infiltration swale or swale-trench system should be integrated into local landscape planning in order to create pleasant scenery. However, this factor is hardly demonstrable.

Social and economic factors

Social acceptance and costs are the final important factors to consider in selecting on-site infiltration measures.

In the five above factors, those in the first three classes concern spatially distributed information on the planning area and are available in most of cases. They can also be evaluated quantitatively or decisively. Factors in last classes are relevant only to actual design and construction of the measures. In the planning stage, only factors in the first three classes are relevant.

CASE STUDY: BASIN OF THE WUPPER RIVER

Introduction

As an end-user within the DayWater Project (Chapter 1), the Water Federation of the River Wupper in Germany, the Wupperverband, is responsible for the river basin management in his area (Figure 7). Therefore the knowledge about quantitative and qualitative processes in the water cycle is essential. To manage the interaction between catchments, sewer treatment plant and water bodies, an integrated approach to water management in the river basin of the Wupper is important. For this purpose, an extensive assessment of rainfall-runoff-processes in the urban sub catchments of the chosen river basin was needed, but as important is a useful system to assess potentials for decentral stormwater management in the catchment area as a whole.

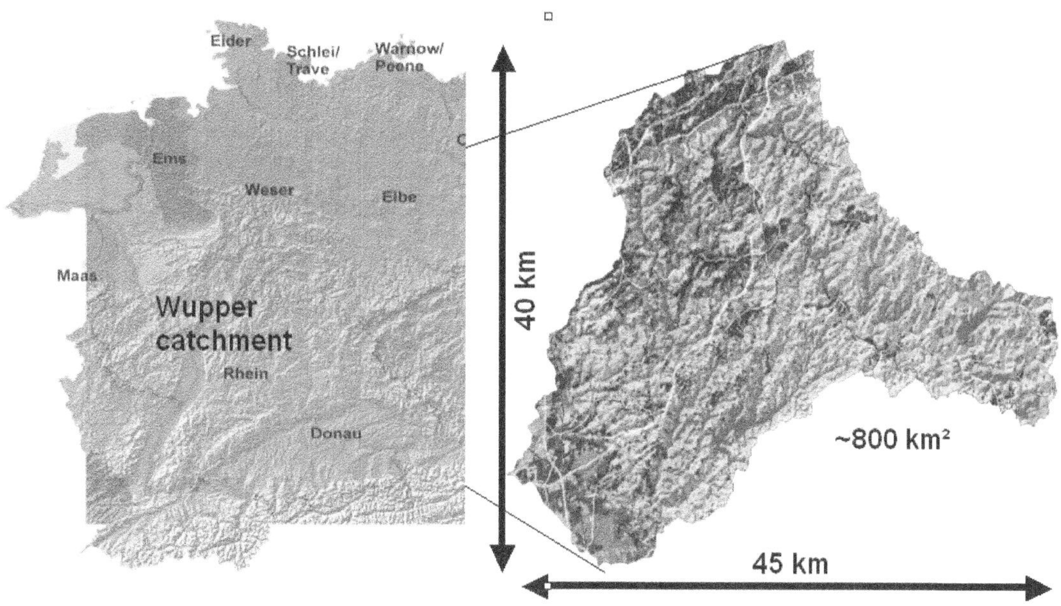

Figure 7: Catchment of the Wupper River.

The assessed part of the Wupper river basin covers an area of 40,100 hectares and consists of six urban runoff catchments. For each of these urban catchments a rainfall-runoff model was set up with the application STORM (Chapter 14). On the basis of the output, decisions could be made concerning measures to reduce or eliminate negative environmental effects of CSOs by possible use of decentral BMPs. Since the basin is very wide, a knowledge-based decision support tool for a first screening of the potential for on-site measures was required. Therefore the Flext spatial decision support tool was applied to the river basin.

Application of the ES in combination with GIS platform

The expert knowledge had to be put into the ES, whereas at the same time relevant information for the area was put into a GIS application. The ES was applied to the GIS application, after which maps were generated, showing potentials for suitable BMPs. The decision rules applicable for the Wupper River basin, were based on information for Hydrology, Urban drainage, Economy, Agriculture and Environmental protection. The results of the testing on the River Basin of the River Wupper can be seen in figures 8 and 9.

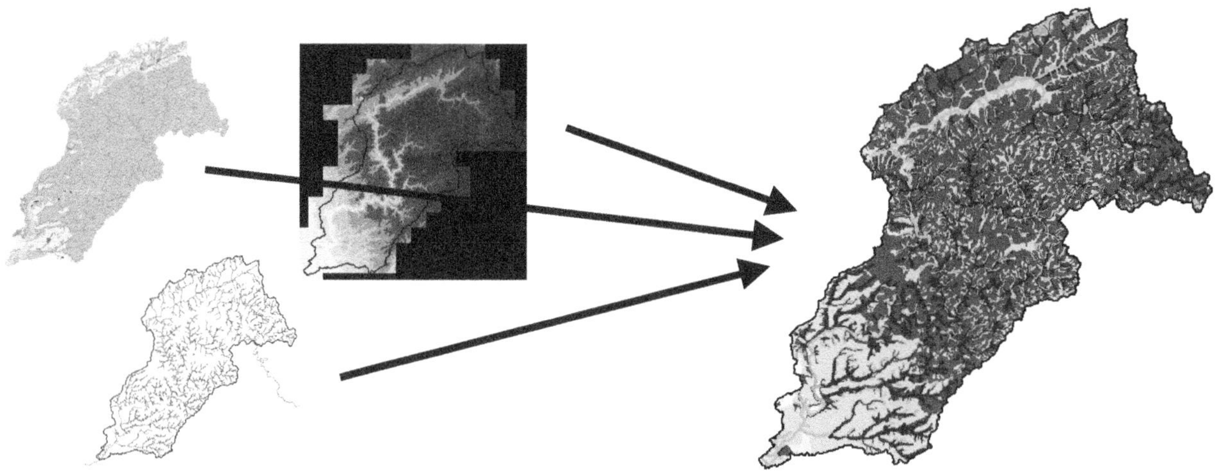

Figure 8: Application of ES and GIS to the River Basin of the Wupper River.

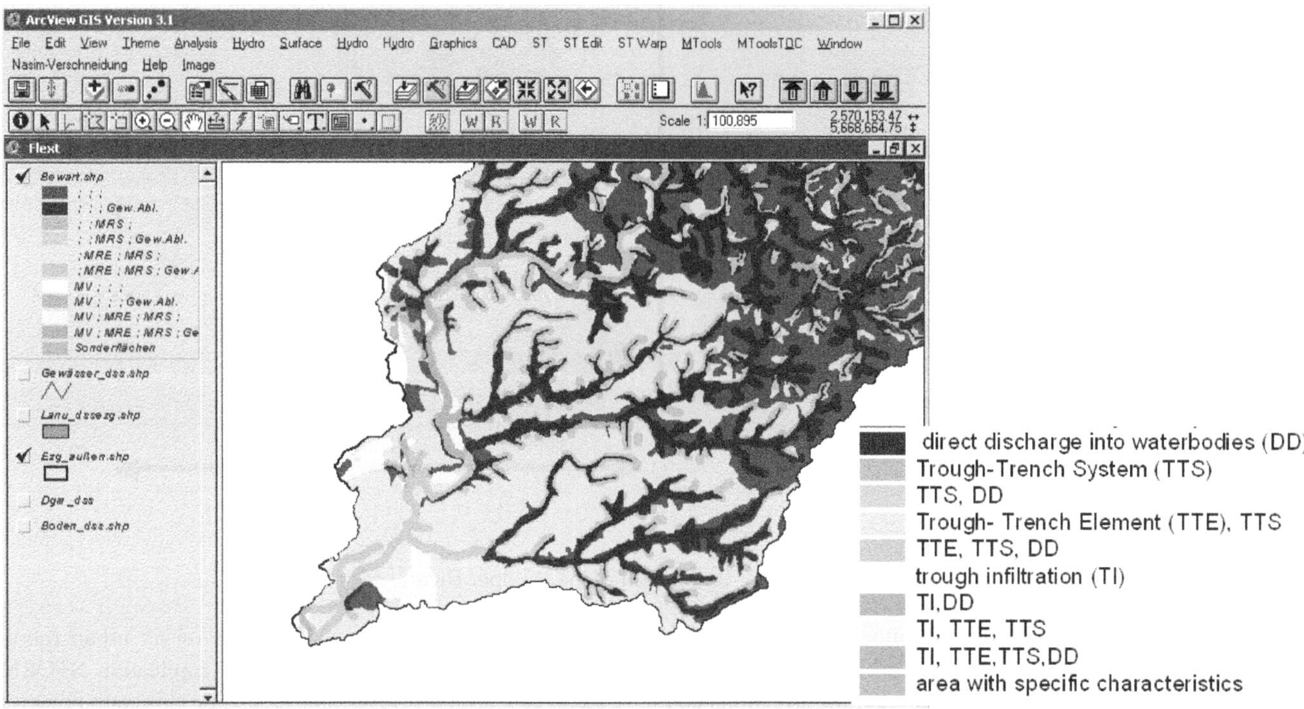

Figure 9: Specification of the assessed on-site BMP potentials near Leverkusen.

DISCUSSION AND CONCLUSION

Further application of Flext

The developed problem-specific expert systems can be applied as a DSS in other projects of planning of the on-site stormwater infiltration. The pre-defined knowledge may be necessary to be slightly modified in order to meet the projects specific demands, such as due to lack of data in certain factors, introducing new factors or simply the change of the different critical values of some constraints on certain measures. However, the possibility of modification of the pre-defined knowledge is already addressed in this transparent and opened system. In fact, this tool is to applied by respective researcher fellows to building ES or DSS for the analyses of decentralised water management measures on agricultural land and measures with considerations for environmental protection and preventive flood control. In general, the integrated model (Flext + GIS) can be used for developing GIS-based rules systems for totally different problem domain.

A simple important application may be the development of an intelligent end of a DSS in which the complicated decision-making procedure can be formulated as decision trees or networks of nodes and the different professional models or functions necessary in relevant steps can be activated in the corresponding rules of a formulated knowledge base. The advantages of using Flext in this case are:

- Complicated decision-making procedures can be clearly formulated as structured trees or networks,
- For the dialog in each decision-making procedure, an informative graphic interface can be designed,
- Flext is easily integrated into a GIS platform, if necessary.

Limitations and further extension

The present version of Flext lacks of the possibility of dealing with uncertainties, which is, in fact, unavoidable in practice when only an estimation instead of accurate measuring value for some constraint factor is available. For example, in the developed system for evaluation of on-site stormwater infiltration, the answer to "Whether the area to be connected is heavy polluted? is likely to be "maybe". In this case, a factor indicating the probability of a more clear or accurate answer (such as answer "yes" with a probability of 0.6 and "no" with 0.4) can be introduced and the inference engine should be extended, considering this factor in its pattern-matching mechanism. In fact, in a full fuzzy system, not only the data or facts but also rules may be attached with probability factors. Users may explicitly specify the probability factors to each input data during the execution of a developed system at the human-computer dialog. It can also be calculated by the system itself according to a pre-defined model for a collection of pre-defined vague vocabulary (such as "maybe", "likely", "hardly", "probably", etc.). This calculation is usually called defuzzlation. Apart from the ability to process uncertainty, it may also be necessary to add some self-learning ability to Flext knowledge formulator, although it is now already very simple and easy understandable.

REFERENCES

Beecham, S., Kandasamy, J., Khiadani, M. and Trinh, D. (2005) Modelling On-Site Detention on a Catchment-Wide Basis, *Urban Water*, 23 – 32.

Cawsey, A. (1998) *The Essence of Artificial Intelligence*.

Jackson P. (1999) *Introduction to Expert Systems*, Addison Wesley Longman Ltd.

Lange, N. T. (2003) Einsatz von Methoden des "Soft Computings" in der Hydrologie, Bericht 15, Fachgebiet Wasserbau und Wasserwirtschaft Universität Kaiserslautern

Sieker, F. and Sieker, H. (2002) Naturnahe Regenwasserbewirtschaftung in Siedlungsgebieten – Grundelagen und Anwndungsbeispiele, Neue Entwicklungen – 2. Neu bearbeitete Auflage –Kontakt & Studium Band 508, Renningen-Malmsheim, expert verlag

Sieker H., Bandermann S. and Zimmerman U. (1999) Geographical Information Systems Used as a Tool for Decentral Stormwater Management. *Proc. of 8th Int. Conf. on Urban Storm Drainage*, Sydney, Australia, 30 Aug.-3 Sept. 1999.

Sieker H., Bandermann S. and Deckert A., (2005) Potenziale dezentraler Regenwasserbewirtschaftung im Einzugsgebiet der Berne.

Part 3

Applications of Adaptive Decision Support

Chapter 16

DayWater ADSS: an incentive for action but a source of questions - an end-user account of the prototype development

Claire Beyeler

'SYNDICAT MIXTE MARNE VIVE': AN ORIGINAL LOCAL WATER INSTITUTION TACKLING STORMWATER MANAGEMENT ISSUES

Presentation of the 'Syndicat Marne Vive'

The 'Syndicat mixte Marne Vive' (Living Marne River syndicate) located at the eastern urban part of the Ile de France region covers about a 100 km^2 of the downstream Marne River basin (Figure 1). It is a local government, similar to a district, gathering twelve members (13 municipalities, 2 public bodies) and has the objective to increase the Marne water quality in order to protect the (drinking) water resource and to support safe swimming activities. In the first half of the 20th century, this part of the river Marne was well known for week-end leisure activities including swimming, canoeing, dancing, fishing, etc. The syndicate is dedicated to carrying out studies and providing advice to its members to enable these activities to return. However it cannot be directly involved in any public work to achieve this goal and this must remain the responsibility of each member of the syndicate. On the basis of a global diagnosis of the river (including an assessment of pollution flow, water quality, fauna and flora diversity and fragility, hydraulics, sewage systems, uses) a 'Marne Vive Plan' has been elaborated and was approved in May 2000 by its board of elected officials.

The 'Marne Vive Plan', focuses on increasing the water quality, the river biodiversity and its scenery and provides a selection of actions to be followed to achieve this. These range from the construction of sewers and the improvement of local regulations, to assessments of every day practice in local public work departments (Figure 2). The 'Marne Vive Plan' is meant to be an incentive for local governments, a help in the definition and the planning of their work programme. The 'Marne Vive Plan', although rather general in its guidelines for stormwater pollution, aims at studying precisely the different practices followed by each of the members and identifying the priorities of each public work department whilst determining the impact of numerous overflows on the river. To this end a model has been used to assess the major pollution problems.

All the actions mentioned in the 'Marne Vive Plan' need financial support to be launched (Figure 2). For example the diagnosis has been supported by a public grant up to 80%, and the five first years of the 'contrat de bassin Marne Vive' program have been estimated at 52 million euros, the financial aid reaching between 20 to 30% of this amount.

Any tool that can bring help and knowledge to the syndicate or directly to its members is sought because dealing with water management and using innovative techniques is very unevenly developed in local governments. Furthermore managing stormwater appears more clearly today to be the main answer to attaining water of relatively good quality and reach the 'swimming' water quality during summer in the Marne River.

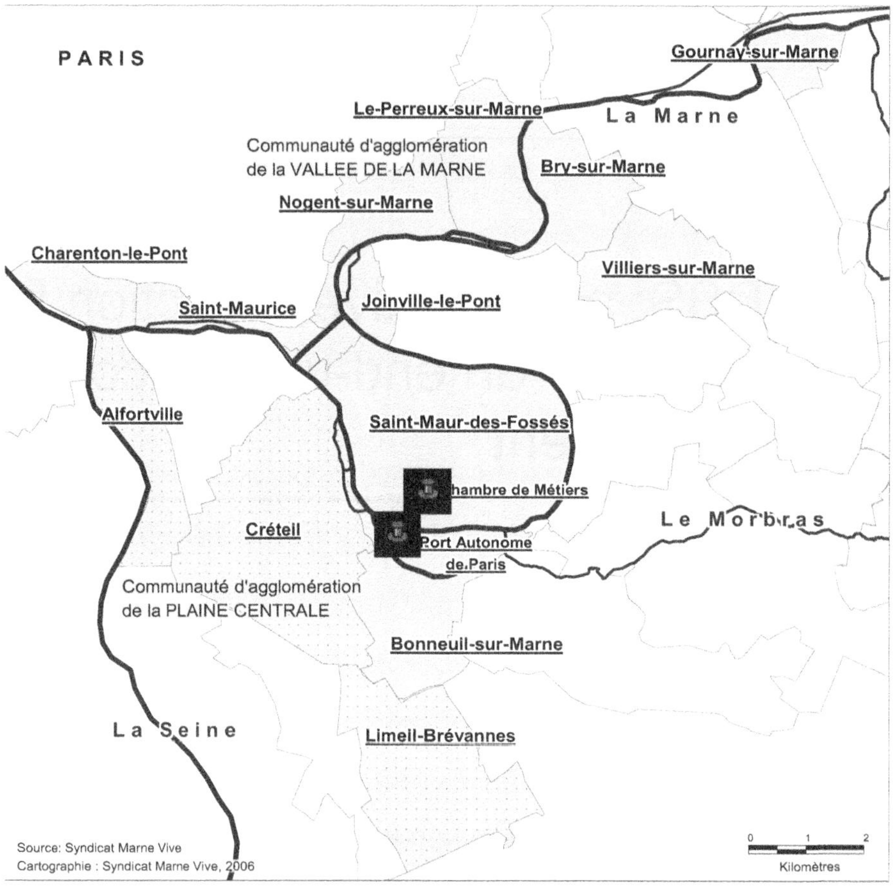

Figure 1: Map of the members of the 'Syndicat Marne Vive' (Living Marne River syndicate) showing, in pink, its 12 member municipalities along the downstream part of the Marne River basin.

The need to manage urban stormwater

In the 'Syndicat Marne Vive' territory, as in many urban areas around Paris (Figure 1):

- stormwater pipes are aging,
- stormwater system hydraulic capacities have reached their limits,
- discharged stormwater is the reason for specific and damaging pollution with regard to receiving ecosystems,
- urban stormwater discharges or overflows are no longer accepted,
- money for investment in the stormwater field is lacking,
- local officials and engineers have to face legal responsibilities (concerning floods and pollution),
- citizens react strongly against flooding, certain types of pollution and tax increases,
- sustainable development issues need to be taken into account.

Basic water management needs to include knowledge of policy, planning, design, implementation, maintenance and finance. Any tool aiming to give help and guidance to storm water management has to address a very wide range of topics and to be site specific.

Within the entire syndicate territory, one of the important needs is to estimate how much stormwater reaches the river. The quantity, quality (pollution flux), frequency and impact on water quality and the ecosystem need to be known, section by section. With this knowledge it will be possible to give priority to some discharge points and to institute control procedure. However, initially, each municipal water department or public work department is trying to reduce the rain water contribution to the system. Some examples are listed below, but to encourage the implementation of these techniques, information and examples are needed:

- infiltration wells have been tested in several parks,

- green roofs have been prescribed in a few new public buildings,
- porous materials have been introduced while renovating street surfaces.
- a new regulation policy has been implemented in two or three towns in order to disconnect individual roof runoff.

Many conferences have been held on water and stormwater management but few examples have been thoroughly analysed and often only the pollution aspects or the general funding questions were addressed. Although some users are aware of the complexity of stormwater management, very few have the required knowledge and skill to confidently defend the use of best management practices (BMPs) *vs.* classical end-of-pipe measures. Moreover the use of BMPs generally requires being involved in the design stage not only with the water and sewer department but also with the urban planning and the road department. In the operating stage further urban departments can still be involved (park, building, street cleaning, etc.). Therefore the need for data, examples and methods is essential.

What help is needed for the engineer, technician or elected official faced with stormwater management problems?

Some of the options available when confronted with a special situation are:

- to register for a training session,
- to phone for known or unknown colleagues who may have faced the same situation,
- to buy specific books and specialist magazines,
- to contract with a consulting firm.

When faced with complex situations, city engineers often opt to contract out their problem with firms that have experience of similar special studies and to be able to present one or more solutions to the stakeholders. However, nowadays in the water management field, local government is keen to keep the process in house in order to develop and establish a local know-how. Therefore, in order to address their problems, they need not only precise and diversified information supported by project examples, but also a thinking aid. This is in order to help them to analyse the different aspects of their problems, identify their needs and finally appreciate a range of solutions.

A stimulating guide is needed, offering different types of data and approaches to the subject. This guide should be aimed at giving a comprehensive overview of a problem but not necessarily providing a ready made solution. The guide should be able to be shared with stakeholders, or at least, the outputs could be shared in order to reach a solution. In this way the end-users would be trained and guided in the achievement of viable solutions.

That's why, when the DayWater project started, the syndicate applied as an end-user (Chapter 1). This decision to participate to a European research program was taken by the syndicate in order to exchange with searchers on the syndicate needs and to create an incentive within its member for facing stormwater management. The letter of intent written at the DayWater project start stated: "If the syndicate is selected as an end-user, it will be considered, for the local governments that have decided to associate in 'Marne Vive' syndicate, as a recognition of their involvement in improving their practices and acting in a more co-ordinated way on the same catchment. It will stimulate them in questioning their planning methods and in searching for new ways of facing urban runoff. What the syndicate expects from the DayWater research program is a help in decision making. Local governments need effective information on stormwater management practices, as well as sets of arguments leading them towards the best source control measures, with a minimum of risks and a sustainable environment."

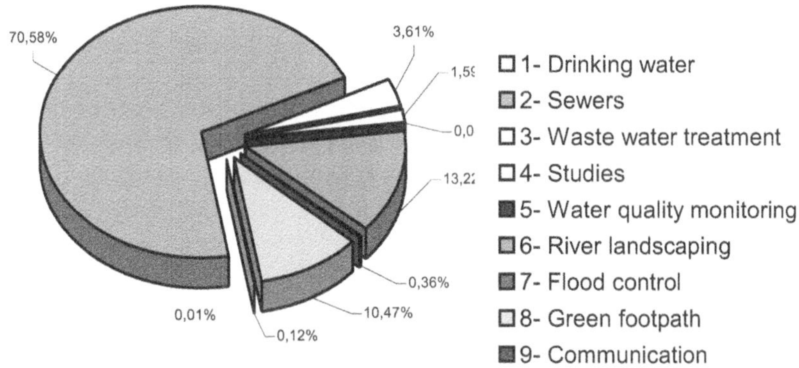

Figure 2: Financial distribution of various actions within the 'Marne Vive Plan'.

A DECISION SUPPORT SYSTEM FOR URBAN STORMWATER SOURCE CONTROL

A decision support system (DSS) is a tool which should be able to offer benefits mentioned above, but it has to be remembered that its user will have many expectations:

- gaining time: the user is usually short of time and is therefore looking forward to rapidly and easily reaching simple to understand information,
- gaining precise and true knowledge and having access to useful updated data,
- benefiting from the experiences of others,
- finding a way to solve a particular problem while sharing the decision with others,
- feeling confident about the implementation of proposed solutions.

The DayWater project is launched: expectations of possible users are high

The problems that stormwater source control DSS users are facing are numerous and complex and as soon as the idea of the stormwater adaptive decision support system (ADSS) was launched, their expectations were high. For them, the expected ADSS should be:

- a tool easy to use by urban stakeholders (engineers, technicians, elected officials, planners, developers, etc.)
- a tool offering methodological advice, case studies, scientific, technical and financial information,
- a sharing of experiences, both good and bad,
- an incentive to think globally,
- a dynamic database,
- an adaptative approach allowing their request to be site specific.

Such ADSS should offer guidance and advice on all aspects of stormwater management, providing access to precise knowledge, useful data, well described and analysed case studies in order to benefit from the experiences of others. The ADSS should seek to answer special problems while sharing the decision with others. It should offer assistance to improve their stormwater management, to promote BMPs and lead to a solution to the raised problems. The ADSS user is expecting a 'Do It Yourself' (DIY) kit, built on a very large sum of knowledge linked together in various ways in order to lead to a solution or to propose solutions that can be discussed with and presented to others.

User comments during the development of the ADSS

First comment: the difficulty to describe one's need or profile

Possible ADSS users have tested and tried out the ADSS components, at different stages of their development. In the first version each user had to identify his need before being given access to any information. This was considered necessary to allow recognition of user's need by the computer system and to assist pathway planning within the system. Hence, the user, in looking for answers and seeking assistance, was being asked very detailed and technical questions that could be seen as a delay and a restricted access to information. It was as if the user was saying "I don't know enough, I don't know how to analyse my situation" and was being answered with "you don't know enough about your problem, sorry we can't help". It was a very strange feeling for a user, knowing what the practical problem is, to be faced with not being able to enter a system which has supposedly been designed to provide access to information. Discussions which were held on this issue rapidly led to a change in the entry structure to the system (Chapter 2):

- a 'free mode' was created enabling the user to cruise through system and to visit and pick up any relevant information and store it in a 'caddy',
- a 'guided tour' was provided also offering the possibility for the user to escape introductory questions by selecting a response "is not important".

This evolution is very satisfactory because the user is being aware of the need, or the possible need of more information, but is not left out of the system when an answer is not immediately obvious. While scrolling the questions, the user understands that the range of solutions provided will vary according to the amount of input given (Chapter 4).

Second comment: facing the complexity

The complexity of stormwater management problems were illustrated by the original 'Hydropolis' drawing (Figure 3): it was unusable as the basis of the ADSS home page. A simplification was essential and, following full discussions, the welcoming page of the ADSS became a city photograph divided into squares each dedicated to a special theme such as 'aspects of water', 'BMP database', 'MCC approach', 'risk and vulnerability', etc. (Figure 4). The complexity has not been transferred to each theme and the links between the different issues are less obvious in such a picture.

If the ADSS user is looking for precise and effective information, he wants to be sure that while browsing through the system no relevant aspects of his problem have been forgotten. While going through BMPs database for example, he has to be reminded (but not asked questions about) the risk related to each technology, their cost, their social aspects, etc. He has to be made aware of the possible complexity of each situation through an overview of the different related aspects. He also has to be invited, through a link, to visit the other themes. When using the 'guided mode', such assistance seems available (Chapter 4).

The multi-criteria analysis of possible techniques or measures (BMPs or alternatives) takes up a side of the complexity, by stressing the importance of the balancing of each aspect and showing that it can lead to very different ranking results. While going through the scoring steps, it gives to the user the range of the complexity. But the challenge is to address the complexity and the diversity of the situation without drowning the user in too many questions or alternatives.

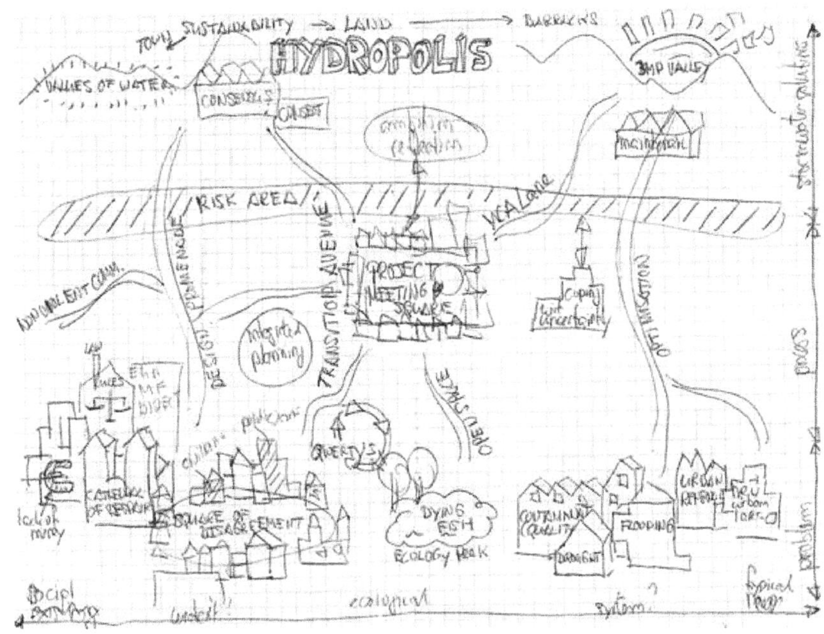

Figure 3: Original drawing prepared by Geldof for the ADSS front page (Chapter 3).

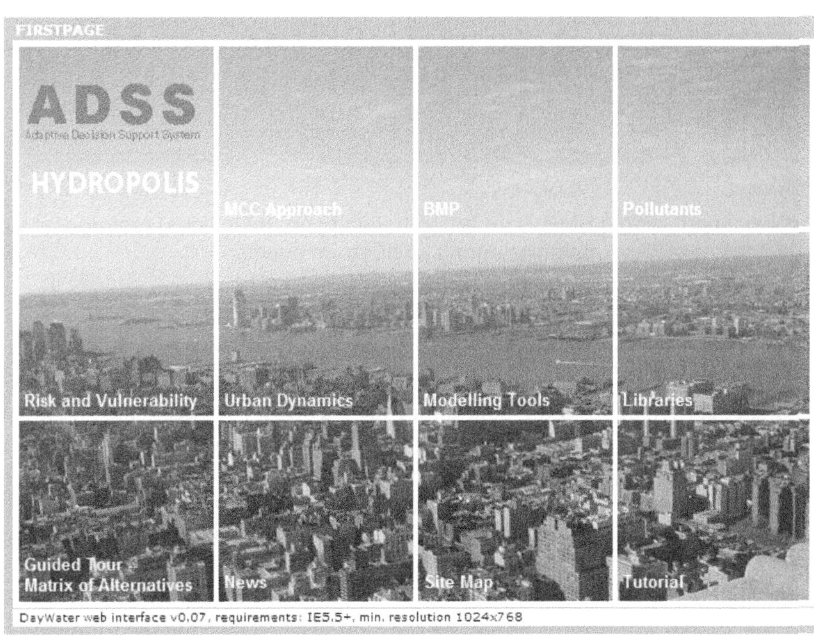

Figure 4: Final city picture for the ADSS front page (Chapter 1).

Scientific knowledge versus user needs and practical experience: two different languages?

The ADSS user is very demanding and wants precise answers to questions. For example he can be faced with the type of problem described below (Figure 5):

- Can I build an infiltration well without polluting the ground water resource?
- Knowing that there is an industrial area nearby, what type of pollutants can we expect and which ones are dangerous to the local stormwater uses?
- Is there a way to measure these pollutants? What is the risk level? What are the guidelines?
- Can a local policy regulate some types of industry in order to forbid a number of pollutants and to make sure that the different infiltration techniques will not generate new sources of pollution?

Wide information exsists in different parts of the ADSS, within several tools: CHIAT, RICH, SEWSYS, STORM, etc. (Chapters 5, 11, 12 and 14). But the difficulty is to guide the user towards the needed information and to help him with his thinking. The user's way of thinking is a result of field experiences and of everyday practice. The scientific approach and thus the organisation and the selection of the information in the ADSS are the result of a scientific thought process. Obviously there is a difference, a gap, between scientific knowledge and practical knowledge. The ADSS has to play the role of a translation tool and a guiding map enabling the user and the scientific to meet.

Once reached, some tools will awaken the curiosity of the user and raise new questions. But he may not find what he is looking for. He will, for example, face the heterogeneous level of scientific knowledge on risk and pollution (Chapters 11 and 12). But the ADSS should give him more confidence than uncertainty in order to give him the incentive to implement a new technology. The gain of knowledge and comprehensive thinking should build his confidence and organise the arguments that he will share while presenting his stormwater source control project to other stakeholders.

Figure 5: ADSS user considering using a BMP, here an infiltration pond: (blue lines) preferential stormwater circulation pathway during small storm events.

CONCLUSION: WHAT IS THE FUTURE OF THE DAYWATER ADSS?

The above account of an user participation to the building of the DayWater ADSS makes it clear that this is a rich and challenging project. But all the users' expectations have not been fulfilled and maybe not all the scientists' aims have been reached. The ADSS needs improvement and a future.

The ADSS is a prototype

The ADSS user needs for a guidance tool and an operational tool. The current DayWater ADSS is a prototype and not yet a fully achieved and operational software. A prototype can be described through the following questions:

- who can use it: skilled or unskilled users? what is the profile of its user? (Chapter 2),
- what is its structure? (Chapter 3),

- which functions are included? (Chapters 2 and 3),
- what are its components? (Chapter 1),
- what are the results of the testing? (Chapters 17 to 19),
- what are the identified improvements to be made?
- what can its future be?

For the final DayWater conference in Paris (3-4 November 2005), the ADSS prototype offered:

- sources of information including databases on BMPs, risk, costs, etc.
- an access to modelling and comparative tools: MCC, STORM, SEWSYS, CHIAT,
- different itineraries to reach information according to the profile of the user.

So at this stage it is clearly a DSS, because it provides access to organised and structured information and offers different navigation possibilities through the system to help decision making. It is clearly adaptative because many options are offered to the user in order to built his own case or to test different situations and also because the structure of the software enables the users the updating of different components (case studies, BMPs, etc.). It is also a training or awareness building system because it opens the mind of the user towards new fields of knowledge.

What improvements could be made in the ADSS?

Hydropolis, the first page of the ADSS system, offers a site map where each component is briefly described in two lines in order to explain its aim: the users have asked for a more user-friendly presentation and utilisation. In fact some information was considered as lacking by the users while the relevant DayWater project partner was able to explain that it was actually present. There are some improvements to be made in order to make the ADSS more legible, to make the navigation through the ADSS easier and to create an invitation to explore it.

For lack of time, the step of testing the system by different types of users has not been totally implemented, although it is a very important step in order to identify the more useful or used components and to improve the quality of the navigation in order move from the present prototype to an actual 'product'. It could also lead to new user profiles and specialised navigation routes, in order to simplify the first contact with the ADSS and to enhance its training or awareness building function.

How can the ADSS be improved and maintained?

This topic has been regularly tackled during the last DayWater conferences but no specific solution has yet been reached. The ADSS prototype is the result of the work of a joint team of academics, public and private institutions. The commercialisation of the ADSS is not possible yet because the prototype still needs improvements but the European Community funding is finished. One solution could rest on the shoulders of the users. The users have shown their enthusiasm for the ADSS but they also have expressed their needs and requirements. To give the ADSS the future they want, they will have to join together, maybe at a national level, in order to develop it. And last but not least, a host is need for the ADSS computer system in order to provide all the computer functions and the software maintenance.

For many core end-users involved in the DayWater project (Chapter 1), the experience has been very rich. All the working conference have provided a chance to express their needs and share with academics their field experience and everyday needs. The ADSS prototype is a 'bridge' between scholarly knowledge and end-user needs but it still lacks a few 'pieces' in order to welcome large traffic!

REFERENCES

Chéron, J. and Puzenat, A. (2004) *Les eaux pluviales: récupération, gestion, réutilisation* [Stormwater: collection, management, reuse], Editions Johanet.

Le Pape, E. (2003) *Le zonage d'assainissement : eaux pluviales* [Water sanitation zones : stormwater], ENTPE Internship Report, CERTU.

Chapter 17

Factors of success and failure in the implementation of urban stormwater source control measures: the Nijmegen municipality experience

Antal H.J. Zuurman and Anton J. Verhoeven

INTRODUCTION

There are many ways the municipality of Nijmegen (Netherlands) performs a sustainable urban water management. They are all written down in the 'Nijmegen Water plan' (van Beurden *et al.*, 2001) a policy plan in which all the water partners work together in order to solve the problems on water quality and quantity. One strategy to achieve this goal is the disconnection of paved area urban runoff from the sewer. Disconnection itself is a source control measure allowing better water quantity and quality (Chapters 1 and 8).

This paper highlights what Nijmegen municipality learned from the implementation of such urban runoff disconnection. To understand the need of a 'Water plan' for the municipality and the need of disconnection, this paper provides first some background information about the 'Nijmegen Water plan' and it aims. The second part considers urban runoff disconnection in Nijmegen and addresses issues like: "What are possible techniques? What are constraints for application? And finally, what is learned while implementing these techniques in practice?" The third part describes the way Nijmegen municipality communicates to the public on the changes in urban stormwater management. The final part describes briefly what value the Adaptive Decision Support System (ADSS) has for Nijmegen, as DayWater end-user (Chapter 1). It also gives some clues on what Nijmegen can to offer to the DayWater project and to the ADSS development.

WATER PLAN OF NIJMEGEN

Introduction

Since the high water levels in the Rhine and Waal and the local nuisances due to heavy rainfall in the last decade, the perspective on water management has changed (Min. V&W, 1999) (Figure 1). It made clear that more attention has to be paid to retention of stormwater at all scales within catchment areas. Especially in newly built urban areas the challenge is to retain stormwater within the spatial structure i.e. roads, houses, firms, parks, watercourses, ponds, etc. This is also necessary for the existing urban area of Nijmegen.

 DayWater: an Adaptive Decision Support System for Urban Stormwater Management. Edited by Daniel R. Thévenot. ISBN: 9781843391609. Published by IWA Publishing, London, UK.

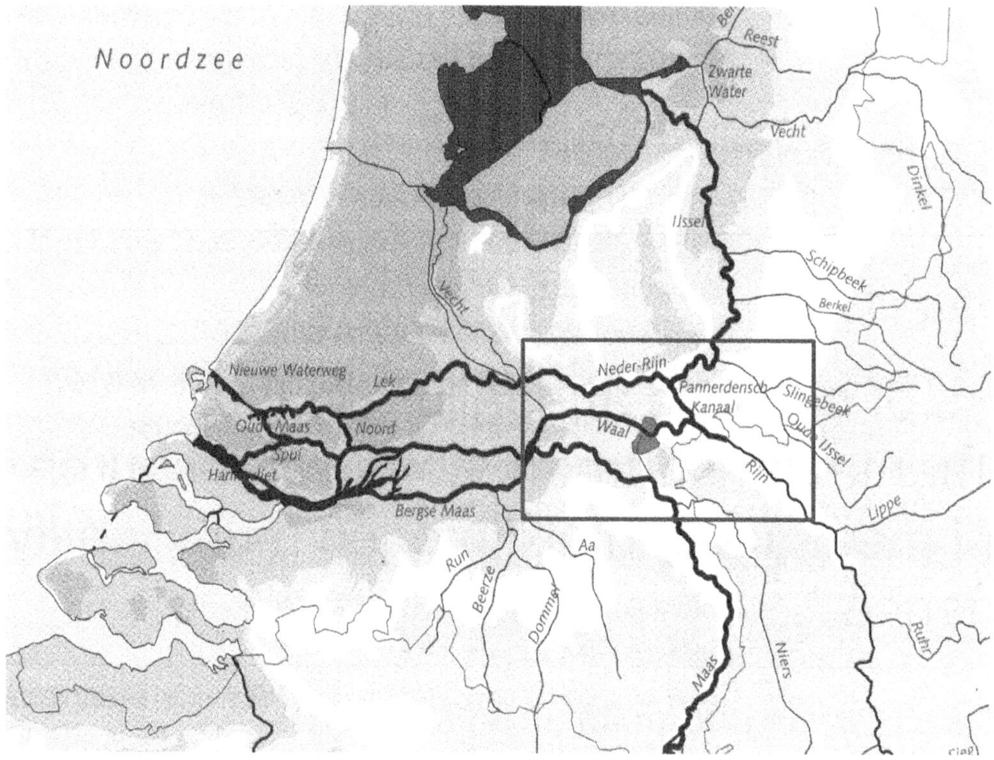

Figure 1: River delta within the Netherlands and situation of Nijmegen city.

To understand the need for the 'Water plan', this section first describes the characteristics of each part of Nijmegen then addresses the water quality issue as one of the main drivers for the 'Water plan'. The final paragraph describes the major goals of the 'Water plan'.

Characteristics Northern part of Nijmegen

North of the river Waal, a whole new development is under construction around a small old village called Lent (Figure2). As usual in The Netherlands, the new development will have a separate sewer system and source control measures will be applied. The spatial planning began there ten years ago with the design of the water system, which was revolutionary at that time. Nowadays several water courses around the neighbourhoods are built. Later on, all the water courses will be combined into one water system and two lakes will be developed. Those lakes have two objectives: retention during the wet season and delivering water during dry season, when the Waal River causes severe drainage to the water courses system. The Waal river is the most important shipping route of North-west Europe. Near Nijmegen the avaialable tranport width is limited and is a bottleneck for the discharge of all the stormwater of the countries upstream Nijmegen.

Characteristics Western part of Nijmegen

In the Western part of the city of Nijmegen (Dukenburg and Lindenholt quarters) retention of stormwater within the existing urban area is the most important issue to be addressed (Figure 2). These quarters were developed more than 20 years ago and have a (dated) system of water courses and ponds that is used for retention and transport of stormwater towards the downstream (rural) area. Next to rainfall the stormwater system in these quarters is fed by both local and regional seepage. The local seepage is caused by the Maas-Waal Canal and fluctuates depending on the water level in the canal. The regional seepage is caused by a more or less constant flow of groundwater from the hills surrounding the city at the Eastern side.

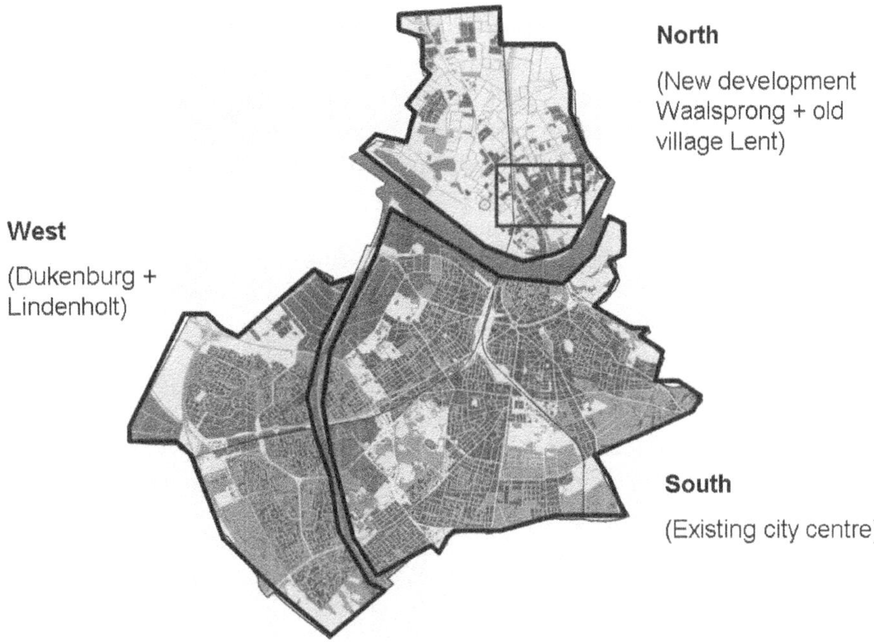

Figure 2: Partition of Nijmegen in characteristical areas.

Characteristics Southern part of Nijmegen

South of the river Waal, the village of Nijmegen lies since ca. 2000 years on the slopes of the hills resulting from the glacier times (Figure 2). This city area has a combined sewer system with some free sewer outlets into the river. The ground water levels vary from 40 m below ground level (East) to 2 m below (West). Ten years ago the municipality council decided to change their water policy.

Instead of taking end-of-pipe measures (building retention basins for runoff storage), the municipality is now disconnecting the paved areas and infiltrate the stormwater into the ground. Where possible, this is done in a visible way (Figure 3). In the shopping centre, several famous fountains (rain water fed) are present and, in the neighbourhoods, many other technical solutions are used.

Figure 3: Girl playing with disconnected stormwater: fountain fed by rain water in the shopping centre.

Water related issues that require a water policy plan

Another issue that influenced the perspective on stormwater management is the receiving water quality. Inventories over the years have shown that the chemical/ecological quality of surface water have been deteriorated in many places. It does not correspond to the international environmental quality standards (EQS), thus creating risks for people, animals and plants. In urban areas sustainability means that polluted wastewater require removal from watercourses and ponds. Through improving the surface water quality, there will be fewer algae and duckweed and the surface waters will present a higher oxygen level, especially during summer.

In order to avoid future generations facing these problems again, special attention has to be paid to reduce or eliminate the urban sources of pollution. This issue is supported by the EU Water Framework Directive which aims to maintain and improve the aquatic environment. In the Netherlands, there is a policy that states that the pollution of surface water (caused by direct or indirect overflow of the sewer systems) (Figure 4) has to be reduced by 50% in 2005 (CIW, 2001). Because the Nijmegen municipality changed its policy from retention devices to disconnection paved area runoff, considering that disconnection is a slow process because of its high cost, a solution had to be found for reaching this objective. The municipality and the water board have built an extra purification step on the wastewater treatment plant (WWTP): taking into account this improvement, we now have until 2019 to disconnect urban runoff from some 200 hectares of paved area.

Figure 4: Combined sewer overflow (CSO).

Main goals of 'Nijmegen Water plan'

The integrated improvement of water systems in existing urban areas can best be done taking into account the European dimension of the river basin. For Nijmegen, this means the North West Europe (NWE) region: raising public awareness and increasing knowledge exchange contribute significantly to raising awareness at the transnational level, i.e. in Belgium, France and Germany and to strengthen the social cohesion in that region. Therefore the city of Nijmegen is since many years involved in several European projects (IRMA Interreg Rhine Meuse Activities), Urban Water, DayWater) and partially involved in projects such as SDF(Sustainable Development of Floodplains) and Freude am Fluss.

At national and regional level these goals are also very important: improvement of the water system means that, at the different phases of the project, various actors have to be involved to create a sustainable water system. Therefore the water partners (municipality of Nijmegen, Water Board Rivierenland, Province of Gelderland, Rijkswaterstaat i.e. the Directorate General for Public Works and Water Management, and Vitens i.e. the drinking water company) decided in 2001 to upgrade the urban water system in the city of Nijmegen and to raise a broad awareness about sustainable water management (Geldof *et al.,* 2003). In the so called 'Nijmegen Water plan' they subscribe to the following goal: "To work together to get a sustainable water chain in the city of Nijmegen, in order to achieve a healthy and resilient water system and a pleasant environment for the citizens at minimal social costs".

A sustainable water-chain is characterised by a fine-tuning between the production and distribution of drinking water, the use of it and afterwards the collecting and purifying of the waste water in the wastewater treatment plant. A healthy water system contributes to a good environment for people, animals and plants. A resilient water system can easily deal with disturbances in quantity (heavy rainfall) and quality (urban runoff, wastewater). One of the main objectives of the 'Water plan' is "Disconnection of paved areas in the old city (combined sewer)". The short and long term disconnection targets are displayed in Figure 5.

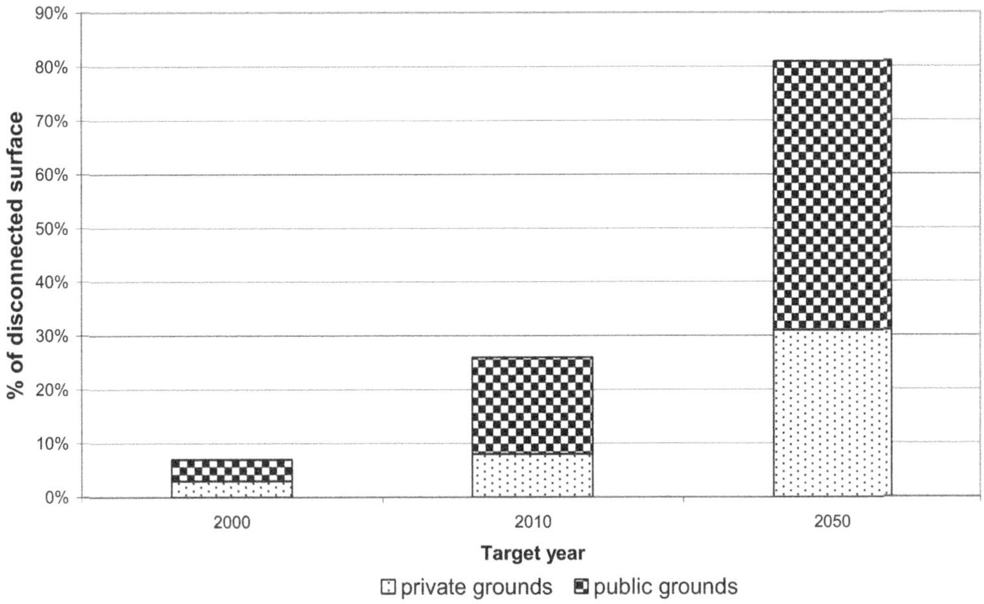

Figure 5: Short to long term urban runoff disconnection targets in "Nijmegen Water plan".

Realising those targets in the past years was very difficult: the main reason was a lack of experience and a lack of projects combining urban runoff disconnection with reconstruction of the area, sewer replacement etc. Combination of activities is a financial constraint because autonomous disconnection is very expensive (Gemeente Nijmegen, 2004). We have now a lot of experience with different techniques and we combine disconnection not only with replacement of sewers but also with building of new roads or new houses. In 2005 the new 'Municipality Sewer Policy Plan' was approved by the city council and even bigger targets were fixed, because of the delay in the past (Witteveen en Bos, 2005). We no longer have targets fixed as a % of the total paved area, but only targets in absolute figures. The development of disconnected surface is displayed in Figure 6.

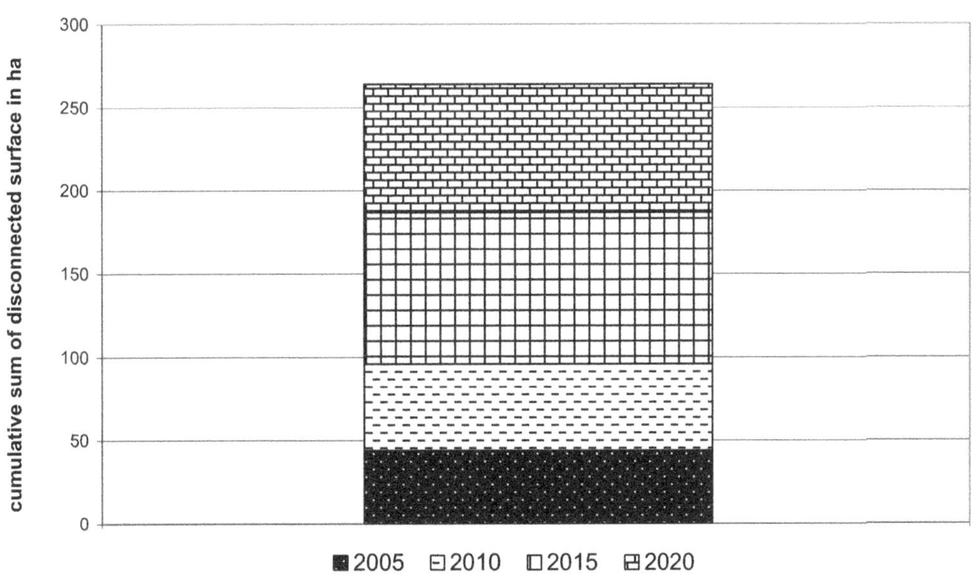

Figure 6: Short to mid term urban runoff disconnection target (ha) in 'Nijmegen Water plan'.

DISCONNECTION OF IMPERVIOUS SURFACE FROM THE SEWER SYSTEM

The main part of this section describes, in three steps, all our experience in urban runoff disconnection. After a short introduction the applied techniques in Nijmegen are presented, followed by an explanation why these systems have been selected. Then in the third part tips & tricks about design, implementation and maintenance are given. The last section emphasizes a case, called 'Voorstadslaan', because a lot of difficulties were faced in this project and the lessons learned are valuable for the DayWater project.

Introduction

The term "Disconnection" should be clear to avoid mistakes in communication about this subject. Disconnection is an abbreviation of disconnection of impervious surfaces from the sewer system (Zuurman,, 1998). In general this means that roof and street runoff no longer enters the sewer system. After the disconnection process an appropriate treatment device is required in order to store or/and infiltrate the stormwater into the ground. When that is not possible, due to high ground water levels, the runoff can also be transported to the surface water.

The treatment techniques can be divided in four main solutions. In decreasing order of sustainability (from source measures to end-of-pipe solutions) (Zuurman, 1998) (Chapter 8):

(1) Avoid stormwater turning into runoff e.g. green roofs, porous paving,
(2) Collect and reuse stormwater for several purposes e.g. flushing toilets, grass watering,
(3) Infiltrate stormwater into the soil e.g. Aquaflow (paving with water passing voids), swales (grass infiltration fields), trenches,
(4) Store stormwater in natural and artificial systems e.g. ponds, concrete basins.

Disconnection techniques in the municipality of Nijmegen

The impervious surfaces consist of roofs mostly situated in private areas and paving in the public area. In Nijmegen six techniques have been applied: five techniques were applied in public areas and one technique only in private ones:

(1) Visual water art projects,
(2) Transport gutters at private area,
(3) Swales i.e. grass infiltration fields,
(4) Porous infiltration pipes (diameter 400, 600 and 800 mm),
(5) Porous paving,
(6) Aquaflow (paving with water passing voids).

Visual water art projects

Nijmegen has several projects in this type, mainly situated in the city shopping centre. These projects use rainwater in an attractive way. Examples shown are a large gutter/waterfall in a steep chopping street and the 'cheater fountains' (Figure 7): the 'cheaters' are called that way because you never know where and when the stormwater will pop up from the ground, this resulting obviously in wet clothing for kids.

Figure 7A: Waterfall at Stikke Hezelstraat.

Figure 7B: 'Cheater fountains' in Koningsplein.

We also create water playgrounds (Figure 8): they all make the public aware of the way Nijmegen disconnects the roofs of the buildings and uses stormwater instead of valuable drinking water. Other public organisations can be informed easily during visits.

Figure 8A: Water playground at day nursery.

Figure 8B: Water playground at public grounds.

Transport gutters at private area

When the Nijmegen municipality replaces broken sewers, we combine this operation with pavement runoff disconnection in the public area. When we design infiltration devices, we give it enough storage for the roads but also for the roof runoff in that area. An employee of the water service desk visits each private owner and asks if the municipality is allowed to create a beautiful gutter in order to disconnect a part of their roof runoff. If they give permission, then within two months such specific gutters are created: stormwater runs through their gardens to our street and is stored there (Figure 9).

Figure 9: Examples of gutters created on private grounds towards public space.

Swales ('wadi' or grass infiltration fields)

In the last ten years, numerous city projects combine green space, in urban planning, with surface infiltration devices. In The Netherlands, such systems are called 'wadi'. This Arabic word refers to the dry rivers in the desert, which once a year swell and flood the area (Zuurman and Hartman, 2001). In the remainder of the year the rivers are dry. Figure 10 shows a possible appliance and Figure 11 displays a cross section of a Dutch swale.

In the same way the Dutch 'wadi' system works: during rainfall they flood and, after a short period, they will be dry again (RIONED, 2006). In the South of Nijmegen a big ecological zone was created and the lowest part functions as a 'wadi'. When the water flows some art works are moving: also here the citizens are made aware of disconnection in a visible way.

Figure 10: Ecological swale in Grootstal (1997).

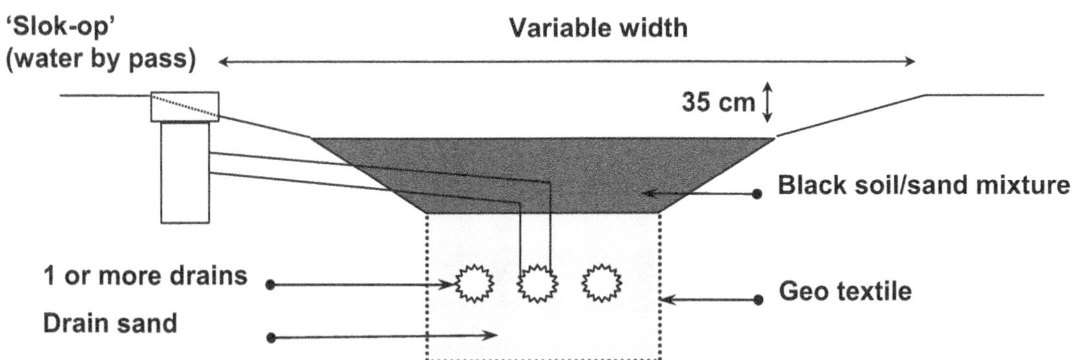

Figure 11: Cross section of a Dutch swale.

In the new settlements of Nijmegen the stormwater is transported in gutters to the swales (Figure 12). In normal times the stormwater will infiltrate and will be purified by the soil. In times of heavy rains, stormwater can overflow to the surrounding surface water.

Figure 12: Recently constructed swale in North Nijmegen.

Porous infiltration pipes

Since 1998, a concrete company developed this product. The main problem used to be the strength of these non reinforced concrete pipes, but the pipes have improved in strength (www.dehamer.nl) (Figure 13). Nijmegen applies porous concrete infiltration pipes since 1999 and has already approximately 6.5 km of such infiltration pipes. Most of the public disconnection of roofs (see above) is done in combination with this kind of infiltration devices.

Figure 13: Porous concrete pipe in testing stage in laboratory.

Porous paving

Porous paving originates from Germany and have over 20 years of experience (Figure 14): porous paving is combined with a layer of drain sand. The main principle is that 60% of the stormwater evaporates just like in natural circumstances. The remainder water is stored in the layer of drain sand and from there added to the groundwater (www.natuurlijksteen.nl) (Figure 13).

Figure 14A: Construction of porous paving.

Figure 14B: Porous paving slab.

Aquaflow/Form pave 2000 (paving with water passing voids)

Aquaflow originates from Pratt (Coventry University): in England they have more than 10 years experience with this system (www.aquaflow.nl) and up to now it is mainly applied in parking lots. A Dutch concrete company redeveloped this system for Dutch applications (Figure 15) and applies it as paving /infiltration device for roads (Figure 16).

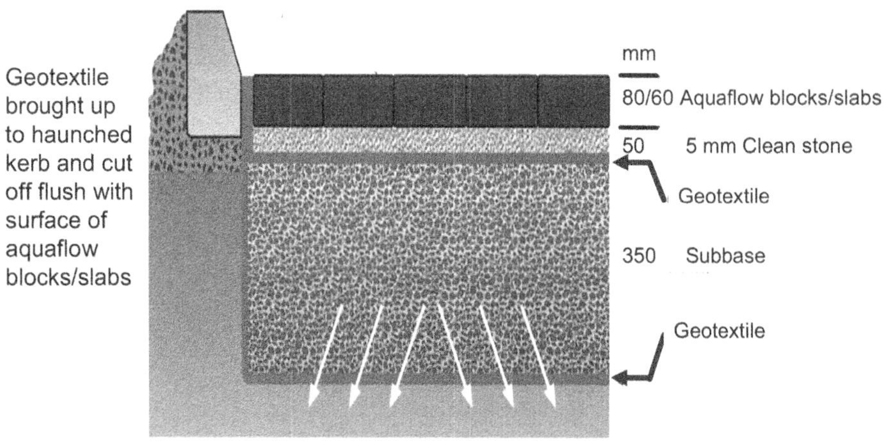

Figure 15: Cross section of Aquaflow for porous road applications.

The top layer consists of stones with water flow through possibilities via the voids: the voids are filled with basalt split (Figure 15). Underneath the paving layer, a layer of broken stone takes care of the storage of the stormwater: from there it infiltrates into the soil or it is drained to surface water (Figure 16).

Figure 16: Construction of porous street paving using Aquaflow.

Constraints for application of urban runoff disconnection techniques

Public awareness

In general the philosophy is simple: if possible a technique should contribute to public awareness of rainwater transport or disconnection. 'Visual water art' projects (Figure 7) and gutters at private ground (Figure 9) are perfect examples. When a system is visible, it is easier to communicate with the citizens about the effects of their behaviour related to stormwater quality. Washing of cars, spoiling chemicals (oil, paint, solvents) and leaving the dogs excrements on the street have all a great negative impact on stormwater quality. The use of pesticides on their private grounds (the municipality does not use them since ten years) does also have a negative impact on the storm and groundwater quality.

Maintenance

The second constraint is that the system is accessible for inspection and maintenance. We want a system that operates now and in the future. Open gutters and infiltration surfaces (as swales, trenches, Aquaflow and porous paving) are therefore preferred to underground devices (e.g. storage chambers). An extra argument is the purifying function of soil (including vegetation) related to the one of pure sand or clay. A swale can be regenerated by replacing the polluted soil layer after 20-30 years. The voids between the Aquaflow pavings are easily cleaned and provide a safe system. Porous paving can be regenerated by a sort of water vacuum cleaner including high pressure water jets.

Porous infiltration tubes are only accessible via observation wells, requiring specific dimensions in the well and in the manhole. The infiltration tubes are cleaned by a rotating wheel and using a 200 bar vacuum to remove the sludge in the pores of the pipe. But it is very difficult to inspect the sand underground the pipes.

Geology and ground water

The third constraint is the part of town where the disconnection takes place. Nijmegen is lying on the slopes of a former glacial 'hill'. On the Northwest side, the ground water tables are high at times of high water levels in the river Waal: therefore it is not recommendable to use here infiltration pipes which lie three meters below ground surface: Aquaflow and porous paving are used here.

On the hillside of town there already several streets where infiltration pipes are used in combination with gutters from private ground. Because of uniformity (for maintenance), we will continue to do so. A new technique that we want to try out here is to let the stormwater flow freely over the streets towards (bigger) central situated infiltration devices. This requires smaller investments and releases money for the monitoring programme. Another possibility is to create filter pits or other stormwater purifying devices.

Paving type

The fourth constraint is the type of road which we want to disconnect. Is it asphalt or brick paving? Since we wish to conduct disconnection along with replacement of the sewer or the road, this is an important issue, since asphalt is much more expensive to replace (Hartman *et al.*, 2002).

How many cars, buses and trucks are passing by? What is the possible danger of pollution which is coming from the roofs of the buildings and from the road? We had the policy not to disconnect runoff from main roads, but new techniques for disconnection are now available together with techniques for stormwater treatment. Therefore our policy is changing.

Tips & tricks

When new systems are introduced within an organisation, they often meet resistance or difficulties. In the past few years, we learned some valuable lessons that are discussed in this section. The working process has been therefore divided into three stages: (i) designing, (ii) implementation and (iii) maintenance. Below is a list of general tips, the 'Voorstadslaan'case being used as an example.

Designing stage

General remarks

- Precise hydrological data is required: mistakes with layers/lenses of clay/loam that lock off drainage towards groundwater should be avoided.
- The water board sets the water levels of surface water: the possible rise of ground water level was not taken into account when infiltration devices were calculated.

- The amount of seepage from different sources was not or not carefully enough calculated.
- Stopping groundwater abstraction causes a rise of the groundwater level that was not taken into account when the infiltration devices where calculated.
- Urban planners design most of stormwater storage locations at highest situated area although water nuisance occurs always at lowest situated areas.
- The permeability of the soil was not tested in the laboratory.
- Wrong type of grass in swale was prescribed whereas the grass type could not bear 24 h flooding periods.
- No escape route was provided for the stormwater in the infiltration devices under extreme climatic conditions and this created water nuisance.
- The designed infiltration devices had no entrance for inspection and maintenance.
- The wrong type of inspection well for infiltration pipe was prescribed thus creating extra space for wastewater storage and the infiltration pipe was sooner polluted.
- The precise technical descriptions for carrying out infiltration devices were not available and the soil became compacted, pores polluted with silt and water nuisance appeared even at implementation stage: this last lesson will be explained in the case below.

In the Voorstadslaan case a high groundwater level is possible and therefore we choose to use Aquaflow in this street instead of the common infiltration pipe technique used at that time in Nijmegen. But since it was an asphalt road and it was not allowed to use bricks there, it was decided to let the stormwater come to the underlying storage via the parking lots. Technically this change of design was possible, but considering what we know now about maintenance of this kind of devices, we will never do that again (Zuurman, 2005a)!

What did we learn from this case?

An 'Aquaflow street' is normally flat and stormwater infiltrates over the whole surface. The design of this road and of the parking lots was like for normal asphalt roads, because the water from the road had to flow to the parking lots, where the Aquaflow is situated. Afterwards, we consider the storage capacity was too small, since we did not plan an escape route for stormwater, during the whole winter there was a water nuisance (Figure 17).

Figure 17: Water nuisance on street next to Aquaflow

Afterwards when the grass infiltration fields remained wet, it appeared that the hydrological data were not precise enough: at some places loam lenses ware available. Also the hydraulic permeability was not enough accurate: at a meter depth, a heavy packed layer of sand was available that blocked somehow the water flow towards the ground water.

Next to the green spaces, the parking lots were constructed: such green space area was expected to function as boarding and alighting strip for the children (Figure 18). The other side of the parking lot was not safe because the parking lots are relatively small and the traffic velocity was too high to get out safe (Zuurman, 2005). Unfortunately the green space was often wet and muddy and did not offer an attractive opportunity to board or alight. This raised a lot of emotions and therefore complaints from the citizens. Afterwards, we will build an extra board and alight strip in the grass to step out dry and several extra street crossings.

Figure 18: Girl forced to alight in the green space.

Implementation stage

General remarks

- Drainage of building sites, before creation of infiltration devices, had not been looked at.
- The soil for swales became compacted by heavy building machines.
- The required height differences for transport of stormwater from the rear of row houses to the front via public space was not created thus yielding water nuisance and ponding at the rear of the row houses.

In the case Voorstadslaan, we did not plan the implementation stage well enough and did not take into account this new infiltration technique. The planning was the same as for a common (not disconnected) road. But we did forget that we had to insure the storage/drainage capacity during the whole implementation stage (Figure 19).

Figure 19: Lack of storage capacity in implementation stage.

In winter time we were not able to finish building the roads and the infiltration fields, although it was incredibly important for getting rid of the stormwater. This led to severe water nuisance and we were forced to communicate further with the citizens about that. Luckily we have our 'Water Service Desk' for that task (see below).

What did we learn from this case?

In winter time, it is not possible to realise the top layer of the asphalt: therefore the road level was lower than the Aquaflow level. The rainwater created ponds upstream the Aquaflow and did not infiltrate. Normally that is not a problem, but because there were no normal swirls, the water could not reach the parking lots and therefore there was a lot of water on the streets (Zuurman, 2005b).

Nearby large new buildings were build and this yielded a lot of heavy traffic and many digging activities. Because this occurred 'upstream' of the runoff disconnection site, sand was transported during storms to the Voorstadslaan and the sand filled up the pores of the parking lots: this also led to a loss of storage capacity.

Stormwater coming from roofs was planned to be stored in the swales (grass infiltration fields) alongside the road. Because of the cold temperatures, the grass did not grow and during the whole winter the inhabitants only saw muddy pools in front of their houses (Figure 20). Now that the grass has grown, there is still too less drainage capacity and therefore the pavement between the gardens and the infiltration fields will be rebuild as porous pavement (Zuurman, 2005a).

Figure 20A: Flooding of green space 'downstream'. Figure 20B: Flooding of green space 'upstream'.

Maintenance stage

General remarks

- No one was responsible for maintenance or responsibility was unclear (Figure 20).
- Required maintenance for each specific device was unknown.
- Too many different runoff disconnection devices were applied, which confuses the maintainer.
- The infiltration device was not maintained since the maintainer was the private land owner.

Figure 21: Lack of maintenance of porous paving: rotten leaves were not removed.

In the Voorstadslaan case we did not foresee the problems that we would get using the parking lots for stormwater drainage instead of the road surface. We also did not foresee the influence of other green aspects (trees, grass used for letting the dogs out) (Zuurman, 2005a).

What did we learn from this case?

Roads can easily be cleaned, but the parking lots are often occupied by many cars and therefore difficult to access. The leaves that felled on the parking lots were not taken away in time and turned into mud which filled up the pores. We now have to clean those pores and fill them up again with new split. Although the design of the grass infiltration fields was done together with the citizens, they now want many street crossings, due to their experience of muddy periods. But mowing the grass becomes impossible: so we have to explain that very clearly.

COMMUNICATION WITH THE PUBLIC

Introduction

In the municipality of Nijmegen communication with the citizens is well developed. It is also one of the main objectives of our 'Water plan': "Raising public awareness by all means" (van Beurden, 2001). This section describes the way Nijmegen communicates to the public about stormwater in order to create public awareness and social support for realisation of projects. First of all creating public awareness is presented, followed by a description of the functioning of the municipal 'Water Service Desk'.

Public awareness

The aim of public awareness is to get the public involved in the transition from a bad surface water quality to a better one: an important strategy for achieving this better surface water quality is the disconnection of runoff from impervious surfaces. Since the public owns half of the connected surface, we need them to achieve our goal of disconnected impervious surfaces. When informing the public, the message will remind only for a short time. When creating a public awareness, it is likely that the memory will last longer. A good example in another field is the movie 'an inconvenient truth' by Al Gore. Public awareness is also required in order to activate the public in disconnecting their own houses.

The municipality of Nijmegen has organized several activities in order to create such public awareness. In this section they are described briefly: (i) water art projects, (ii) garden design competition, (iii) water market and (iv) drawing competition.

Water Art projects

In previous section the 'Visual Water Arts' projects are described (Figure 7): roughly there are two types of 'Visual Water arts' projects. The first type is a device in the public space, especially aimed at the older public. The second type consists of devices for children.

Examples of the first type are the 'Cheaters fountains' (Figure 7B), the 'Water fall' at Stikke Hezelstraat and the 'Water Is' projects (Figure 22). Examples of the second type are the construction of small water playgrounds at 8 day nurseries and the construction of a big water play ground in the Kastanjehof (Figure 8).

The main thought behind this strategy is to make water 'sexy', to show that water is fun. We use the projects to refer to when we communicate to the public which help is needed for runoff disconnection. We show that these devices are fed by disconnected rainwater (Zuurman, 2004).

Figure 22: 'Water Is' project.

Garden design competition

The 'Garden Design' competition was organised by the municipality of Nijmegen, The 'Water Service Desk' of Nijmegen, Municipality of Arnhem, the STOWA foundation, Water board Rivierenland and Water board Rijn en IJssel. The competition was organized for two types of participants. The first group consisted of private house owners that would renovate their gardens. The second group contained professionals like garden architects and green contractors with design affinities (Figure 23).

Figure 23: Example of professional design: leaking water basins looking alike cactus.

The assignment was to design a garden that integrates disconnected rainwater as key element and shows that water is fun and attractive. It had to be a practical design suitable for carrying out. The winner would get 1000 €, for carrying out the design in his garden or the garden that it was designed for. The turnout of the competition was very low, but the presented ideas were good. All ideas were bundled and published. When citizens ask for ideas what to do with rainwater in their garden, this bundle is given to them. These designs are available at http://www.waterbewust.nl/tuinontwerpuitslag.html

Water market

In the past we organised water weeks, water conferences and in 2005 a 'Water market' aiming at making public aware of stormwater disconnection. Such water market that was bringing together water demand and supply. All kinds of organisations (university as well as companies) were there along with our water partners. Schools could visit the exhibition on the boat of 'Rhine water' and there was a drawing competition, together with, of course, a lot of entertainment (Zuurman, 2004) (Figure 24).

Figure 24: 'Water pipe' organ at 'Water market'.

The national campaign 'Holland lives with water' was giving us support (www.nederlandleeftmetwater.nl) (see red-white-blue flags on (Figure 26) Ten thousand people came by that day, because we combined it with the yearly Easter market.

Besides the 'Water pipe' organ as a special attraction (Figure 24), a 'Water pillar' was also opened by the Alderman T. Hirdes (Figure 25): this 'Water pillar' is situated close to the quay and shows the actual water level of the river Waal: it was financed by the Directorate General for Public Works and Water Management.

Figure 25: Opening of the 'Water pillar'.

Drawing competition

As mentioned in the previous section, the drawing competition for secondary schools was an element of the 'Water market' (Figure 26). In the previous weeks many secondary schools in Nijmegen received a package that explained the drawing competition: children had to show in their drawing what would be possible with rainwater around their dwellings or playgrounds, when it is disconnected (Figure 27). The drawing competition was a great success (Zuurman, 2004).

Figure 26: Prize delivery to the winner of the drawing competition.

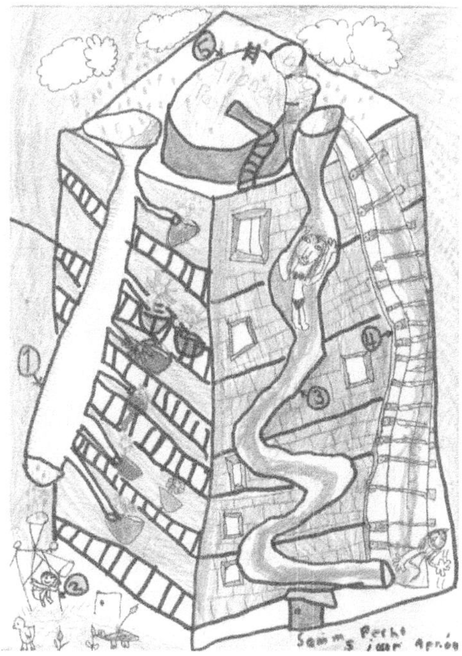

Figure 27: Winning drawing: water slide from roof to garden.

Water Service desk

Another factor for a successful communication with the citizens is the fact that we have an external 'Water Service Desk'. This organisation has a great expertise in communication and education and covers many tasks of the municipality. They are a front office e.g. with a website www.waterbewust.nl, but we use them also as a back office (e.g. developing a biking tour alongside disconnection projects, giving lectures). They also play an important role in the communication with citizens when we want to disconnect private houses together with disconnection of public space (Figure 28). Also in the participation projects for reconstruction of several ponds, they are the project leaders concerning communication. Together with them we make newsletters.

Figure 28: Front cover leaflet of the Water Service Desk.

Besides these activities, the 'Water Service Desk' is also used in runoff disconnection projects. Nijmegen carries out regularly projects in the public space like replacing broken sewers, renovation or reconstruction of road surface and reconstruction of the urban area, etc. When this takes place, disconnection of public and private runoff surface is a key element to all projects. When we disconnect at public space, we try to convince the private owners to disconnect their surface at the same time. For that task, we need the 'Water Service Desk' since they consult the public for the municipality. For all project we follow a strict planning scheme. Table 1 shows the required activities from preparation stage, through realisation to post-processing stage (Zuurman, 2004).

Table 1: Planning scheme for runoff disconnection of private area.

Planning stage	Activities
Preparation	1. Preliminary consultation of the project team 2. Public invitation, information and evening 3. Public invitation, information and evening 4. Inhabitant selection: gutter example 5. Site visit for the gutter construction 6. Construction of an example gutter 7. Letter and leaflet about common disconnection of runoff 8. Site visits of the remaining inhabitants 9. Letter of confirmation of the start of the project
Realisation	Possible letter if the project is delayed
Post-processing	1. Construction of the remaining gutters 2. Second round of visits to inhabitants 3. Second round of gutter construction 4. Letter thanking all participants

The planning schedule works very well and gives us results from 50-100% of disconnection of the private roofs. The scores are higher in streets with a lower social status!! We also carried out this communication strategy in the Voorstadslaan case. Despite the reactions related to the water nuisance, there were still people who did want to have their house roofs disconnected in the second round: that meant that there was still confidence that the municipality would solve the problems. The reason for this confidence was the fact that we kept on talking with the citizens and answering properly their telephone calls, mails and letters. We also organised an extra information evening to discuss the possible solutions (Zuurman, 2004).

The municipality takes this effort because we believe in our projects and we do not hesitate to tell the public when we make mistakes: when you communicate openly and realise the planned solutions, you gain public awareness and confidence about handling stormwater.

CONCLUSION ON THE APPLICATION OF THE ADSS: WHAT CAN BE LEARNED FROM ONE ANOTHER?

This section is very short and describes what, in view of the Nijmegen end-user, is the value of the developed Adaptive Decision Support System (ADSS) within the DayWater project.

From ADSS to Nijmegen

The scale range of application of ADSS may be very different from the scale of the city of Nijmegen (160,000 inhabitants): this means that Nijmegen municipality would probably never fully use all the tools and components of the ADSS. In practice Nijmegen services will use only elements of the ADSS like 'Urban Dynamics' (Chapter 6) and 'BMP database' (Chapter 8). Many other elements are considered to be more useful for specialists. Nevertheless the ADSS is a sort of framework and functions like a checklist or roadmap for the municipality of Nijmegen (Zuurman, 2005b).

From Nijmegen to ADSS

Nijmegen has a lot to offer to the DayWater project and its ADSS prototype. One of the strong points of ADSS is the 'Case study' section (Chapter 5). Theory is nice, but the real value consists of a mixture of practice and theory. When theory is tested and the results are recorded a lot is learned from it. Nijmegen municipality has at least three pilots with stormwater disconnection projects that could serve as case study in this section of the ADSS system (Zuurman, 2005b):

- Voorstadslaan Case study,
- Kronenburgpark Case study,
- Burchtstraat Case study.

As a format for the runoff disconnection project description, the following set-up is proposed:

(1) General description of system,
(2) Cross-sections and pictures,
(3) What did we learn from it?
(4) Constructions costs,
(5) Required maintenance & costs.

The main aim of DayWater ADSS is to provide supportive data that helps the user to think about what is required for an urban water system: most of these data are more or less technical, whereas in the realisation of projects, communication is the main element to achieve the goal. Nijmegen like the rest of the Netherlands is very crowded and the Dutch are there specialist in combining several functions in one area. In achieving this purpose extensive communication is required. Nijmegen municipality has a lot of experience in this field of communication. The last section of this Chapter provides an extended description of the communication strategy which is significant to the DayWater ADSS (Zuurman, 2004).

REFERENCES

CIW Commissie Integraal Waterbeheer (2001) Riooloverstorten - deel 2: eenduidige basisinspanning, nadere uitwerking van de definitie van de basisinspanning, Lelystad.

Geldof, G.D. and Straathof, J. (2003) Interactieve uitvoering geeft meer resultaat, Deventer, Tauw.

Geldof, G.D., Luiten, J., Gijzel en T. and van Beurden, E.A.E.M. (2003) Interactieve Uitvoering in het Waterplan Nijmegen, H2O, 15.

Gemeente Nijmegen, (2004) Evaluatie van het uitvoeringsprogramma van het Waterplan Nijmegen 2001, Nijmegen.

Hartman, E.C., Zuurman A.H.J., Drost, M, and Laarakker, M, (2002) Handreiking afkoppelen voor gemeenten. in H2O, 4.

Min V&W Ministerie van Verkeer en Waterstaat (1999) Vierde Nota Waterhuishouding - regeringsbeslissing, February 1999, Den Haag.

RIONED (2006) Wadi's: Aanbevelingen voor ontwerp, aanleg en beheer, 9, february 2006, Ede.

van Beurden, E.A.E.M. and Geldof, G.D.(2001) Waterplan Nijmegen, Nijmegen,

Witteveen en Bos (2005) Gemeentelijk Rioleringsplan 2005-2009, June 2005, Nijmegen.

Zuurman , A.H.J. and Hartman, E.C. (2001) Handreiking bij een keuze voor behandelmethoden na afkoppelen, Arnhem.

Zuurman, A.H.J. (1998) Drainage, infiltration & water re-use, Research on the feasibility of disconnecting impervious surfaces from sewer systems in urban areas, Hoofddorp.

Zuurman, A.H.J. (2004) Blue transformation of Nijmegen. Proceedings of Urban Water conference, Lille, September 2004.

Zuurman, A.H.J. (2005a) Disconnection devices in Nijmegen, Urban Water site visit, Nijmegen, May 2005.

Zuurman, A.H.J. (2005b) "Disconnection in Nijmegen - Factors of success & failure with implementation of source control measures", Proceedings of the final DayWater conference, Paris, 3-4 November 2005.

Relevant european websites:
http://www.irma-programme.org/
http://www.urban-water.org/cms/
http://www.daywater.org
http://www.sdfproject.nl/
http://www.freudeamfluss.nl

Chapter 18

Road runoff pollution load and sources in the City of Göteborg, Sweden: an example of SEWSYS application

David Jacobsson, Jonas German, Stefan Ahlman and Gilbert Svensson

INTRODUCTION

It is known that stormwater transports large quantities of toxic substances to receiving waters and that stormwater in some watersheds is the most important source of some substances (German *et al.*, 2005). In the U.S. stormwater runoff was the fourth most extensive cause of impairment of the water quality in rivers and the third most extensive source of water quality impairment of lakes (Novotny and Olem, 1994). German and Svensson (2002) investigated metal content in street sediments and concluded that substantial amounts of heavy metals were associated with sediments on streets and that copper was the metal that showed most elevated concentrations according to Swedish environmental quality guidelines. Runoff from streets and highways have been identified as an important source of pollutants and Marsalek *et al.* (1999) showed that highway runoff was more toxic than urban runoff from mixed and residential areas. However, the knowledge of the local conditions is not well described in many cities. The sources of pollutants in stormwater may vary depending on studied pollutant and catchment, but it is known that traffic is one of the more important sources of heavy metals and organic pollutants like polyaromatic hydrocarbons (PAHs).

This project was undertaken in order to increase the knowledge of the pollutant load from road runoff to different receiving waters in the city of Göteborg, Sweden. The stakeholders in this project are the Swedish Road Administration and the City of Göteborg. The City of Göteborg is represented by the Department of Sustainable Water and Waste Management, the Traffic and Public Transport Authority, the Water and Wastewater Works and the Environment and Health Administration. The Swedish Road Administration and the City of Göteborg have both high environmental ambitions. For example, the Road Administration builds sedimentation ponds along new highways and the city of Göteborg has built a sedimentation pond for cleaning road runoff. The city of Göteborg has also put a focus on pollution in surface runoff in the water plan for Göteborg (Vattenplanen). Emphasis must be made on polluted road runoff that can reach the surface water either by an accident or continuously. This is of course of extra concern when the surface water is used as a raw water resource, which is the case for the river Göta älv. There is also a strong political desire to get the Göta älv, which flows through Göteborg, suitable for swimming.

The results from this study will be used to support decisions regarding in which catchments/receiving waters it is most effective to introduce measures on road runoff, depending on load and ecological sensitivity of the receiving water. Since the quality of our receiving waters are our target, the decisions should be done regardless if it is a national or a municipal road.

METHOD

A dynamic substance flow model called SEWSYS has previously been developed for simulations of urban drainage systems (Ahlman and Svensson, 2002; Ahlman and Svensson 2005) (Chapter 14). Substance flow analysis is a method that can be used to form a description of what kinds of substances are present in a system, their sources, pathways and the sinks in which the substances finally reside. SEWSYS is used in this study to calculate the road runoff pollution load and the sources of pollution. GIS analysis is used to provide necessary input data for the model.

Earlier studies have been relying on measurements of highway runoff to characterise pollutant load and concentrations *e.g.* Stotz (1987) and Wu *et al.* (1998). Kayhanian *et al* (2005) used a Geographical Information System (GIS) based program to calculate annual loads from highway runoff using measured concentrations issued from a database and a rational method to calculate runoff. The accuracy of such a method will be dependent on the amount and quality of the measured data. This method does not give any information about the major pollution sources. In the case of SEWSYS individual pollution sources have been considered (Figure 3).

Calculations using GIS

Data on traffic volumes were collected from the local road authorities; data on catchment boundaries were delivered by the water and sewage works and digital maps with road structures from the department of spatial planning. In total 70 catchments were identified. Since some of the catchments have the same receiving water 34 catchments were studied, which can be seen in Figure 1. For each catchment the road area was calculated. The traffic load per square meter could then be calculated for different parts of the road and the traffic load could be allocated to a certain catchment. These calculations were done in MapInfo and were conducted by the consultant firm, GF konsult. One of the 34 investigated sub-catchments is presented in Figure 2.

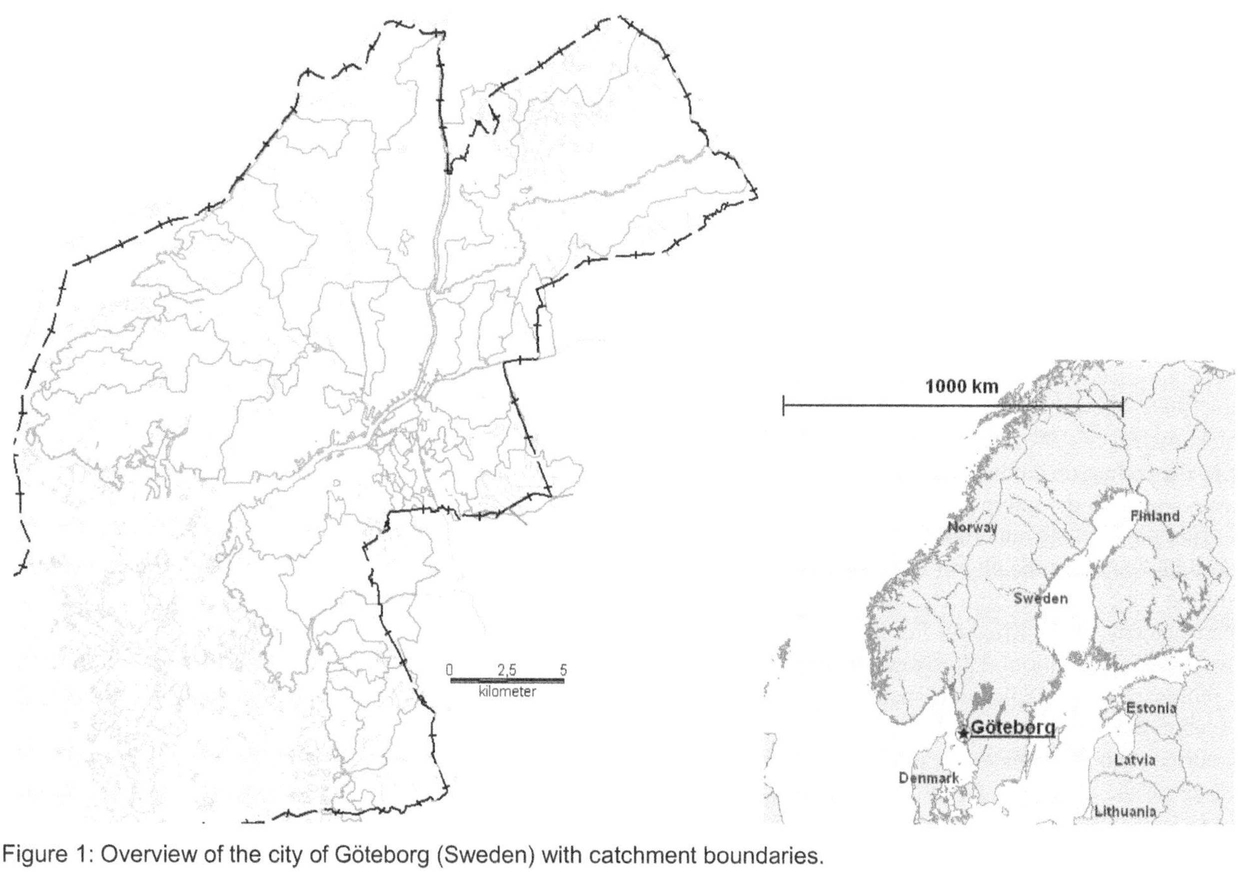

Figure 1: Overview of the city of Göteborg (Sweden) with catchment boundaries.

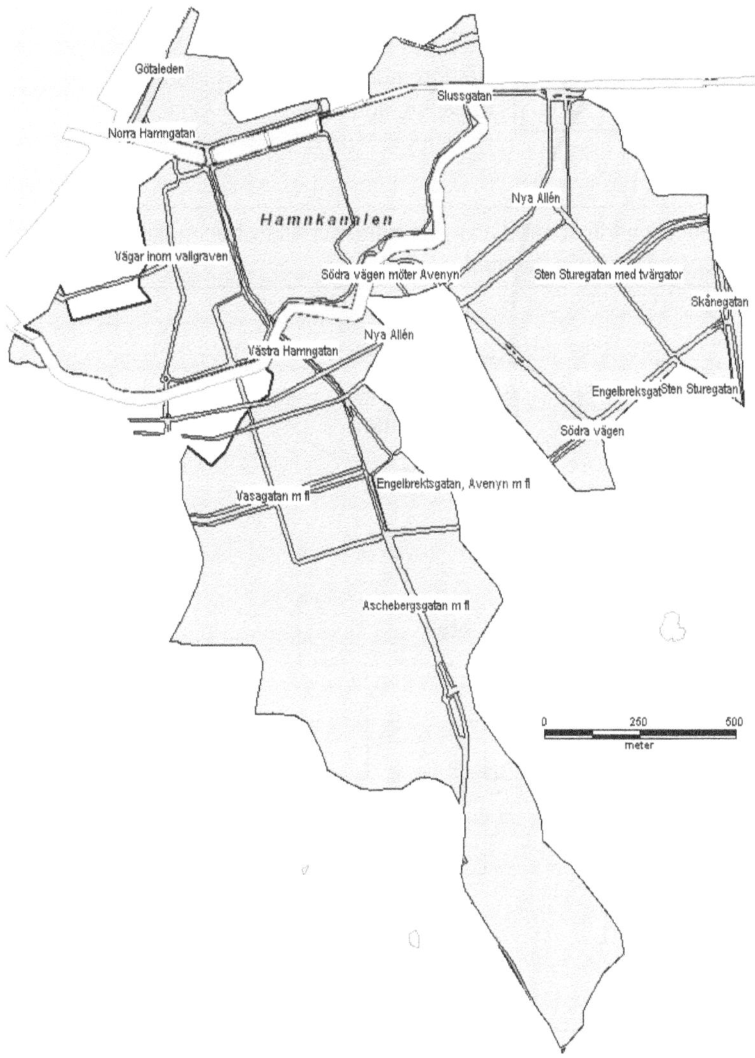

Figure 2: The sub-catchment of Hamnkanalen within the city of Göteborg (Figure 1).

Calculations with SEWSYS

The substances and indicators included in the SEWSYS model used in this study include total amounts of phosphorus, nitrogen, copper, zinc, lead, cadmium, polyaromatic hydrocarbons (PAH) and suspended solids (SS). Each pollutant production is linked to different source parameters such as corrosion rates, atmospheric deposition and other material release factors. The necessary catchment-specific input data for this study are road area and traffic load. Road runoff is not calculated. The results of the model are annual pollution loads and Site Mean Concentrations (SMC). A conceptual description of the generation of pollution and the pathways in SEWSYS is presented (Figure 3).

In order to limit the extent of this study, it was decided to only consider roads and streets that account for about 80% of the traffic load. The total road area studied here is 770 ha and the daily traffic loads on these roads are 8.3 million vehicle-kilometres of which 8% is heavy traffic (trucks, buses). 60% of the total traffic load is performed in sub-catchments presenting a separate sewer system and it is unknown how much of the traffic area that is drained by ditches. The galvanized area of street furniture is assumed to be 2.5% of the total road area: this assumption results from previous studies on the same sites (Ahlman, 2006).

The annual discharges from the 34 sub-catchments were studied for one selected representative year. A time series of successive rain events is used in the simulations with SEWSYS. The rain series (685 mm year^{-1}) relate to year 1926, and gives a statistically good description of a normal hydrological year in the Göteborg region.

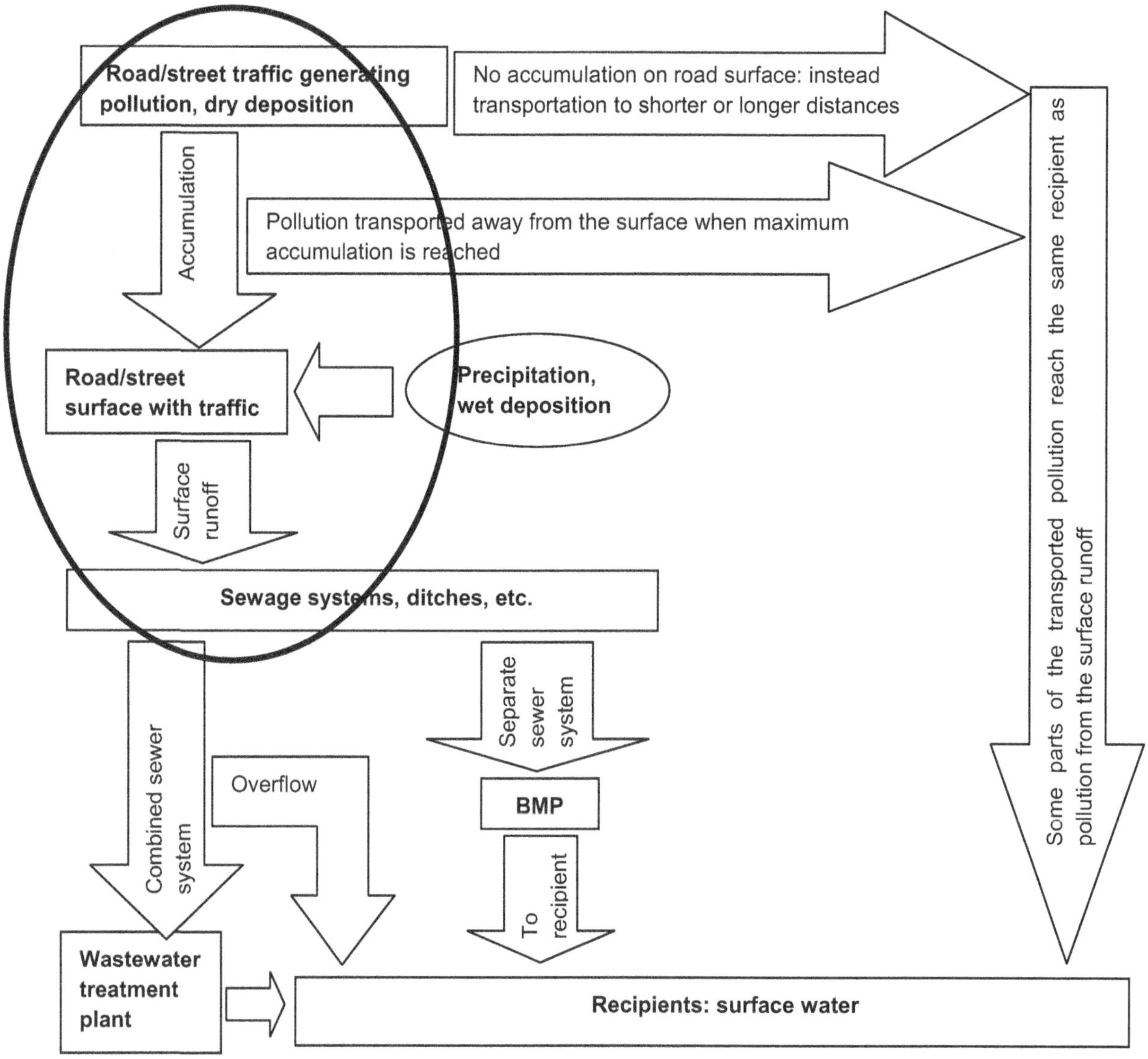

Figure 3: Conceptual description how pollutants are generated in traffic areas and transported to recipient. Processes inside the ellipsis have been modelled with SEWSYS software. Boxes indicate activities, while arrows represent substances flows. There are no correlations with the sizes of the arrows and the amount of pollutions transported. BMP, best management practise.

RESULTS AND DISCUSSION

The total annual loads of pollutants generated from the road traffic and calculated with SEWSYS are presented in Table 1. These annual generated amounts are 5 to 10 times higher than the amounts of pollutants that are transported by surface runoff. This difference results from two steps of SEWSYS calculations: (i) the first step reduces the amounts of pollutants that are available for accumulation to 70% of the generated amount, the other 30% being transported at longer or shorter distances as shown in Figure 3; (ii) the second step is a calculation of the maximum accumulation level for each pollutant: when such maximum accumulation level is reached SEWSYS generates its transport out of the system (Figure 3) (Ahlman, 2006). Because of these two steps in SEWSYS calculations, the annual amounts of pollutants in surface runoff generated from traffic in the city of Göteborg are 480, 1600, 50, 2, 500, 5600, 11 and 1.2 million kg copper, zinc, lead, cadmium, phosphorus, nitrogen, PAH and suspended solids respectively (Table 2). It is unclear how much of the pollutants generated from the traffic that reaches the recipient waters. Some of the pollutants generated from traffic are transported away and deposited on other surfaces such as green areas, parking lots and building. The contribution of pollutants from these surfaces is not simulated in this study.

Table 1: Total annual amount of pollutants generated from traffic, corrosion and dry atmospheric deposition. Wet atmospheric deposition (rain) is not included since it is not available for accumulation. Total generated amounts are the sum of total generated by traffic and by dry deposition, calculated by SEWSYS.

Source	Cu (kg)	Zn (kg)	Pb (kg)	Cd (kg)	P (kg)	N (kg)	PAH (kg)	SS (t)
Brake wear	2100	600	30	0	0	0	0	30
Road wear	160	350	100	1	0	0	30	5600
Tyre wear	100	5800	0	2	470	0	20	400
Exhaust	0	0	0	0	1800	0	0	0
Oil spill	0	40	0	0	0	0	0	0
Zinc corrosion	0	1200	0	0	0	0	0	0
Total generated by traffic	2360	7990	130	3	2270	0	50	6030
Dry atmospheric deposition	20	50	70	1	60	500	0	0
Total generated	2380	8040	200	4	2330	500	50	6030

The five sub-catchments presented in Table 2 accounts for two thirds of the annual pollution load from surface runoff issued from the traffic area. One exception is cadmium for which the five catchments receive 50% of the total pollution load. These annual amounts were compared to the pollutant release by the local wastewater treatment plant (WWTP) (Table 2): this WWTP is connected to 99% of the population of the city of Göteborg and 92% with the other five municipalities. As expected the WWTP is the major contributor of phosphorus and nitrogen to the receiving waters. For copper, the amount in surface runoff from traffic area is one third compared to the wastewater treatment plant effluent. For zinc, lead and SS, surface runoff from traffic area and the WWTP effluent are similar contributors to the receiving waters.

Table 2: SEWSYS assessment of surface annual runoff from traffic area in sub-catchments receiving the highest loads and in Göteborg, compared to the annual discharge from the wastewater treatment plant (Gryaab, 2005).

Sub-catchments/local rivers	Cu (kg)	Zn (kg)	Pb (kg)	Cd (kg)	P (kg)	N (kg)	PAH (kg)	SS (t)
Göta älv upstream Göteborg	66	225	6	0,20	69	650	1,5	170
Göta älv downstream Göteborg	86	290	8	0,28	90	960	1,9	210
Göta älv, upstream raw water inlet	61	219	6	0,20	65	730	1,4	160
Mölndalsån	60	183	5	0,14	60	370	1,3	150
Stora ån	50	157	5	0,16	52	530	1,1	118
Total runoff load in Göteborg	480	1600	50	2	500	5600	11	1200
WWTP effluent	1260	1400	57	40-90	$47*10^3$	$1255*10^3$	-	1100

As can be seen in Table 2, Göta älv River receives most of the road/street pollutants, but Göta älv is also one of the major rivers in Sweden, in terms of annual flow rate. Therefore the expected concentrations of pollutants from the traffic are lower in Göta älv River as compared to other receiving waters with fairly low flows. In Table 3, the local receiving waters with the highest expected concentrations are presented: these concentrations have been calculated by dividing the annual amounts of pollutants in the surface runoff for each of the sub-catchments by the annual flow for each of the receiving rivers; such concentrations of pollutants correspond to an annual average. The result shows that Stora ån is the catchment were the highest concentrations of pollutants generated from traffic can be expected.

Table 3: Annual flow in the receiving waters and expected concentrations of pollutants from traffic (kg/Mm^3=μg/l)

Sub-catchment/local rivers	Annual flow (Mm^3/year)	The proportion of surface runoff compared to the annual flow (%)	Cu (μg/l)	Zn (μg/l)	Pb (μg/l)	Cd (μg/l)	P (μg/l)	N (μg/l)	PAH (μg/l)	SS (mg/l)
Stora ån	6,5	6	8	24	0.7	0,02	8	80	0.16	18
Hamnkanalen	2,2	4	4	13	0.4	0,01	4	56	0.08	9
Osbäcken	2,8	3	3	9	0.3	0,01	3	38	0.06	6
Kvillebäcken	4,3	3	2	9	0.3	0,01	3	44	0.06	6
Göta älv downstream Göteborg	5450		0.02	0.05	0	0	0.02	0.2	0	39
Göta älv upstream Göteborg	5445		0.01	0.04	0	0	0.01	0.1	0	31

As shown in Table 1, SEWSYS allows the identification of the major sources of pollution in road runoff. The distribution between sources contributing to the annual road runoff load is presented in Figure 4. The results shows that brake linings is by far the most important source of copper and tyre wear accounts for 70% of the zinc load. Road wear and dry atmospheric deposition are the major sources of lead. Dry atmospheric deposition is the major source for cadmium and the only source of nitrogen. Road and tyre wears contributes to more than 90% of the total PAHs, whereas road wear is definitely the major source of SS.

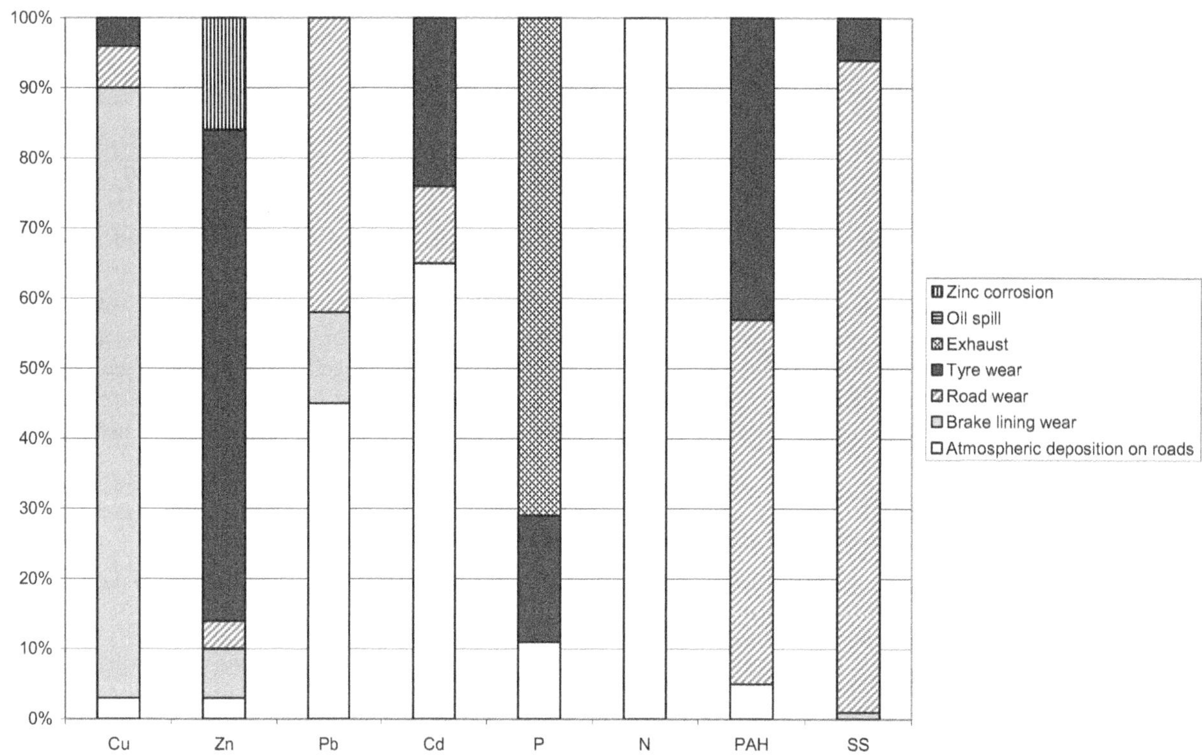

Figure 4: Relative contribution of different sources to the total pollutant load from traffic (%).

According to the used version of SEWSYS prototype, particulate phosphorous in the vehicle exhausts accounts for 70% of the total phosphorous load in road runoff. This result can be questioned since the addition of phosphorous as lubricate and anti oxidation agent has decreased since auto-catalysts were introduced. Instead phosphorous in road wear can be an important source (Kennedy and Gadd, 2003) which has not been included in this work. The importance of road wear as a source of phosphorous also depends on the ballast used in the asphalt.

In order to investigate which roads contribute to most of the pollution, a classification of roads has been done using five classes of the standardised traffic load (vehicle-km.day^{-1}.m^{-2}) (Table 4): within 20% of the total road area, 50% of the vehicle-kilometres are performed in Göteborg. As a result of this, a big proportion of the pollutants are generated from a surface area representing only 20% of the total traffic area. Since nitrogen originates only from atmospheric deposition (Figure 4) the proportions of nitrogen in the different road classes are the same as the traffic area percent in each class.

Table 4: Distribution of the annual traffic runoff loads issued from roads classified according to their traffic load per area.

Standardised traffic load (vehicle-km.day^{-1}.m^{-2})	Vehicle-km/day (%)	Traffic area (%)	Cu (%)	Zn (%)	Pb (%)	Cd (%)	P (%)	N (%)	PAH (%)	SS (%)
>1,75	50	20	40	40	30	30	40	20	40	40
1-1,75	20	20	20	20	20	20	20	20	20	20
0,75-1	15	15	15	15	15	15	15	15	15	15
0,5-0,75	10	20	10	10	15	15	10	20	10	10
0-0,5	5	25	10	10	20	20	10	25	10	10

CONCLUSIONS AND FUTURE WORK

The major conclusions from this ongoing project using SEWSYS model in the city of Göteborg are:

- brake linings are, by far, the most important sources of copper in road runoff, whereas tyre wear accounts for more than half of the zinc load, and road and tyre wear contributes with more than 90% of the total PAHs;
- Five catchments receive up to two thirds of the pollution load in the surface runoff issued from traffic area (roads, streets);
- the receiving waters presenting the highest annual mean concentrations of pollutants, resulting from the surface road runoff, are small rivers with low annual flow;
- for most of the heavy metals and for SS, road surface runoff is equal contributors to the receiving waters as the WWTP;
- 20% of the traffic area, i.e. road surface, generates 50% of the annual road runoff loads;
- it is unclear how much of the pollutants generated from the traffic reaches the receiving rivers, since up to 90% of the pollutants generated from traffic is transported away and deposited on other surfaces, such as green areas, parking lots and buildings; an unknown part of these pollutants will reach receiving waters.

Two separate methods have been used to identify where it is most effective to introduce pollution reduction measures for source control of road runoff. The first approach is to prevent pollution from road runoff to reach one of the five most ecological sensitive receiving rivers in the city of Göteborg. The second approach is to treat all the surface runoff issued from the roads with the highest traffic load. Two examples of pollution reduction measures will be investigated. in order to calculate their comprehensive cost: one example will be suitable for an urban location and one for a rural one. The comprehensive costs of protecting either of the five most ecological sensitive recipients or cleaning the road runoff from roads with the highest traffic load will then be presented to decisions makers.

In the part of the work that has been presented here, SEWSYS have been successfully used for modelling the pollutant load from traffic in an entire city; for the ongoing work further tools from the DayWater project will be used. The BMP catalogue will be used for gathering information on different source control measures (Chapter 8) and the costing tool will be used for estimating and comparing costs for different BMPs (Chapter 10).

REFERENCES

Ahlman, S. (2006) Modelling of Substance Flows in Urban Drainage Systems. PhD thesis, Dept. of Civil and Environmental Engineering, Chalmers University of Technology, Göteborg. ISBN 91-7291-759-8.

Ahlman, S. and Svensson, G. (2002) Modelling substance flows in urban sewer systems using MATLAB/Simulink. In: Proceedings of the Ninth International Conference on: Urban Drainage, E. W. Strecker & W. C. Huber (eds.), 8-13 September 2002, Portland, OR, USA.

Ahlman, S. and Svensson, G. (2005) SEWSYS – a model for simulation of substance flows in urban sewer systems. Urban Water report 2005:11. Available at: http://www.urbanwater.org/file/dyn/00000m/3000m/3065i/UWR%205-11-09-15.pdf (verified August 2007).

German, J., Vikström, M., Svensson, G. and Gustafsson, L.-G. (2005) Integrated stormwater strategies to reduce impact on receiving waters. In: Proceedings of the 10th International Conference on: Urban Drainage, Eva Eriksson, Hülya Genc-Fuhrman, Jes Vollertsen, Anna Ledin, Thorkild Hvitved-Jacobsen, Peter Steen Mikkelsen (eds.), 21-26 August 2005, Copenhagen, Denmark.

German, J. and Svensson, G. (2002) Metal content and particle size distribution of street sediments and street sweeping waste. Water Science and Technology, 46(6), 191-198.

Gryaab (2005) Miljörapport enligt miljöbalken Gryaab 2004. [Environmental report from the regional WWTP for 2004] Gryaab rapport 2005:2. Available at: http://www.gryaab.se/site/admin/actions/upload2/uploads/map6/Miljorapport_Gryaab_2004.pdf, 2005.10.07 [in Swedish] (verified September 2007).

Kayhanian, M., Ha, S. and Stenstrom, M. (2005) Constituents' annual load estimation from highways. In: Proceedings of the 10th International Conference on: Urban Drainage, Eva Eriksson, Hülya Genc-Fuhrman, Jes Vollertsen, Anna Ledin, Thorkild Hvitved-Jacobsen, Peter Steen Mikkelsen (eds.), 21-26 August 2005, Copenhagen, Denmark.

Kennedy, P. and Gadd, J., (2003) Preliminary Examination of Trace Elements in Tyres, Brake Pads and Road Bitumen in New Zealand. Available at: http://www.transport.govt.nz/assets/PDFs/stormwater-inorganic3.pdf (verified September 2007).

Marsalek, J., Rochfort, Q., Brownlee, B., Mayer, T. and Servos, M. (1999) An exploratory study of urban runoff toxicity. Water Science and Technology, 39(12), 33-39.

Novotny, V. and Olem, H. (1994) Water Quality: Prevention, Identification and Management of Diffuse Pollution. New York: Van Nostrand Reinhold. ISBN 0-442-00559-8

Stotz, G. (1987) Investigations of the properties of the surface water run-off from federal highways in FRG. The Science of the Total Environment, 59, 329-337

Wu, J.S., Allan, C.J., Saunders, W.L. and Evett, J. B. (1998) Characterization and pollutant loading estimation for highway runoff. Journal of Environmental Engineering. 122(7), 584-592

Chapter 19

Selection of priority pollutants for Stockholm city - Real life application of CHIAT

Anna Ledin, Eva Eriksson, Cajsa Wahlberg, Arne Jonsson, Anders Baun and Peter Steen Mikkelsen

INTRODUCTION

There are a lot if initiatives for reducing the load of pollutants from urban areas into the aquatic environment. A driving force is the implementation of the European Community Framework Directive (WFD; European Commission, 2000 and 2001), which aims at achieving a good water status in surface, ground and marine waters in 2015. Other driving forces are local goals such as establishing recreation areas in the cities for swimming and fishing, which in turns require improved water quality in most European cities. Among the tools available for the cities to achieve these goals is treatment of both municipal and industrial wastewater in wastewater treatment plants (WWTPs). However, even the most developed WWTPs fail to fully remove some of the heavy metals and some of the xenobiotic organic compounds (XOC). Strategies for handling of stormwater are also on the agendas in many cities. Generally stormwater is mixed with wastewater in the sewage system and treated in the WWTPs, or discharged without any treatment into surface waters. Different kinds of source control options, such as infiltration basins, settling ponds and oil separators are applied in some catchments as a complement to WWTPs or treatment for non-collected stormwater (Chapter 8). Furthermore, the present knowledge regarding the treatment efficiency from these initiatives with respect to heavy metals and XOC is poor. Many cities are now considering other options, besides treatment, for reducing the load of heavy metals and XOC into the receiving water bodies. Available alternatives are regulations and legislation regarding the use of specific compounds or products, voluntary initiatives such as use of products with eco-labels and education with the purpose to change human behaviour and thereby reduce the use of unwanted compounds (Chapter 8).

The heterogenic character of the XOC group as well as the large number of compounds in this group (hundreds of thousands of different types of XOC are present in our society; Ledin *et al.*, 2004) is a major barrier for local authorities to take action. Questions regularly asked are: "How are we going to choose the compounds that should be included in the next monitoring program?", "Which compounds should we optimise the new stormwater treatment facility for?", "Which compounds should we prioritise in our local regulations for what can be discharged into the separate stormwater system?", "What kind of activities should we focus on when we regulate which small industries are allowed to discharge to the WWTP or to the separate stormwater system?", etc.

A tool; **CHIAT** (**C**hemical **H**azard **I**dentification and **A**ssessment **T**ool) has been developed within the DayWater project, in order to supply municipalities and consultants with support when they seek to answer these questions (Chapter 11). The tool includes a number of chemical hazards that need to be taken into account when evaluating the

© 2008 IWA Publishing. *DayWater: an Adaptive Decision Support System for Urban Stormwater Management.* Edited by Daniel R. Thévenot. ISBN: 9781843391609. Published by IWA Publishing, London, UK.

impact from heavy metals and XOC in aquatic systems. These are both acute and long-term effects to living organisms e.g. toxicity, bioaccumulation, carcinogenicity, mutagenicity, reproduction hazards and endocrine disrupting effects, as well as the ability to promote allergic reactions in humans. Bad odour, frothing and colouring are examples of aesthetical problems that also are included (for more details see Chapter 11 in this book or Eriksson, *et al.*, 2005; Ledin *et al.*, 2005 and 2006).

The aim of this study was to select priority pollutants for the City of Stockholm to be applied in their vision to reduce the pollutant load to surface waters in the city of Stockholm (e.g. lake Mälaren and a number of smaller lakes) as well as the Stockholm archipelago. The selection of priority pollutants was performed by applying CHIAT with major focus on XOC.

Specific objectives and frames for the project

The present project was carried out in close collaboration between three partners; The City of Stockholm Environmental and Health Administration, Stockholm Water Co. and Institute of Environment & Resources, Technical University of Denmark (DTU). Initially the partners met and went carefully through all steps in CHIAT and discussed, based on that, the expectations to the project.

It was decided to focus on XOC, and not include heavy metals. The major reason for these priorities is that the City of Stockholm has already made some efforts regarding reduction of discharges of heavy metals (Stockholm's Environmental Programme, 2003) and the lack of a RICH procedure (step 3 'Hazard and problem identification' in Figure 1) for heavy metals.

It was also agreed that no 'Hazard assessment' (step 4 in Figure 1) should be carried out in this project. The City of Stockholm is aiming at selecting the most relevant pollutants to focus on in the coming 10-15 years. It can be assumed that the data needed for carrying out a hazard assessment for these pollutants is not available. This means that the work will be hampered by lack of data, and it was believed that this will bias the outcome from the project.

THE CHIAT METHODOLGY

The procedure applied in this study, CHIAT, is described in detail in Chapter 11 and consists of five steps: (1) source characterisation, (2) recipient, exposure object and criteria identification, (3) hazard and problem identification, (4) hazard assessment and (5) expert judgement (Figure 1) (Eriksson *et al.*, 2005; Ledin *et al.*, 2005 and 2006).

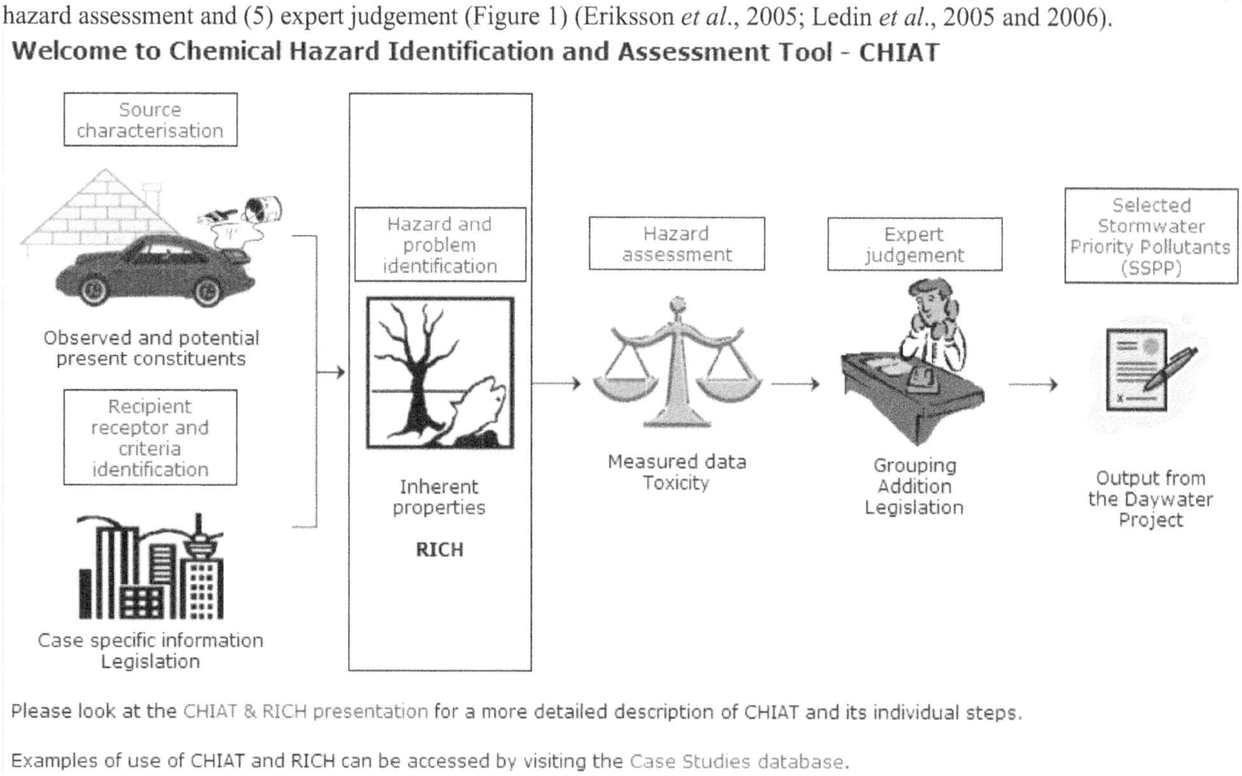

Figure 1: CHIAT: an approach for selecting priority pollutants based on chemical hazard identification and assessment (from Chapter 11).

Source characterisation

The first step aims at collecting a list of potential pollutant. Two literature reviews with somewhat different focuses can be used:

- Searching in the open international literature regarding observations/measurements of XOCs in the relevant types of wastewater: this is generating a list of confirmed pollutants;
- Searching in the open international literature regarding XOCs that potentially could be present in the relevant types of wastewater due to the use of chemical products such as household chemicals (e.g. shampoo, tooth paste, cleaning detergents and washing powders), pharmaceuticals, de-icing agents and pesticides in the gardens, and releases from materials such as construction and road materials, brakes and tires on vehicles: the aim is to identify those pollutants that have not yet been included in monitoring programs.

Lists generated in other projects can be used as a base for new projects (see CHIAT website or Ledin *et al.,* 2005). However, it should be kept in mind that some compounds may have to be added. It could be XOC that are used in this specific catchment like de-icing agents and pesticides which may be relevant for stormwater, or medicines relevant for mixed household wastewater. On the other hand, compounds not used in the catchment do not need to be included and can be removed.

The output from this step is a list with all observed and potentially present pollutants: the 'potential pollutants' (Figure 1).

Recipient, exposure object and criteria identification

The second step aims at framing the specific project/case study in focus. This step includes identification of those different strategies for handling of the selected type of waste- and/or stormwater that should be evaluated in the project. Of concern are the potential human health hazards, technical and aesthetical problems as well as environmental hazards that could occur due to the presence of the XOC in the wastewaters. Use, treatment or discharge scenarios should be investigated in order to identify the potential exposure routes and what or who are exposed (the exposure object). Legislation is reviewed to find e.g. relevant quality and emission standards. Examples of recipients of relevance are water, soil and technical installations. Humans as well as aquatic and terrestrial living organisms are examples of exposure objects. Exposure routes for humans are oral intake, inhalation of aerosols and skin contact. Continuous or pulse emissions need to be taken into account when evaluating discharges to the aquatic environment and applications to soil. The selection of criteria for the hazard assessment is also made in this step, based on the relevant exposure objects (see below). For more details see Chapter 11 or Ledin *et al.* (2005).

The output from this step is a list with 'recipients, exposure objects and criteria for hazard assessment'.

Hazard and problem identification

All identified potential pollutants (first step of the CHIAT procedure; Figure 1) are evaluated in the third step: the hazard and problem identification. The criteria for evaluation are based on environmental fate, short-term aquatic toxicity and long-term effects to living organisms.

The hazard and problem identification is performed according to a ranking methodology titled 'RICH' (Ranking and Identification of Chemical Hazards), which up to now has been developed and tested for XOC (for more details see Chapter 11 and Baun *et al.*, 2006). The methodology consists of a decision tree in which hazardous and problematic compounds are identified. To visualize the sorting of XOC the decision tree can be described as a funnel fitted with several filters. The filters have been set according to specified criteria based on sorption, volatility, persistence to biodegradation, potential for bioaccumulation and aquatic toxicity. There are also one on/off filter for technical/aesthetical problems and a long-term chronic effects-filter considering cancer, mutagenic and reproduction hazards, endocrine disruption effects and allergenic effects. The output is a classification of the compounds in two categories (green and red) depending on their priority as possible pollutants. 'Green' compounds are considered as non-priority pollutants, which mean that these compounds will be excluded from the continuing evaluation with CHIAT. 'Yellow' compounds are passed on to the next filter. These compounds may or may not be priority pollutants depending on the outcome of the following filtration. 'Red' compounds are considered as priority pollutants. The compounds that are identified as hazardous or problematic in this step are listed as 'potential priority pollutants'.

Hazard assessment

The fourth step, the hazard assessment aims at performing an effect-exposure analysis, in order to evaluate if there is an actual hazard under the present circumstances. The selected criteria for a hazard assessment are related to the exposure object. The exposure can be represented by predicted environmental concentrations (PEC), where the values can be based on measured data or model simulations. Evaluation of estimated concentrations for which unacceptable effects are not likely to occur can be done by estimation of predicted no effect concentrations (PNEC). The values can be found in the literature (databases and handbooks) or they can be estimated using toxicity data collected for step 3 applying recommendations by European Commission (2003). Comparison between the PEC and PNEC values are made in order to determine if the compound should be considered as hazardous for organisms in the environment. The pollutants for which PEC/PNEC ratios are above one (1) are classified as priority pollutants. A corresponding evaluation with respect to humans can be performed by e.g. using tolerable daily in-take (TDI) values, to retrieve effect-values.

The compounds that are identified to exceed the criteria selected for causing not wanted effects in this step are listed as 'priority pollutants'.

Expert judgement

The fifth step, the expert judgement, aims at making the final selection of priority pollutants. The expert meeting is not necessarily a single meeting but several meetings and the expert is not a single person but a group of experts with different backgrounds and professions. The selection performed by the experts in these meeting may e.g. aim at reducing the number of compounds due to financial limitations in the specific project. Compounds may be removed based on use patterns in the catchments or grouped based on similarity in structure and fate. In the latter case, an indicator compound may be chosen to represent the whole group. Compounds that are banned can also be removed unless certain reasons exist.

Legislation concerning limit values should be used in order to identify compounds that may need to be added to the list. Such legislation could be drinking water standards as well as environmental quality standards and emission limit values for watercourses, lakes or the sea, reviewed in step 2 (Figure 1). Compounds may also be added if they are priority pollutants present on national/international lists or 'special case' compounds. It should be noted that the selection of priority pollutants performed by the expert group is largely dependent on the purpose of the obtained list, which is illustrated by the results and discussion sections below. The output from this step is a list containing those chemical compounds and other parameters that constitute a hazard after evaluation by the experts: the 'selected priority pollutants'.

RESULTS

Source characterisation

The first step in the CHIAT procedure is to prepare a list of potential pollutants (Figure 1). Information regarding observed and measured XOC, e.g. the confirmed pollutants, was collected by searching in the open international literature. Special focus was on the presence of XOC in different types of wastewaters as well as at different locations in the urban water systems (e.g. outlet from WWTPs). The results showed that:

- 390 XOC have been observed in stormwater,
- 255 XOC have been observed in grey wastewater,
- 221 XOC have been observed in urine,
- 85 XOC have been observed in faeces,
- 410 XOC have been observed in untreated wastewater (WWTP influents),
- 192 XOC have been observed in WWTP sludge,
- 88 XOC have been observed in surface waters downstream from WWTPs.

The search for XOC that may be present, but has not been observed yet, was also carried out in open international literature, supplemented with databases on materials and use statistics. The search resulted in a total of 9,659 compounds, which can be introduced to different types of wastewater due to the use of products (food, household chemicals, personal care products, pharmaceuticals, de-icing agents, pesticides), releases from materials (furniture, clothes, electronic equipment, buildings, roads, vehicles), accidents (transport of goods by tracks including re-loading of goods) and atmospheric deposition. It should, however, be noted that this list can probably be extended if the search for XOC included in products used in the society is continued.

In conclusion, the results from these two literature studies showed that at least 11,000 XOC may be present in different types of wastewater, e.g. they are the potential pollutants.

Recipient, exposure object and criteria identification

Secondly, the recipients and the exposure objects were identified. In this specific case the recipients are all fresh surface water bodies in the City of Stockholm; lake Mälaren as well as a large number of smaller lakes, rivers and creeks. The brackish water in the Stockholm archipelago is also a recipient (Figure 2).

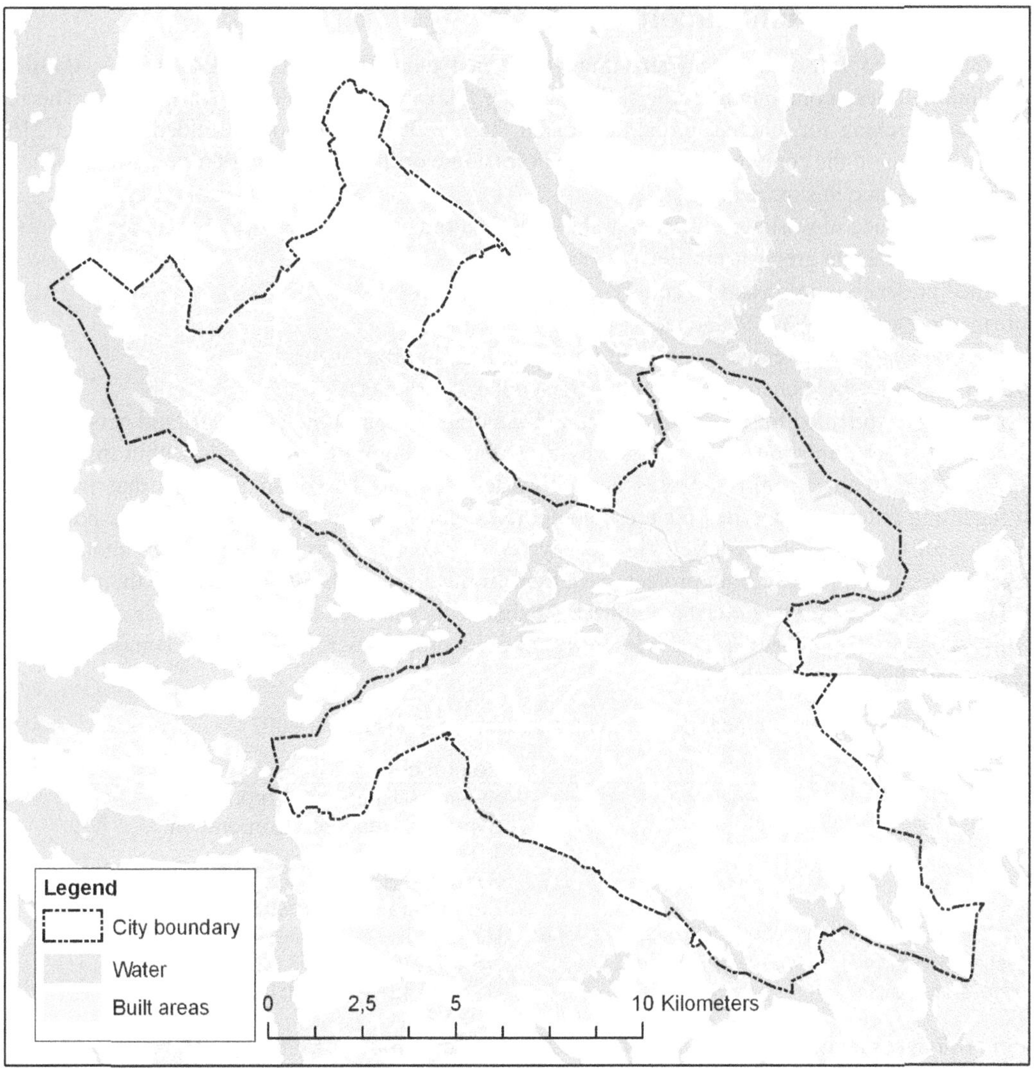

Figure 2: Map of the City of Stockholm and its suburbs (Stockholm City, 2006).

The exposure objects in focus are aquatic living organisms and humans. Aquatic living organisms were chosen due to the direct exposure caused by discharges of XOC to the aquatic environment. Humans were chosen as exposure objects since humans may be exposed due to e.g. swimming in the lakes and estuary, but also because all XOC receiving the recipients have been 'passing by' humans some time during their way from the source (products, construction materials, etc) to the recipient.

No criteria for hazard assessment were selected since step 4 in the CHIAT procedure was excluded in this project. The only identified relevant legislation was the WFD (European Commission, 2001). Furthermore, the following lists with priority pollutants were identified as relevant for this project:

- Swedish Chemicals Agency: 'PRIO' and 'Restricted Substances Database',
- The Chemical Products and Biotechnical Organisms Regulations (KIFS 1998:8),
- OSPAR,
- HELCOM,
- Danish guidelines (Statutory order no. 921, 1996).

The 'PRIO' database is a web-based tool intended to be used to preventively reduce risks to human health and the environment from chemicals. It classifies substances as Phase-out Substances (highest priority) or Risk Reduction Substances (lower priority) The 'Restricted Substances' Database contains information whether a substance or group of substances is restricted according to provisions laid down in regulations issued by the Swedish government or the Swedish Chemicals Agency (www.kemi.se).

Hazard and problem identification

The list generated in step 1 'Source characterisation' contained more than 11,000 XOC. It was, unfortunately, not possible to include all these compounds in the following step 'Hazard and problem identification'. The reason was a limited budget for searching for inherent properties for all these XOC. The partners decided to select 2150 XOC for further investigation, where the compounds should fulfil at least one of the following criteria:

- Confirmed presence in stormwater,
- Confirmed presence in wastewater, either at the inlet or at the outlet of wastewater treatment plants (WWTP),
- Confirmed presence in grey wastewater,
- Confirmed presence in urine and faeces,
- Confirmed presence in WWTP sludge,
- Construction materials potentially contributing to pollution in urban waters,
- Ingredients in household chemicals including personal care products.

The last criterion was fulfilled for too many compounds among the household chemicals and personal care products, with respect to the budget limitations, and that is why a random selection of compounds within this group was made. The selection of XOC with a confirmed presence in wastewater was based on the idea that it is easier to find information regarding inherent properties for these compounds, compared to those XOC that has not been looked for yet. The selection of not-yet-confirmed XOC in stormwater was based on collaboration with another project, where stormwater was in focus, and where both projects benefited from selecting the same XOC for their search for inherent properties. Finally, XOC that are ingredients in household chemicals were selected since there is a lot of attention in the society on this category of compounds. Data regarding inherent properties for these 2150 compounds were collected from a large number of databases (Table 1).

Table 1: List of databases used for compiling inherent properties (verified August 2007).

Database	Reference
Biodeg	Syracuse Research Corporation
Biolog	Syracuse Research Corporation
CCRIS	TOXNET
Chemfate	Syracuse Research Corporation
Chemfinder	CambridgeSoft Corperation
ECOTOX	US EPA
EPISuite (estimation tool)	US EPA (EPIWIN v3.12)
EPISuite (reference data)	US EPA (EPIWIN v3.12)
GENE-TOX	TOXNET
Handbook of environmental Data on Organic Chemicals	Verscheuren K, (1996)
Handbuch Umweltschemikalien	Rippen G (2003)
Hazardous Substances DataBank (HSDB)	TOXNET
IUCLID CD-ROM	European Commission and European Chemicals Bureau
IUCLID on-line	European Chemicals Bureau
KOW	Syracuse Research Corporation
List of unwanted compounds	Danish EPA
N-Class v6.0	Nordic Council of Ministers in collaboration with European Chemicals Bureau,
NIST	NIST Chemistry WebBook
NOVA 2003 datablade	Danish National Environmental Research Institute
NTP	National Toxicology Program
OSPAR	OSPAR Commision for the protection of the Marine Environment of the North-East Atlantic
PAN	PAN Pesticides Database
PhysProp	Syracuse Research Corporation,
SciFinder	Via Dansk Teknisk Videncenter http://www.dtv.dk

It was possible to collect information regarding inherent properties for 1200 (56%) of the 2150 potential pollutants. The results showed that 591 XOC are potential priority pollutants in either water or soil, while 79 turned out to be 'yellow' after the fourth filter (Table 2).

Table 2: Output from step 3 Hazard and problem identification, i.e. the output from RICH.

	Number of XOC
Potential pollutants (from step 1)	2150
Lack of data	950
'Red'	591
'Yellow'	79
Green'	530

Some of the compounds were classified as hazardous due to their slow degradation in the aquatic environment in combination with the ability of bioaccumulate (Figure 3). Others got this classification due to their toxicity in combination with their persistence. The third category contains compounds classified as hazardous due to their long-term toxic effects such as endocrine disrupting effects.

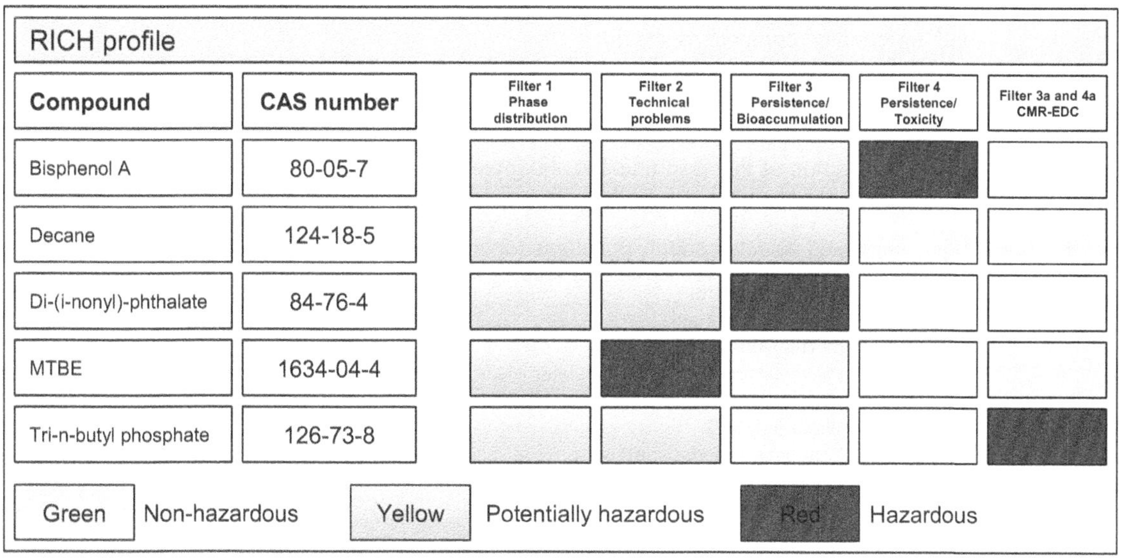

Figure 3: Illustration of the output from RICH for some selected XOC.

Expert judgement

The final step in the CHIAT procedure was carried out at two meetings where representatives from the three partners, as well as invited experts from other universities participated.

At the first meeting, some background information regarding the project was presented. The results from step 3 'Hazard and problem identification', i.e. the output from RICH were presented and discussed, together with a short introduction to the CHIAT procedure. Specific wishes from the City of Stockholm with respect to the expected number of compounds on the final list with selected priority pollutants, was discussed and it was agreed to prepare four lists:

- One list with 'new' highly prioritised selected pollutants,
- One list with 'old' highly prioritised selected pollutants,
- One list with 'new' prioritised selected pollutants, so called 'up-coming' compounds,
- One list with 'old' prioritised selected pollutants, so called 'not yet solved' compounds.

The meeting continued with a discussion regarding criteria for the final selection of priority pollutants. The group agreed to include both water and solid phase compounds in the final selection. They also agreed to exclude those compounds that had been 'yellow' after filter 4, since it was believed that the 'red' compounds constituted a larger potential risk.

The group with experts was divided into smaller groups with 5-6 persons in each group. There was one researcher from DTU in each group. The group received the list with potential priority pollutants from step 3, the so called 'red' compounds. The list contained also information about in which 'filter' these compounds got the classification 'red' (Figure 3), as well as a note regarding in which type of wastewater the compound would be expected to be found. For example: "Compound XX was classified as a water phase compound in filter 1, and was classified as 'red' in filter 3 "bioaccumulation-persistence; XX has been observed in stormwater." The groups realised during these discussions that they needed more information in order to be able to make the final selection. It turned out that information regarding the used quantity of the compound in different kinds of products or the quantity released from materials would be needed. This information was compiled before the second meeting. The list with 'red' compounds was also supplemented with information regarding presence in legislation and priority lists according to the second step 'Identification of recipients, exposure objects and criteria'.

At the second meeting, the list with 'red' compounds contained the following information:

- Compounds name and CAS number,
- 'Filter' causing 'red' in the RICH evaluation,
- Used volume in EU and/or Sweden,
- Presence in WFD,
- Presence on priority lists and databases according to the selection made in CHIAT Step 2,
- Quantity observed in surface water and/or sediment in the Stockholm area.

The final selection of selected priority pollutants for the four lists defined above were performed according to the following: the list with 'new' highly prioritised selected pollutants as well as the list with 'new' prioritised selected pollutants, the so called 'up-coming' compounds should:

- Include High Production Volume Chemicals (HPVC) in EU or Sweden,
- Not include compounds that are on the WFD–list or are included in the Swedish Chemical Inspectorate´s PRIO database.

The group finally made a selection of the most relevant and interesting ones, based on personal knowledge, opinions and preferences, for the short (Table 3) and the long (Table 4) list.

Table 3: List of new highly prioritised selected priority pollutants (short list).

Selected Priority Pollutants
2-Ethylhexyl diphenyl phosphate
5-Chloro-2-methyl-4-isothiazolin-3-one
BHT (2,6-di-tert-butyl-4-methylphenol, Butylated Hydroxytoluene)
DecaBDE (BDE 209)
Di-(i-nonyl)-phthalate (DiNP)
HBCD (hexa-bromo-cyclo-dodecane)
N-metylpyrrolidon
PFOS/PFOA
Tri(2-ethylhexyl)phosphate
Triclosan
Tri-n-butyl phosphate
Triphenyl phosphate

Table 4: 'New' prioritised selected pollutants, so called 'up-coming' compounds long list).

Selected Priority Pollutants
17-beta-Estradiol
1-Dodecanamine, N,N-dimethyl-
1-Nitronaphtalene
2,4,6-Tribromophenol
2,4-Dichlorophenol
2,6-Di-t-butylphenol
2-Ethylhexyl 4-methoxycinnamate
2-Methyl-4-chlorophenol
2-Phenoxyethanol
4-Chlorophenol
4-Nitrophenol
9,10-anthracendion (anthraquinon)
Acetyl hexamethyl tetralin
Benzophenone-12
Benzyl paraben
Benzylbutylphthalate
Beta-Citronellol
Bis(2-chloroethoxy) methane
Bis(2-ethylhexyl) ester decanedioic acid
Chlorhexidine
Cyclohexylthiophthalimid

DCBS (N,N-dicyclohexylbenzothiazole-2-sulfenamide)
Deca BD ethane (Deca-bromo-di-phenyl-ethane)
Di(2-ethylhexyl) adipate
Dihydromyrcenol (2,6-dimethyl-7-octen-2-ol)
Dimethoate
Dimethylamine
Ethyl methane sulfonate
Ethynyl estradiol
Fluoroquinolones (e.g. cipofloxacin, norfloxacin)
Hexamidine
Hexylcinnemaldehyde
HHCB (Galaxolide)
Isofenpos
Isophoron (3,5,5–Trimethyl–2–cyclohexen–1–one)
Isopropyl Myristate
Malathion
Metolachlor
Mono 2-ethylhexyl phthalate
Sorbitan oleate
Tert-butyl phenol
Tetrabromobisphenol A (TBBPA)
Tonalid (AHTN)
Triclocarban
Zinkpyrithion

The list with 'old highly prioritised selected pollutants':

- Should include High Production Volume Chemicals (HPVC) in EU or Sweden,
- Should also include compounds that are on the WFD–list or are included in the Swedish Chemical Inspectorate´s PRIO database as Phase-out Substances,
- Should not include banned compounds or other compounds that of other reasons are not used in Stockholm,
- Was further reduced by selecting representatives for groups of compounds with similar chemical and physical properties.

The group selected, after some discussion, 10 compounds according to these criteria and once again based on personal knowledge, opinions and preferences (Table 5).

Table 5: List of old highly prioritised selected priority pollutants.

Selected Priority Pollutants
4-Chloro-3-methyl-phenol (p-Chloro-m-cresol)
Bisphenol A (4,4-isopropylidene diphenol)
Carbendazim
Chlorothalanil
Diclofluanid
Diphenylamine
Glutaraldehyde (1,5-Pentandial)
Kloramin T (Benzenesulfonamide, N-chloro-4-methyl-, sodium salt)
Limonene (Cyclohexene, 1-methyl-4-(1-methylethenyl)-, (R)-)
Stearyl; N,N-Dimethyl-N-octadecyl-1-octadecanaminium, Chloride

The list with "old" prioritised selected pollutants, so called 'not yet solved' compounds.

- Should include High Production Volume Chemicals (HPVC) in EU or Sweden,
- Should also include compounds that are included in the Swedish Chemical Inspectorate´s PRIO database as Priority Risk Reduction Substances, but not as Phase-out Substances,
- Should not include banned compounds or other compounds that of other reasons are not used in Stockholm,
- Was further reduced by selecting representatives for groups of compounds with similar chemical and physical properties
- Was also reduced based on use patterns, e.g. compounds that are exposed to humans during their use.

The group selected, after some discussion, 29 compounds according to these criteria and once again personal knowledge, opinions and preferences (Table 6).

Table 6: 'Old' prioritised selected pollutants, so called 'not yet solved' compounds.

Selected Priority Pollutants
1,2-Ethandiamine
2-Chloroethanol
2-Propenoic acid, 1,6-Hexanediyl ester
4-Chloro-3-methyl-phenol (p-Chloro-m-cresol)
Acetochlor
Aclonifen
Acrolein
Bentazone
Bisphenol-A-diglycidylether
Cyclododecane
Diazinon
Diethyltriamine
Fenthion
Glyoxal
HBCD (hexabromocyclododecane)
Iprodione
Linuron
MBT, 2(3H)Benzothiazolethione
MBTS (2,2´-dibensotiasyldisulfide)
Metamitron
Methabenzthiazuron
Methylene diphenyl diisocyanate/Diphenyl-4,4´-methandiisocyanate
Methylmethacrylat
Parathion
Tetrabutyl tin
Tetraetylenpentamin
Thiram (TMTD)
Triethylenetetramine
ZDMC (Naftocit Di 4) Ziram

DISCUSSION

A relatively large number of potential pollutants was identified in the first step of the CHIAT procedure (11,000). This is probably only a fraction of the total number of compounds that could be released to the aquatic environment due to activities in urban areas, since several estimates have been made suggesting that the number of anthropogenic organic compounds exceeds 100,000. The work in this study also clearly shows that the number of observed compounds (1449) is just a very small fraction of those compounds that could be present. These figures illustrates that approaches like the CHIAT tool are needed in order to help researchers, authorities, etc to select the most relevant compounds to focus on.

The need for selecting of 2150 XOC out of in total 11,000 for further evaluation with RICH is limiting the objectivity aimed at by applying CHIAT. It was selected in the present study to focus on XOC with confirmed presence in wastewater and stormwater, as well as on two categories of the not-yet-confirmed XOC; stormwater and households chemicals. Unfortunately, these kinds of priorities have to be made, due to budget limitations, and this is of course influencing the final output. However, by applying CHIAT the decisions and selections made are documented and can easily be found by the user of the list with selected priority pollutants. Furthermore, the database with inherent properties increases when projects are accomplished and will with time include more and more compounds.

The work carried out in the third step of the CHIAT procedure shows that one of the major limitations with this approach is the need for relatively large amounts of data related to the potential pollutants inherent properties. It is time consuming to find the data in databases and other scientific literature and in many cases are data lacking. It should, however be noted that one of the criteria for the order of the 'filters' in the RICH procedure was to reduce this problem (Baun, et al 2006), with a need for relatively easily accessible data in the first filter (volatilisation, sorption, water solubility). It should, however, be noted that the data compiled up to now is kept in a database and can be used in coming studies.

The hazard assessment (step 4) was not included, since it was assumed that the environmental concentrations needed for carrying out a hazard assessment, are not available. This means that the work would have required that only XOC that have been quantified in aquatic systems and where the data can be extrapolated to the waters in Stockholm could have been included on the final list. This request was not in consistence with the aim of identifying the most relevant compounds to focus on the next 10-15 years.

Some of the members participating to the expert judgement meetings mentioned during the evaluation of the meetings that they had felt frustrated during this part of the study. The frustration was probably related to the shift from a scientifically based evaluation of the potential pollutants in step 3 to a more subjective based selection in step 5. It was realised by the working group that this shift has to be discussed at the beginning of the expert judgement meetings, in order to reduce the frustration and thereby make sure that the experts really express their opinions during the discussions and selection procedure. The frustration could also be related to the aim of the lists, which is challenging. The frustration has not been observed to this extent in other projects where CHIAT has been used and where the aim e.g. has been to make a monitoring program.

CONCLUSIONS

The following conclusions can be drawn from the present study:

- The CHIAT tool in combination with the RICH-procedure are two very useful tools for selection of priority pollutants.
- The selection of priority pollutants can to a large extent be performed based on scientific knowledge. However, the final selection will be based on more or less subjective opinions.
- The major limitations to this approach are the relatively large amount of data regarding inherent properties that is needed, the difficulties to find data and finally the lack of data for some compounds.

FUTURE RESEARCH NEEDS

The following topics have been identified as relevant in order to continue that work initiated in this project concerning pollutants in the City of Stockholm:

- The database compiled in the first step containing 11,000 potential pollutants need to be extended with more information regarding potential sources, release/emission factors, production and use statistics, etc.
- Evaluation of more compounds according to the 'RICH' procedure needs to be conducted: 2150 compounds were selected out of a total of 11,000. Of these 2150, data were found for 1200 compounds. It is therefore suggested that a project with focus on evaluation of more of these almost 9,000 compounds is initiated.
- Development of tools for carrying out step 4, Hazard assessment, i.e. exposure and effects analysis, appears important.
- Substance flow analysis is necessary in order to generate data for input in the hazard assessment.
- Development of a Web page (Stockholm's Environmental barometer) presenting the results obtained in this study, with possibilities for adding more results when other similar studies have been carried out.

REFERENCES

Baun, A., Eriksson, E., Aabling, T., Ledin, A. and Mikkelsen, P.S. (2006) A Methodology for Ranking and Hazard Identification of Xenobiotic Organic Compounds in Urban Stormwater. Science of the Total Environment, 370(1), 29-38.

CambridgeSoft Corperation (2005) Chemfinder.com database. 100 CambridgePark Drive, Cambridge, MA 02140 USA. http://chemfinder.cambridgesoft.com (verified August 2007)

CCRIS (2005) Chemical Carcinogenesis Research Information System (CCRIS) via TOXNET, http://toxnet.nlm.nih.gov/ (verified August 2007)

CHIAT website: http://CHIAT.er.dtu.dk (verified August 2007).

Danish EPA (2005) The list of unwanted compounds, http://www.mst.dk/Kemikalier/Stoflister+og+databaser/Listen+over+farlige+stoffer/ (in Danish) (verified August 2007)

Danish National Environmental Research Institute (2005) NOVA 2003 datablade, (in Danish)

Danish Statutory order no. 921 (1996). Statutory Order No. 921 of 8 October 1996 on quality standards for surface waters and emission standards for certain dangerous substances discharged to watercourses, lakes or the sea (replaced by no. 1669 of 2006) http://www.retsinfo.dk (in Danish) (verified August 2007)

Eriksson, E., Baun, A., Mikkelsen, P.S., and Ledin, A. (2005) Chemical hazard identification and assessment tool for evaluation of stormwater priority pollutants. Water Science and Technology, 51(2) 47-55.

European Commission (2000) Water Framework Directive, 2000/60/EC.

European Commission (2001) Establishing the list of priority substances in the field of water policy and amending Directive 2000/60/EC. Available: http://europa.eu.int/eur-lex/pri/en/oj/dat/2001/l_331/l_33120011215en00010005.pdf (verified August 2007).

European Commission (2003) Technical Guidance Document in support of Commission Directive 93/67/EEC on Risk Assessment for new notified substances, Commission Regulation (EC) No 1488/94 on Risk Assessment for existing substances and Directive 98/8/EC of the European Parliament and of the Council concerning the placing of biocidal products on the market. Bruxelles, Belgium

GENE-TOX (2005) Genetic Toxicology Data Bank (GENE-TOX) via TOXNET, http://www.toxnet.nlm.nih.gov (verified August 2007)
HELCOM (2005) Helsinki Commission Baltic Marine Environment Commission, List of valid HELCOM Recommendations. http://www.helcom.fi/Recommendations/en_GB/valid/ (verfied August 2007)
HSDB (2007) Hazardous Substances Data Bank (HSDB) via TOXNET, http://www.toxnet.nlm.nih.gov (verified August 2007)
IUCLID CD-ROM (2000) European Commission, Joint research centre. Institute for health and consumer protection. European Chemicals Bureau. 2nd edition.
IUCLID on-line (2005) On-line version of International Uniform Chemical Information Data Base on High Production Volume Chemicals http://ecb.jrc.it/esis/index.php?PGM=dat (verified August 2007)
Ledin A., Auffarth K., Boe-Hansen R., Eriksson E., Albrechtsen H.-J., Baun A. & Mikkelsen P.S. (2004) Use of rainwater collected from non-permeable surfaces - Identification of parameters relevant for a monitoring programme. Miljøstyrelsen, København. Økologisk Byfornyelse og Spildevandsrensning, 48 2004. pp. 1-120. (in Danish) http://www.er.dtu.dk/publications/fulltext/2004/MR2002-072.pdf (verified August 2007).
Ledin, A., Aabling, T., Baun, A., Eriksson, E. and Mikkelsen, P.S (2005) CHIAT - Chemical Hazard Identification and Assessment Tool. En metodik för utvärdering av kemiska risker i samband med hantering av dag- óch avloppsvatten. VA-Forsk, Svenskt Vatten AB, Stockholm. VA-Forsk rapport, 2005-09.(in Swedish) http://www.dataforlaget.net/~vav/filer/VA-Forsk_2005-09.pdf (verified August 2007).
Ledin, A., Eriksson, E., Mikkelsen, P.S. and Baun, A. (2006) CHIAT: Chemical hazard identification and assessment tool for selection of priority pollutants. In: Strategic Planning of Sustainable Urban Water Systems, Malmqvist, P-A., Heinicke, G. , Kärrman, E. Stenström, T.A. and Svensson, G. IWA Publishing, London, UK.
Mikkelsen, P.S., Baun, A. and Ledin, A. (2001) Risk assessment of stormwater contaminants following discharge to soil, groundwater or surface water. In: Advances in Urban Stormwater and Agricultural Runoff Source Controls, J. Marsalek et al. (eds.), 69-80. Kluwer Academic Publishers.
National Toxicology Program (2005) Study results and research projects, http://ntp-server.niehs.nih.gov (verified August 2007)
NIST Chemistry WebBook (2003) NIST Standard Reference Database Number 69 - March, 2003 Release http://webbook.nist.gov/chemistry/ (verified August 2007)
Nordic Council of Ministers in collaboration with European Chemicals Bureau (2005) The N-CLASS Database on Environmental Hazard Classification version 6.0, http://apps.kemi.se/nclass/default.asp (verified August 2007)
OSPAR Comission for the protection of the marine environment of the northeast Atlantic (2005), Hazardous substances strategy. http://www.ospar.org/eng/html/welcome.html (verified August 2007)
PAN Pesticides Database (2005) Chemical search, http://www.pesticideinfo.org (verified August 2007)
Rippen G (2003) Handbuch Umweltschemikalien 5. Auflage. Ecomed Verlagsgeschellschaft, Landberg/Lech. Germany (in German)
SciFinder (2005) Accessed via Dansk Teknisk Videncenter, http://www.dtv.dk (verified August 2007).
Stockholm City (2006) Stockholm City online maps (in Swedish) available http://www.map.stockholm.se/kartago/ (verified September 2007)
Stockholm's Environmental barometer (2007) http://www.miljobarometern.stockholm.se/ (predominantly in Swedish)
Stockholm´s Environmental Programme (2003) ISBN 91 88018 93 8.
Swedish Chemicals Inspectorate (2005) PRIO a tool for risk reduction of chemicals. http://www.kemi.se/templates/PRIOEngframes____4144.aspx (verified August 2007)
Swedish Statutory Order no 1998:8 (1998) (the Chemical Products and Biotechnical Organisms Regulations, KIFS 1998:8), http://www.kemi.se/upload/Forfattningar/KIFS/K98_8.pdf (verified August 2007)
Syracuse Research Corporation (2005) BIODEG database, http://www.syrres.com/esc/biodeg.htm (verified August 2007)
Syracuse Research Corporation (2005) BIOLOG, http://www.syrres.com/esc/biolog.htm (verified August 2007)
Syracuse Research Corporation (2005) CHEMFATE, http://www.syrres.com/esc/chemfate.htm (verified August 2007)
Syracuse Research Corporation (2005) Interactive LogKow (KowWin) Demo, http://www.syrres.com/esc/kowdemo.htm (verified August 2007)
Syracuse Research Corporation (2005) Interactive PhysProp Database Demo, http://www.syrres.com/esc/physdemo.htm (verified August 2007)
US EPA (2000) EPISuite (estimation tool) (EPIWIN v3.12) http://www.epa.gov/opptintr/exposure/pubs/episuitedl.htm (verfied August 2007)
US EPA (2000) EPISuite (reference data) (EPIWIN v3.12) http://www.epa.gov/opptintr/exposure/pubs/episuitedl.htm (verified August 2007)
US EPA (2005) ECOTOX updated Oct. 2005, http://www.epa.gov/ecotox/ (verified an updated version (4.0) August 2007)
Verscheuren K, (1996) Handbook of environmental Data on Organic Chemicals, 3rd Edition, van Nostrand Reinhold co. Dordrecht, Holland.

Part 4

Related research in stormwater management

Chapter 20

Application of a battery of biotests for toxicity characterization of urban stormwater

Anders Baun, Martin Seidl, Lian Scholes, Gudrun Aldheimer, Eva Eriksson, D. Michael Revitt and Jean-Marie Mouchel

INTRODUCTION

The environmental impacts of urban stormwater and road runoff have gained increasing interest during the last decade (e.g. Maltby *et al.*, 1995a,b; Marsalek *et al.*, 1999), an interest reinforced by the implementation of the EU Water Framework Directive with its clear statements on the need to tackle non-point sources of pollution (EU, 2000) (Chapters 12 and 24). Although several studies dealing with toxicity assessment of urban wet-weather discharges have been published (reviewed by Marsalek *et al.*, 1999), the toxic impact of urban runoff is not well described in the literature. Wet-weather discharges can contain a complex combination of environmental pollutants and hence a chemical characterisation may not suffice in explaining toxic potential (Ellis, 2000). As an alternative monitoring strategy, the use of a battery of toxicity tests has been recommended as a more appropriate tool for screening and assessing the potential receiving water impacts of stormwater discharges (Marsalek *et al.*, 1999; Christensen *et al.*, 2006), however, only a few such studies have been carried out. It is within this context that the current study, conducted within the DayWater project (Chapter 1), aims to assess the order of magnitude of toxic effects of chemical constituents in water samples on a variety of biota (rather than specifically determine *in situ* toxicity) to enable the applicability of biotests as indicators for toxicity and hazard ranking.

Runoff samples were collected following rainfall events at four separate urban sites: Nantes in France, Wuppertal in Germany and Luleå and Stockholm in Sweden. Each sample was subjected to a battery of biotests involving bacteria, algae and macroinvertebrates as whole samples as well as additionally being tested following a simplified Toxicity Identification Evaluation (TIE) procedure to investigate the contribution of metals and organic chemicals to total toxicity. In the current chapter, results from water samples collected at the Stockholm site are presented as an example of the data obtained and discussions focus on the application of these results in vulnerability assessment. The results from water samples collected at all four sites have been published in Scholes *et al.* (2007).

MATERIALS AND METHODS

Field site, sampling and chemical characterization

Runoff samples were collected at the inlet to the SORBUS stormwater treatment facility located beneath the elevated motorway Essingeleden on the southwest side of Lilla Essingen island (Stockholm, Sweden) (Figure 1). The SORBUS treatment facility consists of a combination of a sedimentation pond and filters and receives runoff from the motorway, motorway bridge and nearby streets, representing a total catchment area of 15,000 m^2 with asphalt and roof tiles acting as the main contributing surface material types. Following treatment, the flow is discharged into Lake Mälaren (Vägverket, 2006). Details of the catchments, sub-divided into the urban and motorway areas, are presented in Table 1.

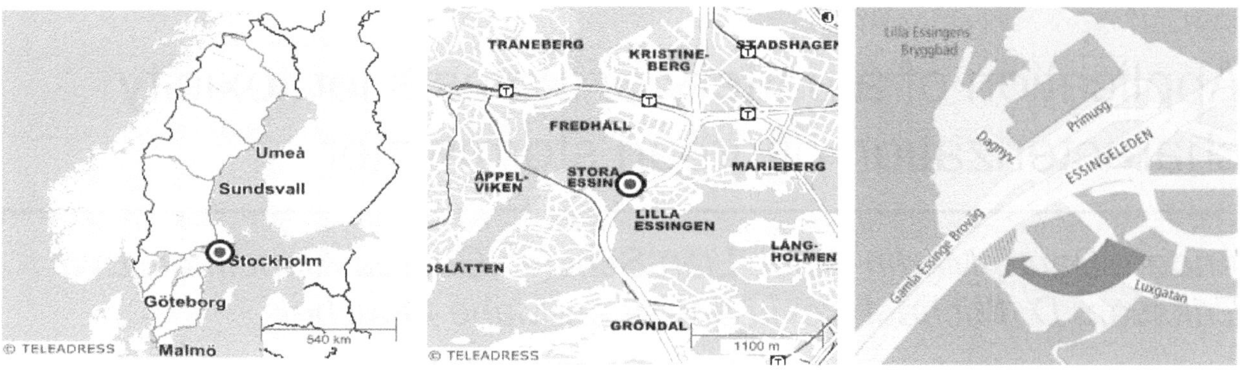

Figure 1: Maps of Sweden, the central parts of Stockholm (The Swedish Yellow Pages, 2004), and Lilla Essingen (Vägverket, 2003)

Table 1: Surface types in contact with the stormwater at Lilla Essingen.

Site	Catchment area	Description	Materials
Essingeleden (motorway)	5,000 m^2	Motorway with 120,000-130,000 vehicles/day	Asphalt
Lilla Essingen (urban area)	10,000 m^2	6 houses	Tiled roofing
		1 parking area	Asphalt
		Motorway	Asphalt
		Urban streets	Asphalt
		Green area	Grass

The Swedish National Road Administration (Vägverket) is responsible of the construction and maintenance of the facility whereas Stockholm Water Company (Stockholm Vatten AB), in co-operation with Vägverket are responsible monitoring the treatment facility. Runoff samples were collected by flow proportional sampling during storm events. Samples were automatically collected from a descending well located prior to the pond from May 2004 to July 2004. The auto-sampler was flow-activated and could collect a maximum of 250 ml/90 s. Water depth, velocity and temperature were automatically recorded, together with rainfall data. All samples were collected by Stockholm Water in 500 ml glass bottles. General water chemistry parameters (total organic carbon, ammonium, nitrate, total nitrogen, ortho-phosphate, total phosphorus, electrical conductivity, suspended solids, and pH) and the total concentrations of heavy metals (Cr, Cd, Cu, Ni, Pb, Zn) were determined by Stockholm Waters own accredited laboratory.

Whole sample tests

Samples were frozen within 24 hours of collection. Prior to testing, frozen samples were left to defrost overnight at 25°C and the sample homogenised by vigorous shaking or mixing. 100 ml of homogenised sample were transferred to settling cylinder and left to settle for a period of 1 hour. 90 ml of overlying sample were used for subsequent testing. The physico-chemical parameters pH, conductivity and suspended solids (SS) where determined on the raw sample. The water hardness was determined on samples for which suspended solids were settled by gravity for 1 hour prior to analysis.

Toxicity tests

The bacterial tests were conducted using the Microtox system which employs *Vibrio fischeri* as the test organism with phenol as a standard reference material. Samples were analysed using the 15 minute whole effluent test which involves the dissolution of 0.2 g sodium chloride in 10 ml of sample before testing, avoiding the need to dilute the sample to achieve the required level of salinity. All samples were tested in duplicate.

In the algal test, flasks were inoculated to 10^4 cells/ml with a log-phase pre-culture of *Pseudokirchneriella subcapitata* (formerly known as *Selenastrum capricornutum).* Mini-scale algal growth inhibition tests (Arensberg *et al.*, 1995), with 4 ml test medium in 20 ml glass vials, were conducted according to the ISO-standard (ISO, 1989). Potassium dichromate was used as a reference compound. The test flasks were incubated on a shaker (100 rpm) in continuous light (90-100 $\mu E/m^2/s$) at 21 ± 2 °C. Samples were withdrawn from each test flask and the controls every 24 hours (*i.e.* 0, 24, 48, and 72 h). In all tests, 5 dilutions (1000, 750, 500, 250 and 125 ml/l) were tested in triplicates and 6 controls were included. Algal growth rates were calculated based on the biomass in the cultures quantified by in-vial acetone extractions as described by Mayer *et al.* (1997). The tests were conducted at pH 8.0 ± 0.3 with typical control growth rates of 1.6-1.8 d^{-1} during the 72 hours of incubation.

The rotifer test was performed using *Brachionus calyciflorus* as test organism and was carried out as described in the French standard method NF T90377 (AFNOR 2000) and as described in the manual of Rotoxkit (Creasel Ltd 2004, Belgium). In each test, three concentrations (999, 666 and 333 ml/l) were used in eight replicates with two controls and the EPA standard medium hard water (SMHW) was used for dilution. The exposure took place in 48 multiwell polystyrene plates with individual volumes of 1 ml. During the test, the rotifers were fed with an algal suspension of $2x10^6$ cells/ml prepared by centrifugation from fresh culture of *Pseudokirchneriella subcapitata*. The algae were grown in the ASM1 medium and washed prior to the utilisation. After 48 hours of incubation in the dark at 25°C on a gyratory shaker (120 rpm) the numbers of rotifers alive and dead were recorded. The reproduction rate was estimated as a mean of logarithm of the number of animals at the end of the test, with the initial number corrected for the number of dead animals.

Toxicity identification evaluation procedures

A number of samples from each site were also subjected to a simplified Toxicity Identification Evaluation (TIE) procedure to evaluate the influence of metals on the total toxicity of the samples (US-EPA, 1991). For this purpose the rotifer test with *Brachionus calyciflorus* was used. To determine the contribution of metals, biotests were carried out before and after the addition of EDTA (Seidl *et al.,* 1998). Heavy metals are chelated by EDTA, with the resulting EDTA–heavy metals chelate being non-toxic. The association between EDTA and heavy metals is much stronger than that between EDTA and Ca^{++} and Mg^{++}. Calculation of metals speciation by chemical equilibrium model MINEQL+ (Schescher *et al.,* 1992) has demonstrated that that 95% of heavy metals are chelated by the addition of EDTA at a concentration corresponding to 1/10th of the total hardness (on a molar basis) of the samples and that, at this level, the residual concentrations of EDTA remain below toxicity thresholds (US-EPA 2005). Hence this approach (the addition of EDTA at a concentration $1/10^{th}$ of the determined total hardness) was adopted in this study giving roughly a level of 10^{-5} M EDTA in the final sample.

RESULTS AND DISCUSSION

Sample characteristics

Table 2 presents the chemical characteristics of samples collected at the site in Stockholm (Sweden). Sample STO1 had a very high content of suspended solids (13,000 mg/l), together with an elevated concentration of organic carbon (1300 mg/l). In general the samples STO1 and STO2 appear to be the most contaminated in relation to a variety of parameters with, for example, the highest concentrations of TOC, ammonium, Zn, Cr, Cu, Pb, and Ni typically being reported in these samples.

Table 2: Chemical characteristics of urban runoff samples collected in Stockholm, Sweden from May to July 2004.

	STO1	STO2	STO3	STO4	STO5
Date	24-05-2004	08-06-2004	11-06-2004	28-06-2004	05-07-2004
pH	7.6	7.2	7.5	7.6	7.1
Suspended Solids (mg/l)	13000	1750	780	210	150
Total Organic Carbon (mg/l)	1300	160	80	31	49
NH_4-N (µg/l)	1200	200	47	180	92
NO_2/NO_3-N (µg/l)	110	540	370	690	213
Tot-N (µg/l)	1800	2700	1550	1900	1600
PO_4-P (µg/l)	66	12	7	15	6
Tot-P (µg/l)	>8000	1400	930	140	280
Conductivity (mS m^{-1})	155	22.0	16.9	29.9	22.2
Zn (µg/l)	12000	3800	2400	500	450
Cr (µg/l)	540	98	55	27	14
Cu (µg/l)	1900	320	170	52	62
Cd (µg/l)	8.5	1.0	1.1	0.3	0.3
Pb (µg/l)	900	110	120	44	26
Ni (µg/l)	290	58	32	12	10

Toxicity of stormwater samples

Results of the Microtox tests found that undiluted samples STO1, STO3 and STO4 showed 10-30% inhibition, whereas sample STO2 did not show any significant response. For sample STO5, it was possible to obtain a full concentration response curve (Figure 2) and the concentrations at which 20% (EC20) and 50% (EC50) of the bacteria were affected is estimated to be 173 ml/l $[128;227]_{95\%}$ and 388 ml/l $[324;461]_{95\%}$. Thus, this sample had to be diluted 2.6 times to reduce toxicity from 100% to 50% inhibition of light emission, and 5.8 times to reduce toxicity from 100% to 10%. This sample was not toxic towards algae (data not shown) and rotifers (Figure 4).

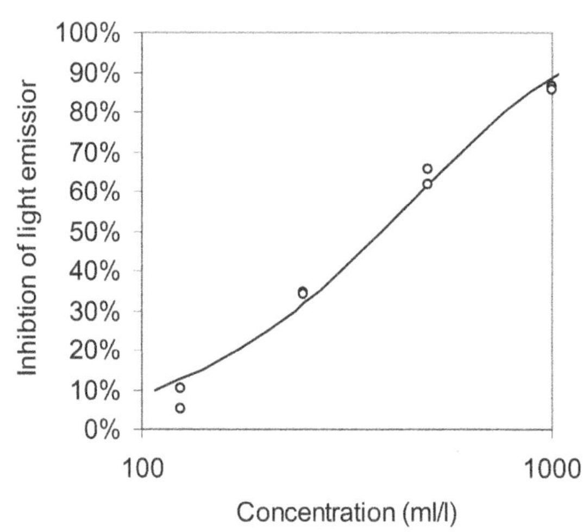

Figure 2: Toxicity of the urban stormwater sample STO5 (Stockholm, Sweden) in the Microtox test after 15 min incubation. Test organism: *Vibrio fischeri*. The concentration unit ml/l refers to the volume of sample added to the medium of the toxicity test.

In algal tests the highest toxic response was obtained with sample STO2 in which the algal growth rate was reduced by 50% in undiluted samples (Figure 3). For samples STO1, STO3, STO4 and STO5, growth rate inhibitions below 15% were observed in undiluted samples. However, due to the high content of suspended solids in STO1 (Table 2) this sample was filtered after settling, as interference in fluorescence measurements due to suspended solids was observed in preliminary algal tests.

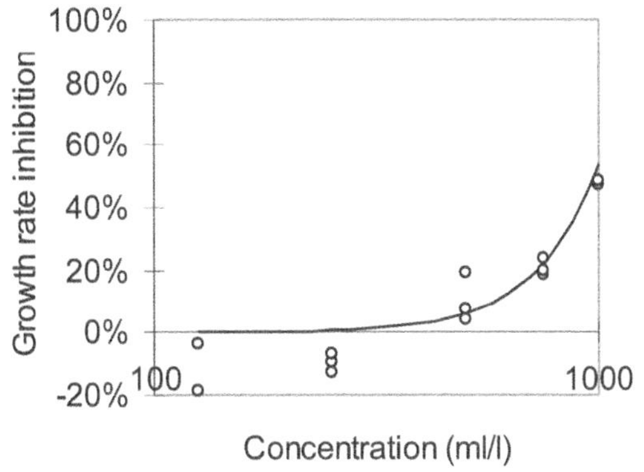

Figure 3: Toxicity of the urban stormwater sample STO2 (Stockholm, Sweden) in the algal test after 48 h incubation. Test organism: *Pseudokirchneriella subcapitata*. The concentration unit ml/l refers to the volume of sample added to the medium of the toxicity test.

Figure 4 illustrates the inhibition of the reproduction of the rotifers *Brachionus calyciflorus* after exposure to undiluted stormwater samples for 48 hours. The stormwater samples STO1 and STO2 had significant impact on the growth of rotifers, *i.e.* 100 and 70% inhibition, respectively (Figure 4, white bars). Samples STO3 and STO5 did not cause any decrease in the reproduction of *Brachionus calyciflorus*, whereas STO4 had a slightly stimulating effect.

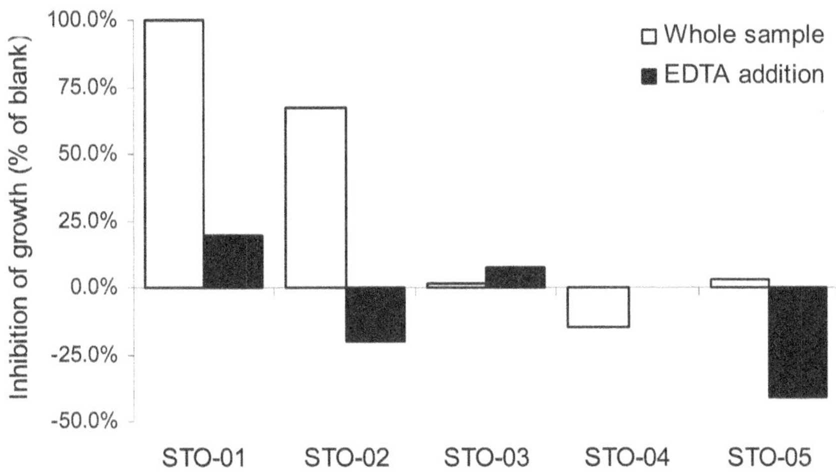

Figure 4: Toxicity of undiluted stormwater samples (Stockholm, Sweden) towards rotifers after 48 h incubation. Samples are tested as whole samples and after addition of EDTA. Test organism: *Brachionus calyciflorus*. A 10% change may be considered as significative.

The total hardness of STO1 was about 1 mM whereas the samples STO2 to STO5 were about 0.5 mM. This means that the final EDTA concentration used in the tests was 0.1 mM respectively 0.05 mM. The addition of EDTA lowered the toxicity of undiluted samples towards *Brachionus calyciflorus* (Figure 4). This effect was observed especially for samples STO1 and STO2 (samples previously reported as toxic), lowering the toxicity by up to 80%. It can be noted that the highest concentrations of Zn, Cr, Cu, and Ni were also reported for these samples (Table 2). From the heavy metal contents we would also expect that STO3 would be toxic, however no toxic effects were observed. The negative inhibition is in fact growth stimulation and was observed in 30% of the TIE-procedures. The mechanism was not elucidated. The addition of EDTA is known to chelate metals to form non-toxic complexes (Campbell *et al.*, 2002) and the used TIE-procedure thereby suggests heavy metals as a key component in the overall sample toxicity.

CONCLUSION: APPLICATION OF BIOTESTS FOR VULNERABILITY ASSESSMENT OF URBAN STORMWATER

As could be expected from the chemical analysis, the most toxic samples for rotifers were those containing the highest level of heavy metals, i.e. STO1 and STO2, which contained 12 and 3.8 mg/l of zinc and 1.9 and 0.32 mg/l of copper, respectively, and also reported high concentrations of all other metals analysed for (Table 2). Considering the high copper concentrations in both STO1 and STO2 and the known high toxicity of copper towards algae (Radix *et al.*, 2000), it may seem surprising that only sample STO2 showed significant toxicity towards *Pseudkirchneriella subcapitata*. However, the total concentration of metals does not reflect the bioavailable concentrations as the speciation of metals may lower the bioavailable fraction significantly. For algal tests of STO1 the additional filtration mentioned above possibly reduced the toxicity of the whole sample. These findings underline that total concentrations of heavy metals cannot be used for prediction of sample toxicity due to complexation and sorption of metals.

The sample STO1 was found to be toxic towards both rotifers and bacteria, but no significant effects on algal growth rates were observed. Rotifer and algal tests identified STO2 as being toxic (Figure 3 and Figure 4), but this sample did not cause inhibition in the bacterial test. The Microtox test was, however, the only one detecting toxicity in STO5 (Figure 2), and for this sample none of the chemicals measurements shown in Table 2 indicate that STO5 should be more toxic than, for example, STO3 or STO4. The reason for using three toxicity tests with different organisms and different toxicological endpoints on all samples was to increase possibilities of detecting toxicity. Due to the limited experiences with toxicity testing of stormwater samples no assumptions as to which organisms may be the most sensitive can be made prior to testing. The results shown in the present study clearly demonstrate the feasibility and value of applying a testing strategy in which a battery of biotests is included. No single organism was found to be the most sensitive in these tests of complex stormwater samples.

In the present study, chemical analysis revealed that samples STO1 and STO2 were highly contaminated with metals and this alone would give rise to concern about the impact of their discharge. Thus, for these two samples it may seem unnecessary to include a battery of toxicity tests in a monitoring programme. However, it should be kept in mind that these samples were also identified by the biotests as being problematic and information on the whole sample toxicity was achieved using the biotest approach. Additionally, the Microtox test also identified STO5 as a potentially problematic sample even though chemical parameters did not indicate this. Moreover, although the results of the TIE tests pointed towards metals as the main contributors to toxicity, there could have been other compounds contributing to total toxicity but due to the limitations of the "chemical window" approach these were not identified. In contrast, the biotest results relate to the toxic effects of the whole samples, information which could not have been achieved by chemical analysis alone.

The most toxic sample would need to be diluted approximately five times to reduce toxicity below the detection limit of the biotests (Figure 2). During a rain event and upon discharge to receiving waters the stormwater will certainly be diluted more than five times. Safety factors (or uncertainty factors) of 100-1000 often applied in risk assessments to take interspecies variation into account and to address the problem that laboratory tests and response test organisms do not directly relate to ecosystem vulnerability. In the present case, a sufficient dilution will most probably be obtained even if a safety factor is included in the effect assessment. It must, however, be emphasised that long-term effects may occur due to the fate and distribution of contaminants in stormwater. The metals found in the present study and organic compounds (most often polyaromatic hydrocarbons) found in other studies of stormwater will primarily be bound to suspended solids, which may settle as sediments in retention ponds or be transported to receiving waters during rain events. To evaluate the potential toxic impact of sediment bound contamination, toxicity tests should be carried out on sediments and/or aqueous extracts hereof along with chemical analysis to identify priority pollutants.

In conclusion, the traditional approach to evaluating the hazards of chemical constituents in stormwater samples is to compare concentrations of single compounds with water quality criteria, if such exist for the compounds identified (by no means always the case). However, the complex composition of stormwater makes estimations of toxicity based on such chemical analysis alone very difficult. It may in some cases lead to an overestimation or underestimation of toxicity, a response demonstrated for stormwater and various other kinds of complex environmental samples by a variety of researchers including Baun *et al.* (2004), Christensen *et al.* (2006), and Kildeby *et al.* (2004). Alternatively the use of a test battery supplies valuable information on the whole sample toxicity, whilst being unable to provide specific information on causes of toxicity. Hence, in terms of a vulnerability assessment (Chapter 12), the use of a test battery approach in combination with chemical analysis, as demonstrated through this study, can provide a more complete overview of the ecological status of runoff than either approach alone e.g. in impact assessment of the structural and non-structural BMPs described in Chapters 8, 9 and 12 (Scholes *et al.,* 2007).

REFERENCES

AFNOR (2000) Détermination de la toxicité chronique vis-à-vis de Brachionus calyciflorus en 48 h [Chronic toxicity assessment towards Brachionus calyciflorus during 48 h]. NF T90-377. AFNOR, Paris, France.

Arensberg P., Hemmingsen, V. and Nyholm, N. (1995) A miniscale algal toxicity test. Chemosphere 30, 2103-2115.

Baun, A., Ledin, A., Reitzel, L.A., Jonsson, S, Bjerg, P.L. and Christensen, T.H. (2004) Xenobiotic organic compounds in leachates from ten Danish MSW landfills – chemical analysis and toxicity tests. Water Research 38, 3845-3858.

Christensen, A.M., Nakajima, F. and Baun, A. (2006) Toxicity of water and sediment in a small urban stream (Store Vejleå, Denmark). Environ. Pollut. 144 (2), 621-625.

Ellis, J.B. (2000) Risk assessment approaches for ecosystem responses to transient pollution events in urban receiving waters. Chemosphere 41, 85-91.

EU (2000) Directive 2000/60/EC of the European Parliament and of the Council of 23 October 2000 establishing a framework for Community action in the field of water policy. 2000/60/EC (23 October 2000).

ISO (1989) Water quality - Fresh water algal growth inhibition test with Scenedesmus subspicatus and Selenastrum capricornutum. ISO 8692. International Organization for Standardization, Geneva, Switzerland.

Kildeby, M.R., Nielsen, J.G., Jensen, L, Ledin, A.. and Baun, A. (2004) Toxicity of xenobiotic compounds in Harrestrup Å, Denmark. Vand & Jord 11, 4-7 (in Danish).

Maltby, L., Forrow, D.M., Boxall, A.B.A., Calow, P. and Betton, C.I. (1995a) The effects of motorway runoff on freshwater ecosystems: 1. Field Study. Environ. Toxicol. Chem. 14, 1079-1092.

Maltby, L., Boxall, A.B.A., Forrow, D.M,. Calow, P and Betton, C.I. (1995b) The effects of motorway runoff on freshwater ecosystems: 2. Identifying major toxicants. Environ. Toxicol. Chem. 14, 1093-1101.

Mayer, P., Cuhel, R.. and Nyholm, N. (1997) A simple in vitro fluorescence method for biomass measurements in algal growth inhibition tests. Water Res. 31, 2525-2531.

Marsalek, J., Rochfort, Q., Mayer, T., Servos, M., Dutka, B. and Brownlee, B. (1999) Toxicity testing for controlling urban wet-weather pollution: advantages and limitations. Urban Water 1, 91-103.

Radix, P., Léonard, M., Papantoniou, C., Roman, G., Saouter, E., Gallotti-Schmitt, S., Thiébaud, H., Vasseur, P. (2000) Comparison of four chronic toxicity tests using algae, bacteria, and invertebrates assessed with sixteen chemicals. Ecotox. Environ. Safety 47, 186-194.

Seidl M., V.Huang & J.M.Mouchel (1998) Toxicity of combined sewer overflows phytoplancton: the role of heavy metals. Environmental Pollution 101:107-116.

Schescher W.D. and D.C.McAvoy (1992) MINEQL+ : a Software Environment for Chemical Equilibrium Modelling. Computer and Environmental Urban Systems 16, 65-76.

Scholes, L., Baun, A., Seidl, M., Eriksson, E., Revitt, M. and Mouchel J.-M. (2007) Assessment of stormwater ecotoxicity using a battery of biotests. In: Highway and Urban Environment Proceedings of the 8th Highway and Urban Environment Symposium Series: Alliance for Global Sustainability Bookseries. Springer, New York, USA.

US-EPA (1991) Methods for aquatic identification evaluations - Phase I, Toxicity characterization procedures. EPA/600/6-91/003. United States Environmental Protection Agency.

US-EPA (2005) ECOTOX database. U.S. Environmental Protection Agency: www.epa.gov/ecotox (verified August 2007).

Vägverket (2003). Nu kommer dagvattnet från Essingeleden att renas med ny prisbelönt teknik [The stormwater from Essingeleden will be treated by a new award-winning technique]. http://www.vv.se/templates/Pressrelease____7584.aspx (in Swedish)

Vägverket (2006) Information on SORBUS treatment facility, Stockholm, Sweden. http://www.ecotechstockholm.se/upload/Sorbus.pdf (verified September 2007).

Chapter 21

Development of a distributed hydrological model for urban catchments

Fabrice Rodriguez and Hervé Andrieu

INTRODUCTION

The rainfall-runoff transformation on urban catchments involves various physical processes. Soil-atmosphere interface processes encompass evapotranspiration and interception by the vegetation. Soil surface processes regroup infiltration of water in the soil, surface runoff and evaporation of surface and soil water. Sub-soil processes consist in the interflow and the drainage of the soil water by trenches and sewer pipes defects. The hydrological response of a catchment is indeed sensitive to these processes: evapotranspiration is considered as an important component of the water budget in urban areas (Grimmond and Oke, 1991); the role of soil in the generation of runoff has been highlighted by recent works (Belhadj *et al.,* 1995; Berthier *et al.,* 2004,). An integrated urban water modelling approach seems to be necessary for a better understanding of the hydrological behaviour of urban catchments, taking into account the physical processes listed before (Jia *et al.,* 2001). A distributed hydrological model, based on urban databanks, has been developed in this aim (Rodriguez *et al.,* 2005). In conjunction with these observations, ongoing urban growth drives local authorities to integrate innovative stormwater source control techniques into a hydrological perspective (Chapter 1). Such techniques, commonly called Best Management Practices (BMPs) (Chapter 8), often favour rainwater infiltration in order to reduce runoff volumes and/or rainwater pollution. To take into account theses new requirements, hydrological modelling practices have to be adapted. Especially, they should focus on a more physically based approach, including a better description of the urban spatial and temporal variability, and a more detailed representation of physical processes at small scale. The available geographical information in cities could be very useful to achieve this aim: urban environment is well-documented, thanks to urban databanks.

The aim of this work is to present a hydrological model based on urban databanks and able to reproduce the spatial and temporal variability of hydrological processes on urban catchments. This distributed model, called MUHDI (Modèle Urbain d'Hydrologie Distribuée), includes a morphological description of the urban environment on a hydrological point of view (Rodriguez *et al.,* 2003). The main objectives of this model are (i) to determine continuously the different components of rainwater fluxes, such as the surface runoff coming from the different land use types and the soil runoff, coming from the infiltration of water into water systems through sewer defects and (ii) to indicate the saturation level and the capacity of soil to infiltrate water on long data series. This paper is devoted to the description of the URBS model. The first section introduces the main descriptive elements resulting from urban databanks. Then, the hydrological processes represented in the model are introduced. The third section is devoted to the application of the model on the urban catchment located at Reze near Nantes (France) and presents the first results.

 DayWater: an Adaptive Decision Support System for Urban Stormwater Management. Edited by Daniel R. Thévenot. ISBN: 9781843391609. Published by IWA Publishing, London, UK.

MORPHOLOGICAL REPRESENTATION OF THE CATCHMENT IN MUHDI

The urban databanks (UDBs) available in cities allow representing an urban catchment through elementary runoff concentrating surfaces and the hydrographic network. This representation is based on the main structural elements of the city, *i.e.* the cadastral parcels and the street network. The urban cadastral map available in many French urban databanks (Figure 1 left) covers the main geographical elements required to describe hydrological processes in an urban catchment: (i) cadastral parcels, houses, street sections and possibly vegetation serving for a 2D description of the urban surface, and (ii) topography, street segments, storm sewers and rivers for the description of the hydrographical network, hereafter designed as Runoff Branching Structure (RBS) (Figure 1 right). The RBS is a vector map of water flow paths along street gutters and inside the sewer network, from each cadastral parcel up to the catchment outlet. Elementary runoff concentrating surfaces are called Urban Hydrological Elements (UHEs), each of them containing one cadastral parcel and its corresponding adjacent street portion. Each UHE is connected to the street and thus to the RBS. The geometrical characteristics of one UHE include surface area, impervious fraction incorporating building and street surface area (covered by trees or not), vegetation fraction, slope and length, the connection point to the RBS and the depth of the drainage network at this point. These characteristics allow for both a surface representation of the UHE and its cross section. This cross section is then decomposed in three land use types, including house, street and natural soil, each of them being possibly covered by trees (Figure 2A). Three vertical profiles corresponding to each land use type are defined for each UHE. Each vertical profile is represented by 3 reservoirs respectively related to the surface, the vadose and the saturated zone of the soil (Figure 2B).

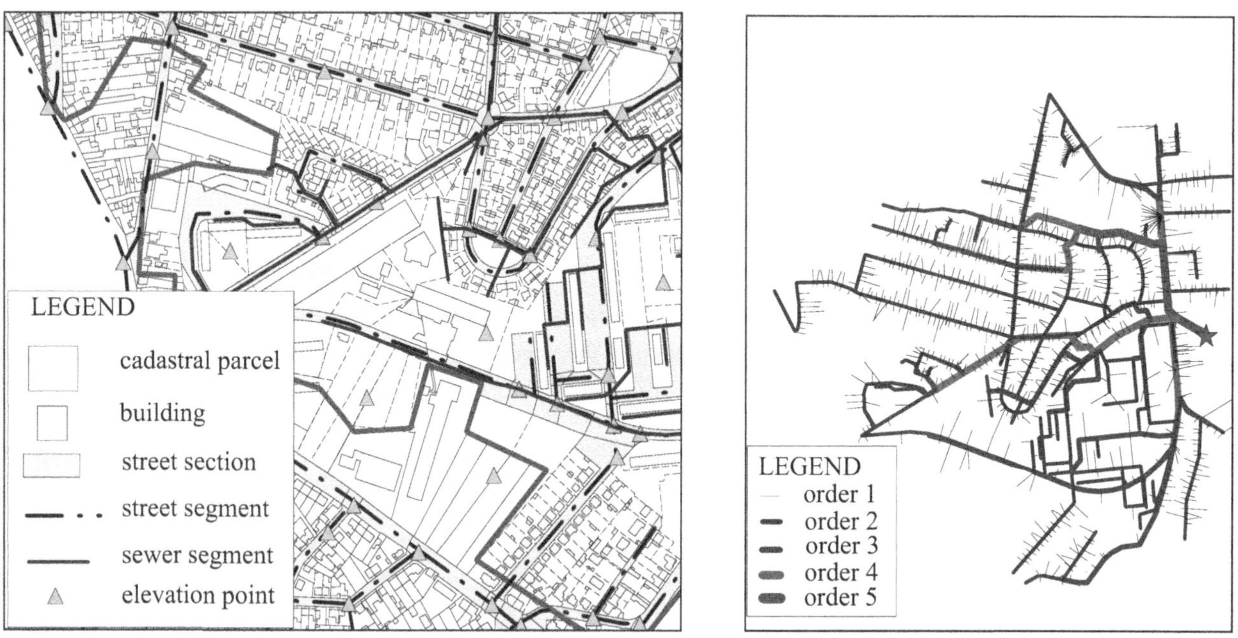

Figure 1: (Left) Main elements of urban databanks (UDBs) for the city of Nantes (France); (right) illustration of the Runoff Branching Structure (RBS) map of an urban catchment, using Strahler order values.

DESCRIPTION OF THE HYDROLOGICAL PROCESSES IN THE MODEL

Physically based hydrological models generally include a deterministic representation of the vadose zone behaviour. The main objectives of the model being to represent all hydrological processes, both for long time series and with short time steps, a simple method has been implemented here. This method assumes a schematic representation of both the soil and the water flow processes, based on a vadose zone reservoir which maximal depth is *a priori* not limited. Each vertical profile contains a vadose zone inside which the water content is considered as uniform. The hydrological processes are described in two steps. First, rainfall interception by trees has been described by using the model of Calder (1977): the tree leaves are represented by a simple reservoir model, supplied by rainfall and drained both by evaporation and by a drainage function occurring when the water storage exceeds a minimum value. Then, water flows through each land use type (here noted with *) have been modelled using the same vertical scheme (Figure 2B):

- The soil surface is represented by a reservoir filled by rainfall P^*_t which falls under possible vegetation, and emptied by evaporation E^*, infiltration I^*_{NSZ} and surface runoff R^*. The evaporation flux E^* is assumed to be proportional to potential evapotranspiration rate and to the water stored in the surface reservoir. The infiltration flux I^*_{NSZ} is controlled by both atmospheric and soil conditions: it depends on the surface storage S^*_S and on the infiltration capacity of the vadose zone. Surface runoff R^* occurs when the surface storage exceeds its maximal capacity S^*_{max}.

- The vadose zone of the soil is represented by a reservoir, filled by infiltration from the surface and by capillary rises I^*_{SZ} from the underlying saturated zone, and drained both downwards by percolation I^*_{SZ} (fluxes are calculated according to Darcy's law) and upwards by transpiration TR^* from the root zone in the natural soil and street profile. As mentioned above, the water content in this vadose zone is assumed to be vertically uniform, and the Brooks and Corey relations (Clapp and Hornberger, 1978) describe the variation of suction head and hydraulic conductivity with the zone water content. The vadose zone is mainly parameterised by the hydraulic conductivity at natural saturation K^*_S which is zero for houses, as no water infiltrates through roofs.
- The level of the saturated zone of the soil is represented in the model. It increases with the contribution of the upperlying vadose zone and decreases both by capillary rise and by direct drainage in sewers if surrounding soil has become saturated. This direct drainage has been estimated by considering that the sewer pipe acts like an ideal drain (Cassan, 1994):

$$I_{sew}(t) = K_S(z_{SZ})\,\Delta t\,\frac{\lambda}{L_i}\left(z_i - z^i_{net} - z_{SZ}\right)^{\mu}$$

where z_{sz} is the saturated zone level, $z_i - z^i_{net}$ the sewer network depth, L_i the UHE length, λ and μ two parameters depending on the type and the status of the drain and the trench containing the drain. This drainage may be separated into 2 components: drainage into the wastewater drainage network and drainage into the rainwater drainage network. We introduced a factor $f_{sew\text{-}rw}$ representing the distribution of these two drainage components, depending on both the presence and the relative positions of drainage networks, and possibly deriving from UDBs. At each time step and for each UHE, the saturated zone level is averaged from the saturated levels encountered in each land use type, allowing the lateral fluxes in the vadose zone.

The runoff at the outlet of one UHE is the sum of the different flow contributions, containing the runoff flow of each land use type and the groundwater drainage:

$$Q(t) = A^{hou}R^{hou}(t) + A^{str}R^{str}(t) + A^{nat}R^{nat}(t) + A^{tot}f_{sew\text{-}rw}I_{sew}(t) = Qhou(t) + Qstr(t) + Qnat(t) + Qsew(t)$$

where A^* represents the surface of each land use type in the UHE, and R^* the runoff produced by each land use type: (hou) houses, (str) street and (nat) natural soil.

The transfer processes are simply modelled using a travel time method based on the estimation of a distributed runoff velocity (Rodriguez *et al.*, 2003).

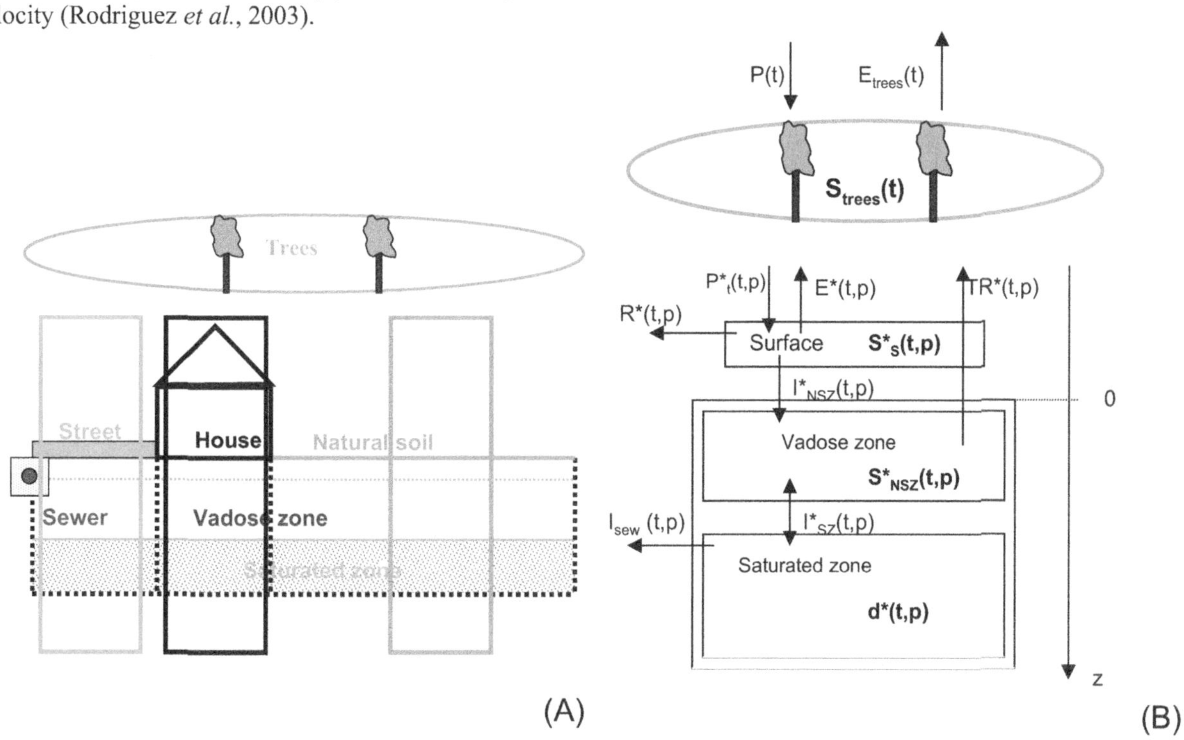

Figure 2: (A) UHE cross section and (B) detail of one vertical profile with associated water fluxes.

APPLICATION OF THE MODEL TO URBAN CATCHMENTS

Principles

This work is aimed at evaluating a production function model able to simulate both the different components of runoff flow and the hydrological behaviour of the subsurface soil. A comparison between observed and measured hydrological variables will serve as a basis for this evaluation. This model has been developed through a realistic representation of hydrological processes in urban catchments, using 15 parameters. Most of them may be estimated from literature review or field measurement and from a previous explicit modelling of this UHE (Berthier *et al.*, 2004). A sensitivity analysis has been conducted in order to compare the influence of the different parameters on the simulation of both runoff flow and saturation level. This analysis shows the importance of the parameter governing direct infiltration into sewers (λ) and of the soil hydraulic conductivity ($K^{nat}{}_S$). No automatic calibration procedure has been implemented in this work.

Case study

This model has been applied to a small urban catchment (5 ha) located in the city of Rezé, near Nantes (West France). The drainage network of this catchment contains separate sewers and its land use consists of single-family housing. The impervious surface connected to the drainage network reaches 37% and the soil is considered as silt clay. This catchment has been continuously monitored since 10 years and has already been used for hydrological modelling tests (Rodriguez *et al.*, 2000; Rodriguez *et al.*, 2003) The data measured contain: (i) the rainfall intensity and the discharge at the outlet from 1991 up to 2002, (ii) the daily water table level through three piezometers installed at two locations of the catchment, from September 1995 to December 1998, (iii) groundwater drainage runoff, measured during three winter periods in wastewater drainage network, providing a good estimation of both rain- and waste-water drainage (Berthier *et al.*, 2004) and (iv) meteorological data such as Penman evapotranspiration, collected from the meteorological station located at 5 km from Rezé. Figure 3A shows the different measurements locations and the catchment morphology.

Application of the model on Rezé catchment

Given the homogenous type of housing, the 70 cadastral parcels of the catchment have been represented by a single UHE which morphological features, resulting from UDBs, are presented on Figure 3B. The response time of this catchment is short, *i.e.* lower than 10 minutes; as this work focuses on the production function processes, no transfer process was taken into account for this application. Continuous simulation has been performed during 10 years using a 1-hour time step. The parameter set used to run the model is listed (Table 1): parameters have been initialised thanks to physical considerations taken either from literature or from *in situ* observations. Only the parameter λ has been adjusted in order to minimize the deviation between observed and simulated volumes.

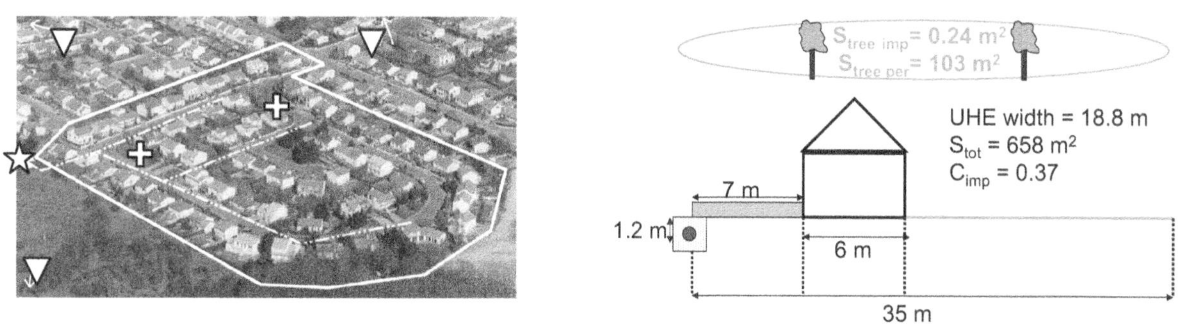

(A) (B)

Figure 3: (A) Rezé catchment: (✚) piezometer, (★) outlet and (▼) rain gauge locations; (B) morphological characteristics of the representative UHE.

Table 1: Parameters of MUHDI model applied to Rezé near Nantes (France).

	Parameter		Natural soil	Street	House
surface	S^*_{max}		5.0	3.5	0.5
	K^*_s		1.3 10^{-5}	7.5 10^{-8}	0
soil	θ_S	0.43			
	Ψ_e	0.20			
	b	5.00			
	M	0.20			
	z_{root}	1.30			
sewer	λ	4.00			
	μ	2.00			
	$f_{sew\ rw}$	0.37			

FIRST RESULTS

All results presented below are dependent on the chosen parameter configuration, and constitute a first attempt to explain the hydrological behaviour encountered on this catchment. Three comparison criteria were used for this model's evaluation: the Nash criterion (C_{Nash}), the correlation coefficient (R^2) and a bias criterion (C_b). The rainfall events have been selected through an automatic procedure and the values of flow rate volumes and flow coefficients have been determined. From 1991 to 2000, the rain event sample contains of 850 events which rainfall intensity exceeds 1 mm/h.

Simulation of a rainfall event

The simulation of a rain event is presented in Figure 4. This graph plots the different contributions to runoff at the outlet of the catchment due to a rainfall event on 26 February, 1996, and demonstrates the significance of commonly neglected processes, such as the runoff produced by natural soil Q_{nat} and the drainage into the sewer system Q_{sew}, both being especially important in winter. Such drainage, not significant at the rain event scale, may represent large volumes at the year scale. Moreover, the contributions of impervious surfaces are quite similar, because these surfaces (roofs and streets) were already wet at the beginning of the event. For this event of 35 mm rain depth, the surface runoff resulting from pervious surfaces is delayed but important, because of the high saturation level, varying from –0.7 to –0.55 m during the event.

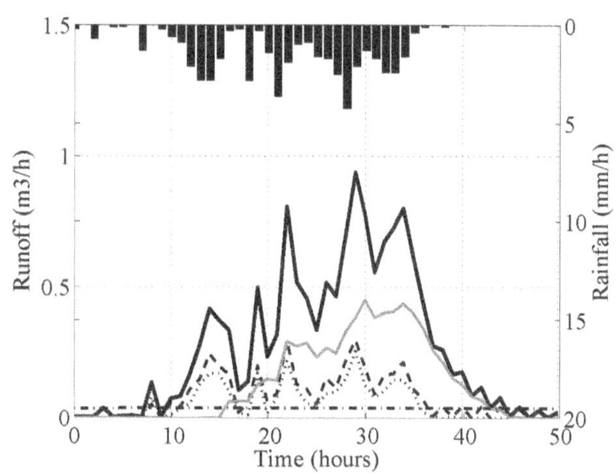

Figure 4: Runoff contributions (m^3/h) simulated with MUHDI model on Rezé catchment using rainfall intensity (mm/h) on 26 February 1996: (— Q) total-, (··· Q_{hou}) houses- (-- Q_{str}) street- (— Q_{nat}) natural- and (- · Q_{sew}) drainage-, - Instantaneous hourly flow rates.

Simulation of runoff flows

Three types of results have been compared. First, the instantaneous hourly flow rates Q have been compared to the observed ones; then, for each rain event, the flow rate volumes V and flow coefficient C_F have been compared to the observed ones (Table 2). There is a good correspondence between simulated flow rate volumes V or instantaneous flow rates Q and the observations: the MUHDI model reproduces half of the observed flow coefficient variability ($R^2 = 0.54$) and the shape of the simulated histogram is close to the observed one, both following a Gaussian distribution.

Moreover, the model allows for the comparison of the main hydrological fluxes for the given analysis period. Figure 5A represents the simulated fluxes occurring in the representative UHE: values are expressed as a percentage of gross rainfall on the whole period and the results depend on the parameter configuration used. However, these simulations show some interesting points. For instance, the total evaporation represents 50% of the gross rainfall and the rain interception by trees is not totally negligible (2.7%). The main contribution to the flow rate encompasses the runoff on streets (7.5%) and the runoff on houses (15.1%) but the direct infiltration into sewers represents more than 10% of the total rainfall. These fluxes are not really validated, because of the lack of appropriate field data, but are in agreement with other experimental results (Hollis and Ovenden, 1988; Belhadj *et al.*, 1995).

Table 2: Comparison criteria for simulations of runoff volumes, flow coefficient and instantaneous flow rates for the whole period.

Parameters	C_b	R^2	C_{Nash}
V	+1.40%	0.95	0.95
C_F	-0.48%	0.54	0.50
Q	+2.02%	0.88	0.86

Simulation of saturation levels

Due to the representation of the catchment with a single UHE, the simulated saturation level is assumed to be uniform: but the piezometric measurements are located on two points on the catchment (Figure 3A). The comparison will consequently focus on the dynamical evolution of saturation levels during two years of available data (Figure 5B). Both the simulated and observed levels increase during winter and decrease during summer. The variation of simulated level agrees with the observations, except for summer 1998 where the model is not able to reproduce the particularly strong and fast decrease of the saturated level; such decrease could be of anthropic origin, for instance, pumping.

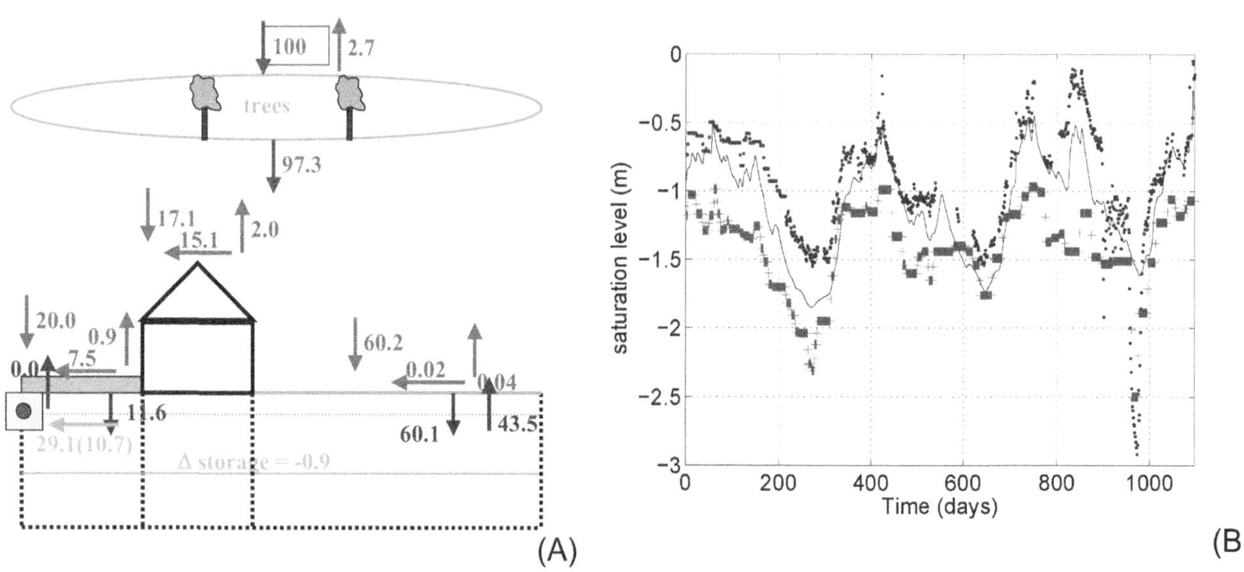

Figure 5: (A) Simulated rainwater fluxes (expressed in percentage of rainfall) on Rezé catchment from 1st January 1991 to 31 December 2000- the arrows corresponds to the fluxes cited in Figure 2B; (B) evolution of (—) MUHDI simulated and observed saturation levels (01/01/1996 – 31/12/1998): (•) PZB and (+) PZH represent the different piezometers

CONCLUSIONS

This first application of MUHDI model to a small urban catchment demonstrates the possibilities offered by this kind of modelling: the Rezé catchment has been described by a single UHE and the main rainwater fluxes have been simulated continuously during 10 years. This model has been evaluated through the comparison of these simulations with observed flow rates during the whole period and with observed saturation levels during two years: it allows the assessment of water fluxes often neglected in urban areas (evapotranspiration, soil infiltration into sewers) and shows that they could be significant, in accordance with previous experimental studies (Berthier *et al.,* 2004). The application of the model to a larger catchment has not been presented here but has highlighted the relevance of distributed modelling (Rodriguez *et al.,* 2005) by showing a spatial representation of most hydrological variables, such as the capacity of soil to store water. This model could contribute to create a development-aid tool for urban catchments, and a possible application could be the representation of innovative stormwater technologies and the determination of their influence on the hydrological behaviour of an urban area. Such technologies favour indeed the infiltration of rain water into the soil, and the consequence of this may be a local increase of the groundwater level (Göbel *et al.,* 2004). The use of such a distributed model, combined with GIS, allows for both (i) the representation of different planning scenarios involving BMPs (Chapter 8) and (ii) the analysis of the influence of these scenarios on the urban water budget at local and catchment scales. This kind of application could be implemented in decision support systems such as Flext (Chapter 15), as a professional simulation model, to take a benefit of a physically based approach of continuous hydrological processes modelling. The modelling of long rainfall data series is indeed necessary to evaluate the impact of infiltration stormwater management on both the decrease of runoff flows and the modification of soil water storage capacity.

REFERENCES

Belhadj, N., Joannis, C. and Raimbault, G. (1995) Modelling of rainfall induced infiltration into separate sewerage. *Water Science and Technology*, **32**(1), 161-168.

Berthier, E., Andrieu, H. and Creutin, J.D. (2004) The role of soil in the generation of urban runoff: development and evaluation of a 2D model, *Journal of Hydrology*, **299**(3-4), 252-266.

Calder, I.R. (1977) A model of transpiration and interception loss from a spruce forest in Plynlimon, Central Wales. *Journal of Hydrology*, **33**, 247-265.

Cassan, M. (1994) Aide-mémoire d'hydraulique souterraine [Underground hydraulics summary book], Presses de l'Ecole Nationale des Ponts et Chaussées.

Clapp, R. and Hornberger, G. (1978) Empirical equations for some soil Hydraulic Properties. *Water Resources Research*, **14**(4), 601-604.

Göbel, P., Stubbe, H., Weinert, M., Zimmermann, J., Fach, S., Dierkes, C., Kories, H., Messer, O., Mertsch, V., Geiger, W.F. and Coldewey, W.G. (2004) Near-natural stormwater management and its effects on the water budget and groundwater surface in urban areas taking account of the hydrogeological conditions, *Journal of Hydrology*, **299**(3-4), 267-283.

Grimmond, C.S.B. and Oke, T.R. (1991) An evapotranspiration-interception model for urban areas, *Water Resources Research*, **27**(7), 1739-1755.

Hollis, G.E. and Ovenden, J.C. (1988) The quantity of stormwater runoff from ten stretches of road, a car park and eight roofs in Hertfordshire, England during 1983. *Hydrological processes*, **2**, 227-243.

Jia, Y., Guangheng, N., Kawahara, Y. and Suetsugi, T. (2001) Development of WEP model and its application to an urban watershed. *Hydrological Processes*, **15**, 2175-2194.

Rodriguez, F., Andrieu, H. and Zech, Y. (2000) Evaluation of a distributed model for urban catchments using a 7 years continuous data series. *Hydrological Processes*, **14**(5), 899-914.

Rodriguez, F., Andrieu, H. and Creutin, J.D. (2003) Surface runoff in urban catchments : morphological identification of unit hydrographs from urban databanks. *Journal of Hydrology*, **283**(1-4), 146-168.

Rodriguez, F., Morena, F. and Andrieu, H. (2005) Development of a distributed hydrological model based on urban databanks - Production Processes of URBS, *Water Science and Technology*, **52**(5), 241–248.

Chapter 22

Stormwater in cold climates

Camilla Westerlund and Maria Viklander

INTRODUCTION

The interest in and knowledge of urban stormwater has increased greatly since the early 1970´s. It was reported already in 1972 by Sartor and Boyd that road runoff was substantially contributing to urban pollution and hypothetical calculations showed that runoff during the first hour of a storm could contribute with considerably more pollution load than would the same city's sanitary sewerage during the same period of time. Within the European Union today, and maybe especially in Scandinavia, access to clean water is taken for granted without a second thought. However, it is not often realized that many of the human activities put a great burden on water quality and quantity. All polluted water, whether polluted by households, industry, agriculture, or traffic returns back, in some way, to the environment and may cause damage to human health or effects on biota, including impacts on vegetation and biodiversity of such aquatic communities such as benthos or fish. The increasing demand for cleaner rivers and lakes, groundwater and coastal beaches was the driving force for the **EU Water Framework Directive** which has the overall aim of reaching, and keeping a "good status" for water within the EU. The protection should include all waters, surface waters and groundwater, and should be achieved before year 2015 (European Commission, 2006). With these increasing demands upon water quality, further research is required in terms of the quantity and quality of stormwater (Chapter 1).

According to Marsalek (1991) and Thorolfsson (2000), the stormwater focus has earlier been on warm or temperate climates and only have small parts been directed towards cold climate perspectives. The importance of understanding the special requirements for stormwater systems in cold climates was pointed out by experts in the field during the international conference on urban drainage and highway runoff in a cold climate, in Riksgränsen, Sweden, March 2003 (Viklander *et al.*, 2003). Further research is required specifically in areas with a cold climate due to the special conditions: (1) storage of precipitation and pollutants in snowpacks during extended time periods, (2) processes in the snowpack influencing the availability of pollutants, and (3) different transport dynamics due to snow and ice. As a result, the runoff quantity and quality will differ compared to snow and-ice free climates (Marsalek, 2003).

In this chapter, processes and influencing factors affecting snowmelt and snowmelt induced runoff in cold climates will be presented. The special requirements concerning the applicability, operation and need of special design of various best management practices (BMPs) in a cold climate will be addressed. This research has been conducted within the framework of the EC funded research project DayWater (Chapter 1).

DEFINITIONS OF COLD CLIMATE

There are many definitions of cold climate areas, but there is no unified definition (Chen *et al.*, 2006). According to the Stormwater BMP design manual for cold climates by Caraco and Claytor (1997), the cold climate identification for

BMP modification, is decided by temperature (the average daily maximum temperature for January, less than 1.7ºC), plant growing season (the number of days between the last freezing day in spring and the first freezing day in autumn, less than 120 days) and depth of snow (greater than 0.9 m), which are data available from National Weather Service or local weather stations. Typically, the most common measure of cold climates is the air temperature during the winter period or in January as the coldest month (UNESCO, 2000; Caraco and Claytor, 1997). In UNESCO´s Urban drainage in specific climates (2000), cold climate is described by a simple definition based on an average historical temperature of winter season less than 0ºC. Another stronger criterion, used in some countries in Europe and North America, is that the region belongs to a cold climate if the return period of January temperature greater than 0 or -1ºC is more than 10 years This means that a warm January with temperatures higher than 0 or -1º C can not occur more than 1 time during 10 years.

It is of importance to realise that cold climate areas could be subdivided into smaller areas with different climatic zones. For example, in Sweden, the winter could be divided into, at least, two different types. One type of winter climate, which is found in the south of Sweden, has a temperature often fluctuating around 0°C. Snowfall is received at precipitation and temperatures below zero, however, the snow is commonly melting instantaneously or shortly after the snowfall. The other type of winter climate is what could be found in the north of Sweden, having long periods with temperatures below 0°C. During this time there is an accumulation of snow and pollutants until spring when the temperature is rising and the snowmelt begins. This melt period is critical when it comes to risks connected to both quality and quantity. The highest concentrations of pollutants and the highest flows are found during snowmelt and rain-on-snow events respectively. The Nordic Road & Transport Research has been dividing Sweden into four different climate areas since the 1970s, using weather statistics (snow depths and January average temperatures) for a longer period, nowadays stretching from 1960 to 1990. Originally, this subdivision was used for traffic safety studies and latter in studies of speed adjustments (Öberg, 2007). Different cold climate classifications could be appropriate depending on the important parameters for the research area. However, for stormwater related research in cold climates, parameters as discussed above e.g., temperature, snow depth and plant growing season, should be appropriate for deciding if an area should belong to cold climate or not.

SNOWMELT PROCESSES AND RUNOFF

Stormwater is affected in many ways by a cold climate, with regards to both physical and chemical aspects (Bengtsson, 1990; Marsalek, 1991; Sansalone, 1996; Novotny, 1999; Oberts *et al.*, 2000). Precipitation in the form of rain immediately runs off or infiltrates, while snow is stored in snow packs in the urban catchment or transported due to different snow handling operations or to snow drifting. The relocation of snow will change the hydrological cycle since the recipient of the snowmelt will be different than the recipient of the precipitation if it had been rain. The resulting snowmelt is dictated by the amount of snow and by climatic variables such as air temperature, wind and solar radiation (Bengtsson, 1990). Furthermore, the snowmelt is largely affected by anthropogenic factors in the urban environment such as de-icing chemicals, solids from anti-skid material, atmospheric deposition, vehicle exhaust, etc..., contaminating the snow and lowering the albedo. Sun exposure, depending on building orientation and topography of the urban area, heating sources, depending on building material and distance to snow etc..., will also affect the snowmelt (Semadeni-Davies and Bengtsson, 1998). An important variable in snowmelt hydrology is the snow water equivalent (SWE) i.e., the depth of water that would theoretically result if the snowpack was instantaneously melted. The SWE determines the water volume available for runoff generation. The SWE can be calculated by multiplying the on-site measured snow density and depth. Another important variable is the percentage of snow covered area (SCA) as it is very inhomogeneous in urban areas (Semadeni-Davies and Matheussen, 2003). The heterogeneity of snow covers and snow covered areas in inner city and residential areas were investigated by e.g. Westerström (1984), Semadeni-Davies (1999b) and Bengtsson and Westerström (1992) in Sweden, Buttle and Xu (1988) and Ho and Valeo (2005) in Canada, Matheussen and Thorolfsson (2001) in Norway. At snowfall in urban areas, the snow is heterogeneously distributed due to turbulence and lee effects of the built environment. Subsequently, the snow is redistributed within the cities due to snow handling practices, removing snow from streets and other impervious areas (Bengtsson and Westerström, 1992). Furthermore, large snow volumes are often transported to central snow deposits outside the cities. Due to these practices and effects, many impervious surfaces are snow free at the beginning of snowmelt. However, Matheussen and Thorolfsson (2001) underlined that it was more the snow handling than the topography which affected the SCA in urban areas. Consequently, snowmelt-controlling variables vary in urban areas; therefore, the snowmelt will change with time and space (Semadeni–Davies, and Bengtsson, 1998; Oberts *et al.*, 2000). During snowmelt, there is usually a larger surface contributing runoff because the frozen ground results in decreased infiltration possibilities (Westerström, 1984; Bengtsson and Westerström, 1992). Research indicates that, in cold climate, the largest runoff events during the year are rain-on-snow events (Westerström, 1984).

STORMWATER QUALITY IN COLD CLIMATES

All urban water discharges are potentially hazardous to the environment; however, discharges in cold-climate regions seem particularly severe. This severity arises from the long winter period where the precipitation falls as snow and stays on the ground for several months, accumulating pollutants till the snow melting period (Figure 1) (Marsalek *et al.*, 2003).

Figure 1: The road "Södra Hamnsleden" in Luleå after snowfall at the beginning of winter (left) and at the end of the snow melt period, after months of pollutant accumulation (right).

Urban snowmelt may have damaging effects on receiving waters, soil, vegetation, animals and micro-organisms. The impact depends both on the properties of the snowmelt, such as concentrations and loads of pollutants, volume, flow, and temperature, as well as the sensitivity of the receiving body. Depending on the concentration and loads of pollutants, the effects can either be short-term, usually due to oxygen-demanding compounds, solids and bacteria, or arise after a long period of time due to heavy metals and nutrients (Malmqvist, 1983). According to Oberts (1982), up to 65% of the annual load of sediment, organic compounds, nutrients and lead can be attributed to spring melt events.

Westerlund *et al.*, (2003), and Westerlund and Viklander (2006), investigated concentrations for total suspended solids (TSS) and heavy metals which showed higher concentrations during snowmelt runoff than during rainfall runoff. Another study, for two sites, that compared seasonal variations was made in mid-Sweden (Bäckström *et al.*, 2003): larger total and dissolved metal concentrations were observed for several of the studied elements (e.g., Cu, Cd, Pb and Zn) in the runoff, during winter compared to runoff during summer. The fraction of particulate concentrations for Pb and Zn was found to increase during winter while other element particulate distribution factors remained constant. The major part of the mass transport of these heavy metals occurred during the winter runoff. Higher loads towards the end of the melt period were also observed for TSS and for total heavy metals (Westerlund *et al.*, 2003). Extreme pollutant loads was observed during the end-of-season melt events when rain falls on snow and flushes soluble and particulate pollutants from the snow banks, as well as from impermeable and frozen or saturated permeable surfaces (Schöndorf and Herrman, 1987). Sansalone *et al.*, (1995) investigated the event mean concentrations (EMCs) for TSS, Cu, Cd, Pb, and Zn at six sites during snowmelt and rainfall runoff: all investigated elements showed higher EMCs for the snowmelt runoff as compared to the rainfall runoff except one site, which had higher concentrations for Cu and Pb during rainfall runoff. Sansalone and Buchberger (1996) again reported higher concentrations for TSS and total heavy metals during melt events than during rainfall events: they explained this observation by the lower flow and the contribution of TSS, and associated metals, eluted from the melting snow bank.

Sources for pollutants in snow and road runoff

Many pollutants are similar to those found for other climatic conditions. However, some of the pollutants are present in larger quantities in a cold climate, as compared to a temperate climate (Oberts, 1994). In Chapter 11, a tool (CHIAT) for selecting priority pollutants in stormwater is presented. As well, in chapter 14, a tool for sources and fluxes of urban stormwater systems. The reason for higher quantities of some pollutants, in a cold climate, relates to the higher energy and fuel usage due to e.g. heating of buildings and cold starts of engines (Malmqvist, 1983): it was found that a cold-vehicle engine produces 2 to 8 times more particles, as compared to a warm engine (Kobriger and Geinopolos, 1984). When lowering ambient temperatures from +20 to -7°C, the emissions of particulates increased approximately 19 times (Ludykar et al, 1999). By lowering the starting temperature from -7 to -20°C, the car engine emission of particles was further accentuated up to 2.5 times. Also, vehicles contribute to snow pollution through motor exhaust and wearing of tyres, brakes, and engines (Hvitved-Jacobssen and Yousef, 1991). There is also a large wear of the pavement due to the use of studded tyres in many areas with cold climates (Baekken, 1994).

Slipperiness control

In North America, it is common to use a "bare road policy", meaning that road salt or de-icing chemicals, such as e.g. sodium chloride (NaCl), calcium chloride ($CaCl_2$), or magnesium chloride ($MgCl_2$), are used to keep the roads clear throughout the whole winter period (Marsalek, 2003). The quality of urban snow is often poor due to these de-icing chemicals, but also due to anti-skid material used on the roads. The primary de-icing chemical used in Sweden is NaCl: the salt is used either as a solid, as a water solution, or mixed with sand or gravel to avoid frozen lumps. During the winter 1990/91 about 210,000 t NaCl were spread on Swedish roads as compared to the winter of 2002/2003, where 260 000 t were used, showing approximately a 20% increase in the use of NaCl (Ojala, L. and Mellqvist, E., 2004). The usage of road salt, as a slipperiness control measure, is more common in the southern parts of Sweden where the temperature is often fluctuating around 0°C, and build-up of ice layers on roads is common in winter time (Reinosdotter and Viklander, 2005). In the north of Sweden temperatures are often lower and the de-icers do not work efficiently at such temperatures. Another difference is that the build up of pollutants is more significant for the winter climates, with long, cold periods, since the freezing-thawing cycles are not as frequent. When this accumulated snow is starting to melt, there are usually high concentrations of pollutants and, later during the melt period, there are large quantities of water and often high loads of pollutants (Oberts, 1994).

De-icing chemicals in the form of NaCl is clearly harmful to the environment, destroying vegetation, disturbing ecosystems and polluting groundwater, etc... The presence of NaCl will shift the partitioning between dissolved and particle-bound heavy metals towards the dissolved phase, which usually is more toxic and hazardous to the environment. Warren and Zimmerman (1994) found that an increase in chloride will increase the dissolved phase of Cd, Cu and Zn. In addition, cyanide concentrations can sometimes be high in snowmelt runoff because of cyanide added to salt to prevent clumping. Furthermore, in cold climate areas where salt is employed, corrosion of building material, such as copper in roofs and galvanized steel used in vehicles, etc..., is increased and contributes highly to the level of pollution (Malmqvist, 1983). According to Malmqvist (1994), 25% of all Zn in stormwater comes from corrosion in areas where Zn is used as a building material. For Cu, the corresponding value can be up to 65-75%. Air- borne residues in urban snow come from long-distance and local sources which differ greatly depending on local activities such as industrial plant operations and solid waste incineration.

According to Marsalek (2003) environmental benefits can be made by smart use of de-icers and by BMP adaptation to allow chloride dilution and passage at low concentrations, instead of accumulating chloride, and reducing or preventing chloride discharge to sensitive surface water and groundwater.

The use of anti-skid material is also a large pollutant source in areas with a cold climate, this material settles in the stormwater pipes, causing partial and sometimes total blockage of the system. Also, the sediments dissipate energy to stay in suspension, reducing the cross sectional area or increasing the frictional losses. Subsequently, their presence will lead to loss of hydraulic capacity and provoke surface flooding or, where combined systems are used, premature operation of CSOs. According to Thorolfsson, 2003, there was a large increase in solids removed from a stormwater pipe after a year of accumulation during 2001, as compared to a year of accumulation one decade earlier. This is due to a greater usage of sand as a skid control material for public safety reason.

Additionally, the sediment adsorbs pollutants which will be transported to receiving waters through the stormwater sewer or during CSO operations. During a melt period Westerlund *et al.* (2003) found that particles larger than 1.6 μm could be associated with the highest concentrations of total metals (Cu, Cd, Ni, Pb and Zn). If additional treatment of the runoff is required during the melt period, a reduction of the TSS phase is most efficient inducing a reduction of high heavy metal concentrations. However, many studies indicates that more than 50% of heavy metal loads are to be found in the sandy fractions (larger than 0.125 mm) (Ellis and Revitt 1982; Stone and Marsalek, 1996; Sansalone and Tribouillard, 1999; German and Svensson, 2002). During the rain period, the correlations between total concentrations of heavy metals and TSS and of different particle sizes were not as significant as for the melt period. The lower correlations are explained by the lower percentage of particulate-bound metals during the rain period, as compared to the melt period: this is due to the lower amount of available particulate material during the rain period, providing less adsorption surfaces for heavy metals (Westerlund *et al.*, 2003; Sansalone and Buchberger, 1996). Cold climate areas, where snow handling strategies exist, will be influenced by the choice of strategy. As discussed in the previous section, the relocation of snow will change the hydrological cycle, since the recipient of the snowmelt will be different. Obviously, this will also apply to the pollutants accumulated in the snow. Furthermore, the choice of local snow deposit as an alternative to central snow deposits might decrease air emissions (Reinosdotter *et al.*, 2003).

Transport of stormwater pollutants

The transport of urban pollutants is also affected by different climatic conditions; therefore, a good understanding of hydrology is needed to address the transport of associated pollutants correctly (Oberts *et al.*, 2000). Rainfall-induced runoff often results in a quick hydraulic response, with rapidly ascending flows and high peaks and volumes. Such rainfall events have a high intensity and can transport both soluble and pollutants associated with solids. It was found by Sartor and Boyd (1972) that contaminant materials on a surface is removed by two simultaneous mechanisms: (1) soluble fractions are present in solutions where the impacting raindrops and horizontal sheet-flow provide good mixing continuously replenishing solvent; (2) particulate matter is detached from the surface through the impact of falling drops, so called rain drop erosion; when detached, the repeated impact of adjacent drops will maintain reasonably large particles in pseudo-suspension. Snowmelt, however, is generally a slower process, which can transport pollutant-associated solids of varying size, depending on the energy of the event. This transport can be dictated by the amount of available pollutants or by the transport capacity of the flows (Oberts *et al.*, 2000).

Pollutant accumulation in a snow pack and its release from it are dynamic processes, including pollutant influx through deposition and precipitation, and pollutant release through intermittent melt-events, rain-on-snow events or the final snowmelt. During every snowmelt, pollutants in the snowpack are released (Figure 2). The initial stages of snowmelt are generally slow and the first melt exerts high pollution concentrations of soluble pollutants, but not a high pollution load. This phenomenon is frequently referred to as a "first flush". As the snowmelt continues, the dissolved pollutants are washed out, leaving particulates and associated pollutants behind. Medium and coarse particles usually stay behind until the snowpack is entirely depleted (Viklander, 1997), or until the occurrence of a high intensity rain-on-snow event with enough energy for transport of these particulates (Westerlund *et al.*, 2003).

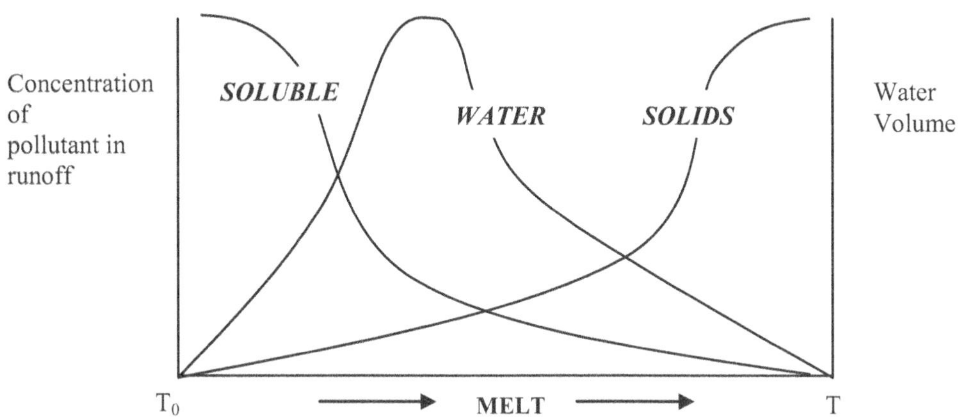

Figure 2: Snowpack release of soluble pollutants, water and solids (total suspended solids and large particles) from initiation of melt (T_0) to end of melt (T) (Oberts et al. 2000).

STORMWATER SOURCE CONTROL IN A COLD CLIMATE

In cold climates, snowmelt or rain-on-snow events might be the biggest event of the year, both in terms of volume and pollutant loads and concentrations. Therefore, these events rather than storm events, which are commonly used elsewhere in the world, might be appropriate in the design of BMPs for both quantity and quality issues. During snowmelt, the occurrence of large volumes of melt water together with large pollutant loads coincides with frozen grounds and reduced infiltration capacity, frozen pipes and inlets/outlets, ice-layers, low biological activity and changed water chemistry (Oberts, 2003). Due to the described cold climate conditions, the effectiveness of BMPs is often lowered and needs special considerations concerning design criteria, performance, maintenance and operation. In the stormwater BMP design for cold climates (Caraco and Claytor, 1997), the following general issues have been pointed out: (1) pipe freezing, (2) ice formations, (3) reduced biological activity, (4) reduced settling velocities, (5) reduced oxygen levels in bottom sediments, (6) reduced soil infiltration, (7) frost heave, (8) short plant growing season, (9) high runoff volumes and pollutant loadings during snowmelt and (10) environmental problems associated with de-icing chemicals, gritting and snow management.

Structural BMPs

Swales/Filter strips

These types of open channels are common BMPs in cold climates. An advantage of swales, in a cold climate, is their capacity to become potential snow deposit areas (Caraco and Claytor, 1997) and of conveying melt water during the snow melt period (Lindwall and Hogland, 1981). However, the snowmelt conveyance can decrease in case of ice blockages at swale inlets and outlets. Furthermore, in a cold climate, the infiltration rate can be very limited and therefore decrease the effectiveness of a swale (Caraco and Claytor, 1997). Stormwater pollution control in grassed swales is controlled by sedimentation, adsorption and infiltration. However, in a cold climate, the positive effects of vegetation, such as flow retardation and pollutant uptake, is decreased during snowmelt. It was reported by Bäckström (2003) that snow-covered swales, from three different sites, indicate removal rates in the 78-99% range for TSS and total concentrations of Cu, Pb and Zn. Dissolved Cu and Zn were instead washed out from the swale with the snowmelt, this process being registered as a negative removal rate. This is in accordance with observations of Viklander (1997) where most of the particle bound pollutants in snow deposits ended up as gravel below the snow layer and the dissolved pollutants were released with the snowmelt. On the contrary, 30-40% of dissolved Pb was retained which was explained by the higher adsorbance of Pb as compared to Zn and Cu. It was also observed that the removal efficiency of a swale depends on the incoming pollutant concentrations: high concentrations gave a high removal rate, mainly due to sedimentation of particles, whereas low incoming concentrations yielded release pollutants (Bäckström, 2003). Considering that swales are widely used in cold climates, the documented knowledge of the performance is rather confined.

Infiltration systems

Infiltration systems for stormwater, such as percolation basins, may have ice-clogging problems inside the facility and in surrounding soils, due to frost penetration during winter and, as a consequence, the infiltration capacity can be reduced during the snow melt period (Caraco and Claytor, 1997; Stenmark , 1992). Also, ice formations in pipes and inlets/outlets and frost heave might lead to floodings and over-flows (Stenmark, 1992; Lindwall and Hogland, 1981). Thus infiltration systems in cold climates should be designed for a higher infiltration rate to compensate for such clogging and reduced infiltration rates. For example Caraco and Claytor (1997) are suggesting not using infiltration systems in areas with permafrost. The performance of a percolation basin during the snow melt period was examined in Luleå during 1990 (Stenmark, 1991): factors such as the temperature distribution inside the basin and the hydraulic conductivity of the soil were investigated both through measurements and calculations. The temperature distribution and water levels were measured by placing probes and observation pipes throughout the basin, both vertically and horizontally. Furthermore, a laboratory study of a percolation basin model, in a climate room, was conducted to investigate the risk of a total ice blockage in the infiltration pipe or in the macadam layer. According to Stenmark, the study indicated a small risk of a total ice blockage, in the percolation basin, during the snow melt period, for minimum temperature of -15°C inside the basin. For insulation and decreased frost penetration, it was recommended to place the percolation basin in areas with vegetation and undisturbed snow layer, i.e., not affected by snow removal practices (Stenmark, 1992, 1995)

According to Marsalek (2003) stormwater infiltration facilities that receive chloride-laden stormwater could contribute to groundwater contamination and metal leaching, and the permeability could be reduced by chloride interference with soil structures. The risk could be mitigated by controlling the chloride use in exposed areas, focusing on infiltration of chloride–free, clean runoff (e.g., from roofs), and restricting infiltration practices to the areas with limited uses of groundwater.

Porous pavement/asphalt

In a study in a housing area in Luleå during 1995–1996, measurements of ground temperature, frost penetration depth, frost heave, groundwater levels and runoff, were performed for two different road sections, one porous and one impermeable (Bäckström, 1999). In this study, it was found that the snowmelt runoff volumes could be reduced by 50-60% by using a porous pavement (Figure 3), which also was found to be more resistant to freezing compared to conventional, impermeable pavement. The reason for such higher resistance was the higher water content in the underlying soil giving an increase in latent heat. Also, Hogland and Wahlman (1990) reported that porous pavements are more resistant to freezing and to frost actions, than conventional roads. The thawing process was also found to be faster for a porous pavement due to the snowmelt infiltration. Indications that roads with porous asphalt become free from snow and ice earlier than conventional roads was observed during several inspections in different parts of Sweden (Hogland and Wahlman, 1990). There were tendencies of more frost heave resistance for porous pavement, as well as decreased frost penetration depth, and shorter length of frost period. Therefore, the risk of frost damage was concluded to be lower for roads with porous paving, which was explained by the absence of stormwater pipes and manholes in the porous pavement. Overall, the Luleå study results indicated good performance during cold climate conditions. However, on streets where the friction material applied in winter was smaller than 2-4 mm, porous pavement was not recommended since the pavement could clog (Bäckström, 1999). It should also be remembered that the Luleå study was performed in a housing area where there was no sanding nor salting operations during winter. According to the BMP manual by Caraco and Claytor (1997) the use of porous pavement should be cautious in a cold climate, due to problems of clogging, the need of regular maintenance and to damages during snow removal operations. The risk of groundwater contamination through chloride-laden stormwater has to be considered and monitoring the groundwater for chlorides is recommended.

Figure 3: Comparison of porous pavement (left) and a nearby impermeable pavement during the snowmelt period (right).

Ponds and wetlands

Ponds and wetlands are wide-spread BMPs and despite their reduced efficiency in cold climate regions, ponds are regarded as the most reliable BMP (Caraco and Claytor, 1997; Oberts 1994). The reduced winter performance of ponds and wetlands is described in a detailed manner by Oberts (1994) and Oberts *et al.* (2000). Ice-layers during winter formed on ponds and wetlands will decrease the effective volume of the pond, thus lower the treatment efficiency. The first snowmelt will enter underneath the ice and create turbulence, whereas already settled particles might resuspend (Figure 4).

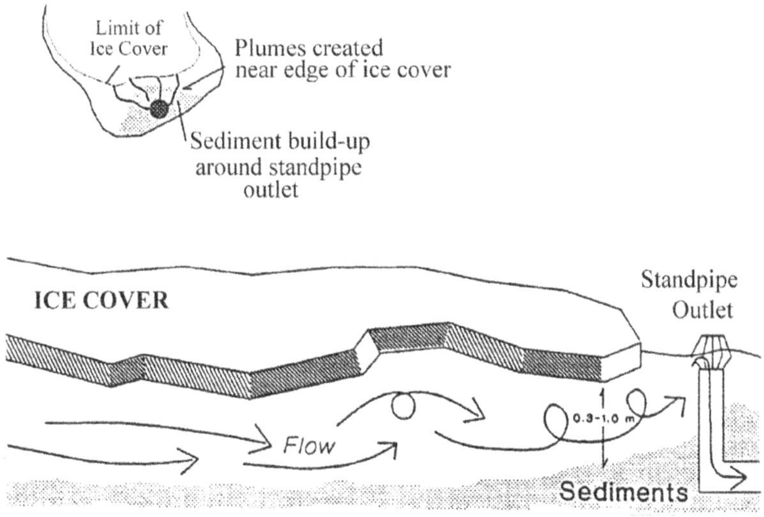

Figure 4: Schematic figure of flow under partial ice cover, either seen from above (top) or from the pond side (bottom) (Oberts, 1990).

However, it was shown in a study by Marsalek *et al.* (2003) that for a stormwater pond with an ice-layer of 0.3-0.5 m, the near bottom velocity during a snowmelt event was not high enough for scouring the bottom sediment. When the volume under the ice is full, additional snowmelt will be forced above the ice, which will further reduce the treatment capability due to the minimal settling depth above the ice. Furthermore, particles settled on the ice will easily be resuspended at the next melt/rain event or melt of the ice-cover (Figure 4). Settling velocity of a particle is controlled by particle size, but also by the shape, e.g. roundness and sphericity, and density of the particles, as well as the viscosity and density of the fluid. Consequently, the higher viscosity of the water due to lower temperatures in winter, results in a lower settling velocity of the particles. The temperature effect on particle settling velocities was discussed by Jokela and Bacon (1990). The ice-cover and decomposition of organic matter hinder aeration of the water-column, leading to anoxic pond conditions (Pettersson, 1996; Oberts *et al.*, 2000) and an increase of nutrients and metals due to the release from bottom sediments (Striegl, 1987). Finally, the vegetation and biological activity for reduction of e.g. nutrients is reduced during the winter period.

The heavy metal (Cd, Cu, Pb and Zn) treatment efficiency of a stormwater pond, with a surface area of 530 m^2 and a depth of 1.2 m, was studied during a rainfall event and a snowmelt event respectively in 1997 and 1998 (Pettersson, 1999). During the rainfall event, the total heavy metal effluent concentrations were rather low at the beginning of the event but increased towards the end, most probably due to the pond volume which was half the size of the runoff volume. The total heavy metal effluent concentrations during the snowmelt event were constantly high throughout the event, as compared to the rainfall event. The conductivity was 10 times higher for the snowmelt event, as compared to the rainfall event, due to the usage of road salt, the main source of high conductivity in urban runoff. It was also concluded that more Cu and Zn were in the dissolved phase during the snowmelt event, most probably due to the higher ionic force of the solution. On the contrary, this was not the case for Pb and Cd, which had higher percentage in the dissolved phase during the rainfall event. According to Marsalek (2003) chloride in ponds have a number of environmental implications such as a strong stratification, inhibiting vertical mixing and aeration of bottom sediments. The lack of oxygen and high concentrations of chloride may lead to release of chemicals from the bottom sediment. Finally, the high chloride concentrations may cause toxic effects on the pond biota.

Another study was made on the performance of a stormwater pond, at Bäckaslöv in the south of Sweden, during summer 1997 and winter/spring 2003 (Semadeni-Davies, 2006). Summer sampling, during 1997, showed a good treatment performance, while the winter of 2003 showed tendencies of lower removal efficiency for some of the measured pollutants. A seasonal comparison showed the same efficiency during summer and winter/spring for Cd (75%), and Cu (49%), while removal of Pb, Zn, and TSS dropped from 79%, 81%, and 80% in summer to 42%, 48%, and 49% in winter/spring. It was also pointed out that the removal efficiency varied greatly between different events.

Oberts *et al.* (1989) report on the effectiveness of 4 different detention ponds in Minnesota during both snowmelt and rainfall (Table 1). The data collected show upon a wide range of pond removal rates and a marked reduction in the ability of a pond to treat snowmelt induced runoff, as compared to rainfall induced runoff. It should be noted that the Woodbury pond was undersized.

Table 1: Mean pollution removal rates (%) of ponds in treating either snowmelt or rain runoff in Minnesota (Oberts *et al.*, 1989).

Pond	TSS	VSS	TP	DP	COD	TKN	NO_3	TPb
	Percent pollutant removal during snowmelt events (during rainfall events)							
McCarrons (0.97 ha)	58 (91)	58 (95)	41 (78)	38 (57)	50 (90)	22 (88)	13 (60)	40 (85)
Lake Ridge (0.38 ha)	72 (90)	63 (70)	10 (61)	6 (11)	- -	10 (50)	19 (10)	18 (73)
McKnight Basin (2.24 ha)	85 (85)	48 (67)	30 (48)	11 (13)	- -	10 (31)	8 (24)	59 (67)
Woodbury (0.15 ha)	-60 (46)	-46 (32)	-17 (24)	-12 (21)	- -	-27 (14)	4 (18)	-40 (42)

TSS: total suspended solids
VSS: volatile suspended solids
TP: total phosphorous
DP: dissolved phosphorous
COD: chemical oxygen demand
TKN: total Kjeldahl nitrogen
NO_3: nitrate nitrogen
TPb: total lead

Oberts and Osgood (1988), reported from a study on a Minnesota wetland system that pollutants can move through a wetland even when it appeared like the system was operating correctly. Table 2 lists the pollutant removal rates for snowmelt and for the two first rainfall events after snowmelt in a 2.5 ha, six-chambered, wetland treatment system. The wetland outlet was completely frozen during the winter and an accumulation of all melt events and base flow occurred in the final wetland chamber. The snowmelt and settled material stayed in the final wetland until the outlet gradually allowed all of the collected melt water to leave. However, the material had settled on ice layers and was washed away by the first two rainfalls that occurred about one week after the snowpack was entirely depleted. A way of improving the treatment in this situation would have been to keep the outlet open to prevent the formation of an ice layer, and then allowing detention to occur in contact with the wetland soils. The calculated removal rates both for ponds and wetland were based upon the mass inflows and outflows for each event, aggregated over the entire study period.

Table 2: Mean pollution removal rates (%) of wetlands in treating either snowmelt or rain runoff in Minnesota (Oberts and Osgood, 1988)

	TSS	VSS	TP	DP	COD	TKN	NO_3	TPb
Snowmelt	82	78	68	68	74	53	54	71
Two rainfalls after melt	4	15	6	-5	10	22	-20	39

According to Marsalek (2003) the effects of chloride upon wetlands are not well understood but it is unlikely that there would be a significant risk of chloride accumulation due to the low water depths in wetlands.

Oil and grit separators

Oil and grit separators (OGS) are often used for small impervious areas to improve the stormwater quality where oil and petrol spills can be expected, e.g. at gasoline stations, industrial areas, and sometimes at urban transportation areas. Different types of OGS either perform sedimentation of grit and TSS, or flotation of oil and grease. Marsalek (2003) reported that deep storage compartment OGS, designed to prevent resuspension of solids, tend to accumulate dissolved salts and, especially, chlorides. The chemo-stratification could potentially hinder removal of fine suspended solids since the dense bottom layer might become a barrier against incoming stormwater. Furthermore, the chloride interference with trapped sediment and adsorbed heavy metals could allow potential metal leaching. Not many investigations are done on OGS performance focusing on cold climate performance, i.e. during snowmelt events.

Gully pots

Gully pots are maybe the most common BMP. Deposited sediments, and associated pollutants such as heavy metals and organic substances transported by stormwater, will be collected in the grit chambers of the gully pots (Karlsson, 2006). Of interest is the performance of specially designed filters installed in gully-pots for treatment of road runoff. Filter systems have been tested *in-situ* in Sweden in a stormwater management system that was constructed in Västerås, in 1998 (Färm, 2003). The system consisted of a detention pond, the filter system and a constructed wetland and area catchment of 4.3 ha, including a highway with a high traffic density (20,000 vehicles/day). The results showed that the filter system (consisting of calcium silicate rock) was not permeable after the winter period. Water could not pass through the filter and the clogging was probably caused by cementation of the material. As for OGS, not many studies have been done on filter systems in gully pots during snowmelt and thus, require further research.

Non-structural BMPs

Stormwater quality can be amended, not only by structural devices, but also through measures where the source of pollutants can be effectively controlled (Chapter 8). Introduction of non-structural BMPs, such as street sweeping, a different snow handling strategy, gully put cleaning etc., or public education campaigns can reduce the need for treatment in structural BMPs.

Street cleaning

After snowmelt, street sweeping can be an efficient measure for removing sediment, and associated pollutants, before the sediment is washed off by rainfall (German, 2003). According to Ellis and Revitt (1982), street sweeping is particularly efficient in removing particles larger than 250 μm and therefore it might be expected that conventional street sweeping will have a small effect on the reduction of toxic runoff, since most metals have a greater affinity for smaller particles, which will be transported through the sewer system into receiving waters. However, despite that the metal content increases with a decreasing particle size, it was found in many studies that the largest metal loads were mostly associated with midrange to coarser particles, i.e. larger than 0.125 mm (Ellis and Revitt 1982; Stone and Marsalek, 1996; Sansalone and Tribouillard, 1999; German and Svensson, 2002). Stone and Marsalek (1996) found that less than 10% of the total metal load was in the smallest fraction; below 63 μm. Street sweeping is affecting the total available metal loadings as well as the elemental size distributions (Revitt and Ellis, 1980). Latter street sweeping experiences from Anchorage, Alaska, in 1999, showed tendencies of good street sweeping efficiency for particles greater than 100 μm (WMP, 2002). Recent studies from Canada have shown that street sweeping can remove sediments down to a few μm (Marsalek, 2006). According to Breault *et al.* (2005) most of the data related to sweeping efficiencies come from studies done in the late 1970s and in the beginning of 1980s, when street sweepers were nearly all mechanical and inefficient, or from manufacturers' reports, most often published in trade journals. It was concluded that vacuum sweeper was 1.6 to 10 times more efficient, as compared to the mechanical sweeper, for all particle sizes (silt and clay <63 μm; very fine sand <125 μm; fine sand <250 μm; coarse sand < 2mm; ≥2 mm coarse sand) (Figure 5). The sweeper efficiencies are probably a function of total mass availability on street surface and location, with respect to the curb, particle-size distribution, and of variations of the sweeper operation.

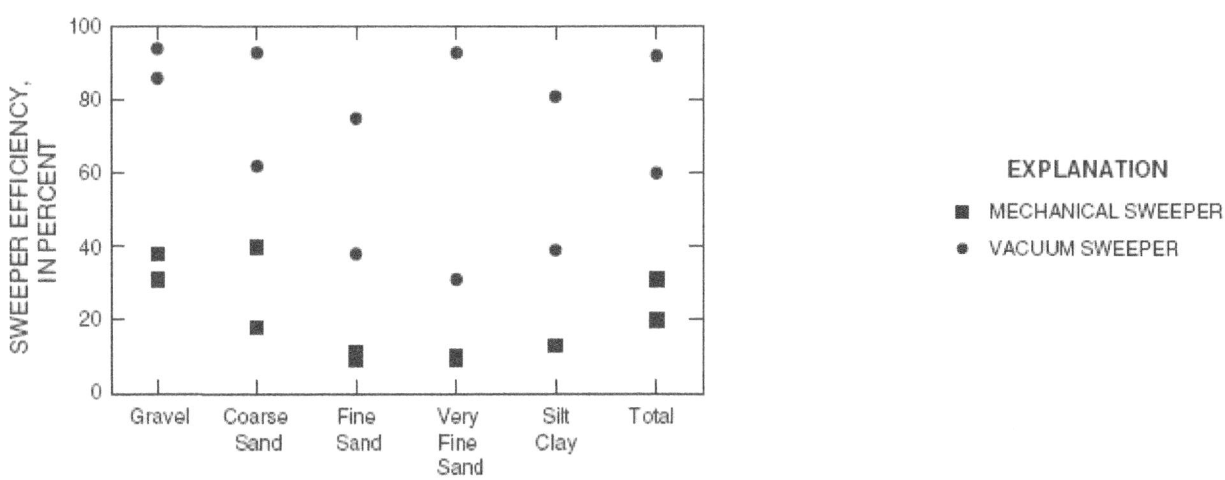

Figure 5: Street-sweeping efficiencies measured for the Pelican series P mechanical sweeper and the Johnston 605 Series vacuum sweeper, New Bedford, Massachusetts (Breault *et al.*, 2005).

To investigate the efficiency of street-sweeping, experiments were performed in Jönköping, Sweden (German and Svensson, 2002). A six week study was conducted: (1) during the first three weeks (sweeping period 1) the street was swept once a week, (2) during the following three weeks (sweeping period 2) the street was swept every day (Table 3). Generally, an increasing frequency of sweeping increased the amount of removed sediment. However, the increased amount of sediment was not proportional to the increase in sweeping effort. The largest loads of metals and sediments were found in the sandy fraction whereas it was concluded that street sweeping with modern sweeping equipment can be an effective pollutant control measure, in terms of removal of heavy metal loadings.

Table 3: Removed sediments and heavy metals in sediment (< 0.25 mm) during the street sweeping test periods (German and Svensson, 2001).

Sweeping period	All (kg)	Sediments < 0.25 mm (kg)	Cr (g)	Ni (g)	Cu (g)	Zn (g)	Pb (g)
1	37.9	6.3	0.4	0.1	1.5	1.2	0.2
2	46.6	10	0.5	0.2	2.5	2.5	0.4
1+2	84.5	16.3	0.9	0.3	4	3.7	0.6

Snow handling practices

For people in winter cities, snow gives joy as well as problems. One of the positive aspects with snow is the possibility for skiing and other outdoor activities; in addition, many people find a snowy city beautiful. Nevertheless, problems such as space limitations, poor accessibility and unsafe road conditions are caused, therefore making it important that the streets are cleared of snow. This is done by the city snow handling practices, which usually consist of removal, slipperiness control, transport, and dumping. These practices achieve the objective of road safety and uninterrupted traffic flow, but cause inconvenience due to the large number of vehicles involved during the ploughing and transportation of the snow. It presents an impact on the environment such as noise and exhausts fumes and, in addition, put great strain on the municipalities' budget. The costs for winter road management are, together with the costs for maintenance of the pavement and street lightning, the largest operation costs for Swedish municipalities. Snow removed from streets and parking areas will be dumped either in water or on land. If the snow is dumped directly in a water course, all the substances are spread in the water. The particle-bound substances settle on the bottom, while the dissolved fractions, to a larger extent as compared to the particle-bound substances, enter the biological systems.

A major part of the city snow must be transported to central snow deposits, which are usually located outside the city centre and housing areas. It is also rather common that snow is dumped into small deposit sites in city centres, the so-called local snow deposits. An increased use of smaller snow deposits inside the city will give a more effective use of the resources. However, the surroundings may be affected in many ways; vegetation may be damaged by the snow weight and the melt water runoff and the plant growing season may be delayed. Under the snow deposit, the soil may be contaminated and the ground temperature will probably be affected. The snowmelt, with its content of pollutants, may cause environmental impacts on close recipients as well as the groundwater. The snowmelt induced runoff may cause damage to buildings due to large volumes of water during a short period. An increased number of local snow deposits will also affect the people. The deposits can be perceived as enrichment in the urban environment or as a problem. How will we handle the risk for personal injures, which can arise if children will use the snow piles as play-yards?

The environment is heavily affected by the polluted snow and choice of snow management. One suggested strategy to improve the environmental effects is to sort the snow in different categories between less polluted snow and heavily polluted snow (Malmqvist, 1985; SEPA, 1983). Heavily polluted snow should be transported to a central snow deposit appropriately located, designed and operated in order to reduce negative environmental effects during the snowmelt period. In contrast, less polluted or clean snow can be dumped at storm water outlets. Consequently, snow-handling procedures can be a good example of non-structural BMPs, under cold climate conditions.

CONCLUSION

The different conditions during a cold climate, as compared to a temperate climate, have a great influence on the quantity and the quality. The concentrations and loads of TSS, particles, and associated pollutants during snowmelt events are high, especially during rain-on-snow events, compared to rain events (Oberts, 1994).

An important factor for pollutant concentrations and loads during snowmelt are the conditions before the occurrence of the melt event i.e. how much material is present due to abrasives or de-icing chemicals, both on the road, and from snowpack elution (Figure 6). Has the temperature been below 0°C during the entire winter, or has intermittent melt events occurred? The number of particles in the snow pack will decrease throughout the melt period as wash-out is continuously taking place. However, the use of anti-skid material, for slippery road conditions, is a common practice in cold-climate regions: they occurr at several occasions throughout the melt period. Therefore, the availability of material might increase during parts of the snowmelt period. Also, of importance is the intensity of the lateral flow that is transporting the particles to the gully pot. This is predominantly important for the larger particles that need a high intensity lateral flow to be transported.

Figure 6: Difference in accumulated material at the same study site in Luleå, Sweden, during snowmelt (left) and rainfall events (right).

The different factors governing the quantity and quality of snowmelt-induced and rainfall-induced runoff respectively suggest different treatment solutions in different climates or during different times of the year.

A comparison was made of heavy metal and TSS concentrations determined in stormwater runoff from a road with 7400 vehicles per day, for both rainfall and snowmelt induced runoff (Westerlund *et al.*, 2003) with concentrations for roads runoff (below 15,000 vehicles per day) calculated by the Swedish Road Administration (SRA). The Luleå study concentrations for heavy metals during the melt period were well in agreement with the suggested min–max range, while the concentrations for the rain period were slightly lower than suggested values. The concentrations for TSS during the melt period were much higher than the suggested concentrations, while the TSS concentrations for the rain period were within the suggested range. The calculated concentrations by the SRA, and by similar institutions in many other countries with a cold climate, could be improved by incorporating values for both rainfall - and snowmelt-induced runoff. As discussed earlier, the differences in content and composition of runoff between snowmelt and rainfall will affect the efficiency of various BMPs. In a study by Bäckström and Viklander (2000) three general indicators (runoff control, pollution control and level of integration) and a fourth indicator specific for cold climate (winter performance) were defined and estimated for a chosen set of stormwater BMPs (Table 4). The study was based on a literature review of stormwater BMPs and the score for each indicator ranged between low (--) to very high potential (++) effect.

Table 4: Positive (+) and negative (-) effects of different stormwater components.

	GENERAL INDICATORS			
STORMWATER BMPs	Runoff control	Pollution control	Level of integration	Winter performance
Combined pipe system	+	-	- -	- -
Separate pipe system (no treatment)	+ +	- -	- -	- -
Stormwater Infiltration	+	+	+ +	-
Percolation basin	+	+	+	+
Porous pavement	+	+ +	+	+ +
Grassed swales and ditches	+	+	+ +	+ +
Wet pond (including wetland)	+ +	+ +	+	+
Dry basin (dry pond)	+	+	+	-
Stormwater reuse (household)	+	-	+	- -
Large-scale snow deposit	+	+	- -	+
Local snow deposit	+/-	-	+ +	+/-

The general indicators were then plotted against the winter performance as seen in Figure 7. From this figure it was then possible to determine which BMPs were more or less suitable in a cold climate. Of course, this information is based upon literature reviews and on the perception of the authors: it should not be seen as a definite answer or solution to the best BMP to choose in a cold climate, but should instead be used as a base for discussion.

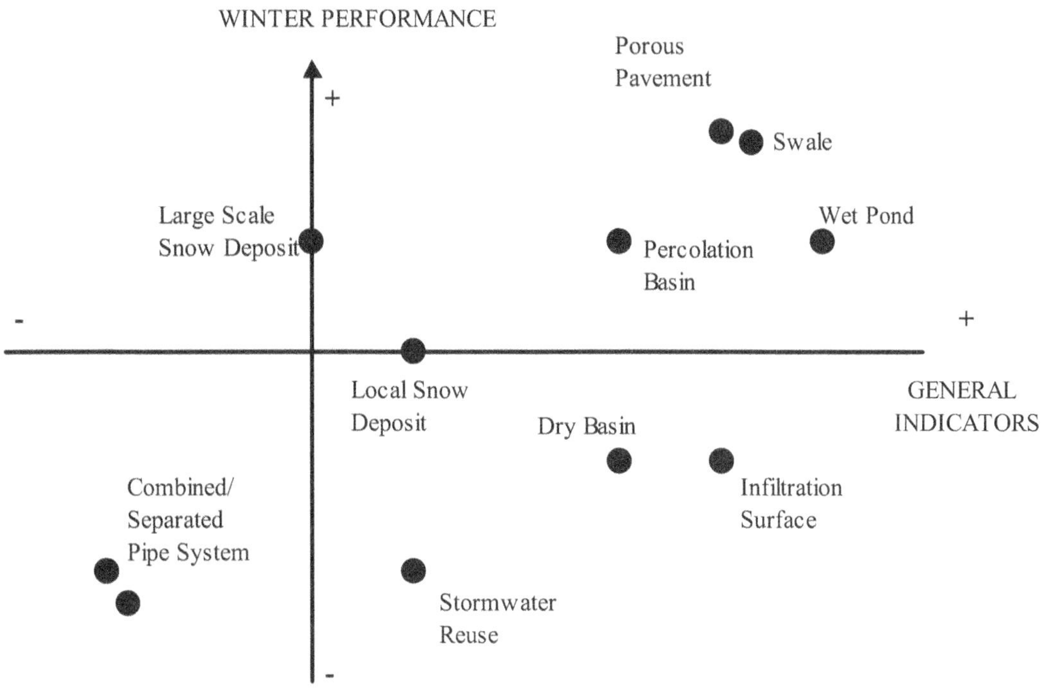

Figure 7: Winter performance vs. general performance indicators (Runoff control + Pollution control + Level of Integration) for different BMPs for stormwater source control in cold climate conditions.

Knowledge about winter performance of BMPs is essential for decision making when dealing with stormwater treatment, snow handling, and street sweeping. Many BMPs developed for controlling flow and pollutants have showed reduced efficiencies or not work at all in winter, during snowy and icy conditions. Still, there is a good understanding of the processes behind the pollutant accumulation in and transport from snow. However, further research is needed on assessing snowmelt impact on receiving water bodies, modifying existing BMPs, and modelling of urban snowmelt quantity and quality (Oberts *et al.,* 2000). To decrease impacts from road runoff in different climatic regions, it is therefore important to understand the dynamics of pollutants and then use this knowledge to predict and prevent environmental damage.

This work was conducted within the DayWater project framework (Chapter 1). Although parts of it are mentioned in the BMP catalogue (Chapter 8), no specific ADSS component was developed for its valorisation.

REFERENCES

Baekken, T. (1994) *Trafikkforurenset snö I Oslo. (Snow polluted by traffic in Oslo)*, NIVAR-report, O-94047, Oslo, Norway. ISBN 82-577-2603-6. (In Norwegian)

Bengtsson, L. (1984) Modeling snowmelt induced runoff with short time resolution. *Proceedings of the Third International Conference on Urban Storm Drainage*, vol. 1, Analysis and Design of Stormwater Systems, Göteborg, Sweden, pp. 305-314.

Bengtsson, L. (1990) Urban Snow Hydrology. *Proceedings of Urban Hydrology Under Wintry Conditions* Narvik, Norway, March 19-21, 1990.

Bengtsson, L. and Westerström, G. (1992) Urban snowmelt and runoff in northern Sweden. *Hydrological Sciences journal*, **37**(3): 263-275.

Breault, R.F., Smith, K.P. and Sorenson, J.R. (2005) *Residential street-dirt accumulation rates and chemical composition and removal efficiences by mechanical- and vacuum type sweepers*, New Bedford, Massachusetts, 2003-04. Scientific investigations report 2005-5184, U.S. Geological Survey.

Buttle, J.M.and Xu, F. (1988) Snowmelt runoff in suburban environments. *Nordic Hydrology*. **18**, 19-40.

Bäckström, M. (1999) Porous pavement in a cold climate. Licentiate thesis, 1999:21 ISSN 1402-1757 Luleå University of Technology, Luleå, Sweden.

Bäckström, M. (2003) Grassed swales for stormwaterpollution control during rain and snowmelt. *Water Sci. Technol.* **48**(9), 123-132.

Bäckström, M., Nilsson, U., Håkansson, K., Allard, B. and Karlsson, S. (2003) Speciation of heavy metals in road runoff and roadside total deposition. *Water, Air and Soil Pollution*. **147**, 343-366.

Caraco, D. and R. Claytor. (1997) *Stormwater BMP Design Supplement for Cold Climates*. Center for Watershed Protection. Ellicott City, MD.

Chen, R., Kang, E., Ji, X. Yang, J, Yang, Y. (2006) Cold regions in China. *Cold regions science and technology*. **45**, 95-102.

Ellis. B, and Revitt, M. (1982). Incidence of heavy metals in street surface sediments: Solubility and grain size studies. Water, Air and Soil Pollution.**17,** 87-100.

European Commission (2006) *The EU Water Framework Directive – integrated river basin management for Europe.* http://ec.europa.eu/environment/water/water-framework/index_en.html.

Färm, C. (2003) Constructed filters and detention ponds for metal reduction in stormwater. PhD thesis 2003:04, ISBN 91-88834-40-9. Department of public technology, Mälardalens högskola, Västerås, Sweden.

German, J. and Svensson, G. (2001) Street sweeping as a pollutant control measure. *Proc. of the 4th international conference on Innovative technologies in urban drainage.* Lyon, France, June 25-27, 2001, pp.383-390.

German, J. (2003) Reducing stormwater pollution –performance of retention ponds and street sweeping. PhD thesis 2003:06, ISBN 91-7291-316-9, Department of Water, Environment, Transport, Chalmers University of Technology, Gothenburg, Sweden.

German, J. and Svensson, G. (2002) Metal content and particle size distribution of street sediments and street sweeping waste. *Water, Science and Technology*, **46**(6-7), 191-198.

Ho, C.L.I. and Valeo, C. (2005) Observations of urban snow properties in Calgary, Canada. *Hydrol. Process.* **19**, 459–473.

Hogland, W. and Wahlman, T. (1990) *Enhetsöverbyggnad, Hydrologiska och Vägtekniska egenskaper.* R90:1990, Byggforskningsrådet. ISBN 91-540-5264-5 (In Swedish).

Hvitved-Jacobsen, T. and Yousef, Y.A. (1991) Highway runoff quality, Environmental Impacts and control, In *Highway Pollution*, Editors:Hamilton, R.S., Harrison, R.M., Elsvier, Netherlands. ISBN 0-444-88188-3.

Karlsson, K. (2006) Pathways of pollutants in stormwater systems. Licentiate Thesis 2006:05. ISSN 1402-1757. Department of Architecture and Infrastructure, Luleå University of Technology, Luleå, Sweden..

Kobriger, N.K. and Geinopolos, A. (1984) *Sources and migration of runoff pollutants.* Vol. III, Research report, U.S. Department of Transportation.

Jokela, J.B. and Bacon T.R. (1990) Design of urban sediment basins in Anchorage. In: *Cold regions Hydrology and Hydraulics.* Ed. by Crissman, R.D. American Society of Civil Engineers, New York, 761-789. ISBN: 0-87262-773-X. 3.

Lindwall and Hogland (1981) *Operational aspects on stormwater inifiltrations.* Report R14:1981, Swedish Council for Building Research. ISBN 91-540-3435-3. (In Swedish)

Ludykar, D., Westerholm, R. and Almén, J. (1999) Cold start emissions at 22ºC, -7 ºC and -20 ºC ambient temperatures from a three-way catalyst (TWC) car: regulated and unregulated exhaust components. *The Science of the Total Environment*, **235**, 65-69.

Malmqvist, P.-A. (1983) Urban stormwater pollutant sources. An analysis of inflows and outflows of nitrogen. phosphorus. lead. zinc and copper in urban areas. Dissertation Series. Dept. of Sanitary Engineering. Chalmers University of Technology. Gothenburg. Sweden. ISBN 91-7032-106-X.

Malmqvist, P-A. (1985) Environmental effects of snow disposal in Sweden. *Proc. of Workshop Minimizing the Environmental impact of the Disposal of Snow from Urban Areas*, Junc 11-12, 1984. Report EPS 2/UP/1, October 1985. ISBN 0-662-14390-6

Malmqvist, P.-A. (1994) *Stormwater characteristics.* Report 1994-11, Va-forsk, Swedish Water and Wastewater Association (VAV) (In Swedish).

Marsalek, J. (1991) Urban drainage in cold climate: Problems, solutions and research needs, *Int. Conf. Urban Drainage and New Technologies (UDT91)*, Elsevier Applied Science, Dubrovnik, Yugoslavia, pp. 299–308.

Marsalek, J. (2003) Road salts in urban stormwater: an emerging issue in stormwater management in cold climate. *Water Sci. Technol.* **48**(9), 61-70.

Marsalek, J., Oberts, G., Exall, K and Viklander, M. (2003) Review of Operation of Urban Drainage Systems In Cold Weather: Water Quality considerations. *Water Sci. Technol*, 48(9), 11-20.

Marsalek, J. (2006) Project chief, Urban Water Management Project, and Research Scientist at the National Water Research Institute, Burlington, Ontario. Personal communication, 2006-11-10.

Matheussen, B.V. and Thorolfsson, S.T. (2001) Snow surveys in Risvollan – Norway. *World Water & Environmental Resources Congress, Urban Drainage Modelling Symposium*, Orlando, Florida, USA, 20-24 May 2001, (ASCE, EWRI).

Novotny, V., Smith, D.W., Kuemmel, D.A., Mastriano, J. and Bartosová, A. (1999) *Urban and Highway Snowmelt: Minimizing the Impact on Receiving Water.* Project 94-IRM-2, Water Environment Research Foundation (WERF), Alexandria, USA.

Öberg, G. Head of research, VTI, the Swedish National Road and Transport Research Institute. Personal communication, 2007-01-16.

Oberts, G.L. (1982) *Water resources management: nonpoint source pollution.* Metropolitan Council, Publ. No. 10-82-016, St. Paul, MN.

Oberts, G.L. (1990) Design Considerations for Management of Urban Runoff in Wintry Conditions. *Proceedings of Urban Hydrology Under Wintry Conditions*, Narvik, Norway, March 19-21, 1990.

Oberts, G.L. (1994) Influence of Snowmelt Dynamics on Stormwater Runoff Quality. *Watershed Protection Techniques*, **1**(2), 55-61.

Oberts, G.L. (2003) Cold climate BMPs: solving the management puzzle. *Water Sci. Technol.* **48**(9), 21-32.

Oberts, G.L. and R.A. Osgood. (1988) *Lake McCarrons wetland treatment system: Final report. on function of wetland treatment system and impacts on Lake McCarrons.* Metropolitan Council, Publ.No. 590-88-095, St. Paul, MN.

Oberts, G.L., P.J. Wotzka and J.A. Hartsoe (1989) *The water quality performance of select urban runoff treatment systems.* Metropolitan Council Publ. No.590-89-062a. St. Paul, MN.

Oberts, G.L., Marsalek, J. and Viklander, M. (2000) Review of water quality impacts of winter operation of urban drainage. *Water Qual. Res.* **35**(4), 781-808.

Ojala, L. and Mellqvist, E. (2004) Vägsalt-användning och påverkan på grundvattnet.SGU-rapport-2004-13. (In Swedish)

Pettersson, T.J.R. (1996) Pollution Reduction in Stormwater Detention Pond - Field study and modelling experiments. Licentiate Thesis, Department of Sanitary Engineering, Chalmers University of Technology, Gothemburg, Sweden.

Pettersson, T.J.R. (1999) Stormwater Ponds for Pollution Reduction. PhD thesis 14, Department of Sanitary Engineering, Chalmers University of Technology, Gothemburg, Sweden.

Reinosdotter, K. and Viklander, M. (2005) A comparison of snow quality in two Swedish municipalities – Luleå and Sundsvall. *Water, Air, and Soil Pollution.* **167**(1-4), 3-16.

Reinosdotter, K., Viklander, M. and Malmqvist, P-A. (2003) Comparison of the effects of using local and central snow deposits: a case study in Luleå. *Water, Science and Technology*, **48**(9), 71-79.
Revitt, D.M. and Ellis, J. B. (1980) Rain Water Leachates of Heavy Metals in Road Surface Sediments. *Water Research* **14**(10), 1403-1407,
Sansalone, J.J. and Buchberger, S.C. (1996) *Characterization of Metals and Solids in Urban Highway Winter Snow and Spring Rainfall-Runoff.* Transportation Research Record 1523: 147-159, TRB, National Research Council, Washington, D.C.
Sansalone, J.J., Buchberger, S.C. and Koechling, M. (1995) Correlations Between Heavy Metals and Suspended Solids in Highway Runoff: Implications for Control Strategies. Transportation Research Record (1483), TRB, National Research Council, Washington, D.C.
Sansalone, J.J. and Tribouillard, T. (1999) Variation in characteristics of abraded roadway particles as a function of particle size. Transportation Research Record (1690), TRB, National Research Council, Washington, D.C.
Sartor, J.D. and Boyd, G.B. (1972) Water Pollution aspects of street surface contaminants. EPA-R2-72-081. U.S. Environmental Protection Agency, Washington, D.C. 20460.
Schöndorf, T. and Herrman, R. (1987) Transport and chemodynamics of organic micropollutants and ions during snowmelt. *Nordic Hydrology*, **18**, 254-278.
Semadeni-Davies, A. F. and Bengtsson, L. (1998) Snowmelt sensitivity to radiation in the urban environment. *Hydrological Sciences Journal* **43**(1), 67-89.
Semadeni-Davies, A. F. (1999b) Snow-cover heterogeneity in Luleå, Sweden. *Urban Water* **1**, 39-47.
Semadeni-Davies, A. F., and Matheussen, B.V. (2003) Progress in urban snow hydrology (1990-2003): from Narvik to Riksgränsen. *Proc. 1st International Conference on Urban Drainage and Highway Runoff in Cold Climate*, 25-27 March 2003, Riksgränsen, Sweden, pp 327-337.
Semadeni-Davies, A. (2006) Winter performance of an urban stormwater pond in southern Sweden. *Hydrological Processes*, **20**(1), 165-182.
SEPA (1983) *Trafikens inverkan på dagvatten,* [The influence of traffic on stormwater], The Swedish Environmental Protection Board, Report 1722 (In Swedish).
Stenmark, C. (1991) The function of a percolation basin in a cold climate. *Proc. Int. Conf. UDT'91.* Dubrovnik.Yugoslavia. 17-21 June.
Stenmark, C. (1992) Local disposal of stormwater in cold climate. Licentiate thesis 1992:24 L. Luleå University of Technology. ISSN 0280-8242.
Stenmark, C. (1995) An alternative road construction for stormwater management in cold climates. *Water Science and Technology* **32**(1), 79–84.
Stone, M. and Marsalek, J. (1996) Trace metal composition and speciation in street sediment: Sault Ste.Marie, Canada. *Water, Air and Soil Pollution* **87,** 149-169.
Striegl, R.G. (1987) Suspended sediment and metals removal from urban runoff by a small lake. *Wat. Res. Bull.* **23,** 985-996.
Thorolfsson, S.T. (2000) Specific problems in urban drainage in cold climate. Urban Drainage in Cold Climates. Urban Drainage in Specific Climates, Vol II: Sægrov, S., Milina, J. and Thorolfsson, S. T. (eds.)Maksimovic, C.(Chief ed.). UNESCO IHP-V / Technical Documents in Hydrology / No. 40, Vol. II, pp. 120–147
Thorolfsson, S.T. (2003) Problems in urban drainage in cold climate-experiences in north European Atlantic regions. Proceedings of 1st international conference on urban drainage and highway runoff in cold climate, 25-27 March, Riksgränsen, Sweden, pp. 247-257.
UNESCO (2000) *Urban drainage in cold climate.* Urban drainage in specific climates, Vol II: Saegrov, S., Milina, J., and Thorolfsson, S.T. (Eds), Maksimovic, C. (Chief Ed.). UNESCO IHP-V/ Technical Documents in Hydrology No 40, Vol II.
Viklander. M. (1997) Snow quality in urban areas. Ph.D thesis, Luleå University of Technology, Luleå, Sweden.
Viklander, M., Marsalek, J., Malmquist, P-A., Watt, W.E. (2003) Urban drainage and highway runoff in cold climates:conference overview. *Water, Science and Technology*, **48**(9), 1-10.
Warren, L.A. and A.P. Zimmerman. (1994) The influence of temperature and NaCl in cadmium, copper and zinc partitioning among suspended particulate and dissolved phases in an urban river. *Water Res.* **28**(9), 1921-1931.
Westerlund, C., and Viklander, M.,(2006) Particles and Associated Metals in Road Runoff during Snowmelt and Rainfall. *The Science of the Total Environment*, **362**(1-3), 143-156.
Westerlund, C., Viklander, M., and Bäckström, M. (2003). Seasonal variations in road runoff quality in Luleå, Sweden. Water, *Science and Technology*, **48**(9), 93-101.
Westerström, G. (1984) Snowmelt runoff from Porsön residential area, Luleå, Sweden. *Proceedings of the Third International Conference on Urban Storm Drainage.* vol. 1, Analysis and Design of Stormwater Systems, Göteborg, Sweden, pp. 315-323.
WMP (2002). Anchorage OGS and street sweeping as storm water controls: performance analysis. Municipality of Anchorage watershed management program, Document no: WMP APr02002, November 2002.

Part 5

Related Decision Support Systems

Chapter 23

The PETUS project: a step to fill the gap between theory and practice of urban sustainability

Emmanuel Lardieg, Marcel Miramond and Elham Alfakih

THE PETUS PROJECT: PRESENTATION

The background

Prior to PETUS (2005), partners worked together for about four years as part of a COST Action C8 (2004) which dealt with urban infrastructure and sustainability assessing a range of case studies carried out by local authorities and other government organisations. The main objective of the action was to develop better solutions for sustainable urban infrastructure by exchanging experiences on and by promoting the diffusion of best practices concerning sustainable urban infrastructure. The secondary objective was to achieve a common understanding on sustainability and related criteria for decision-making in urban planning and design dealing with infrastructure. This action of the COST C8 was more an exploratory action than a research one, its overall goal being to establish a state-of-the-art related to practices and theories on urban infrastructure. During the time of conclusions of the COST Action C8 (2004), two particular points were raised by the different partners:

- The first underlying pillar of the PETUS project was the following observation: urban actors often hesitate between the theory of sustainable development, provided by international texts and charters ("Brundtland" Report, UNESCO convention, Rio convention, European directives, national legislation, etc.) and the difficulty of putting the concept into action. The theory and practice of sustainability both exist. Examples of urban infrastructure projects which managed to be successful in incorporating environmental, economic and social elements exist. However points common to the study of the practices and the exploration of theoretical documents are not obvious. What is a sustainable project? This gap between theory and practice should be understood in terms of a lack, *i.e.* a lack of tools to put principles into practice.
- Every urban infrastructure from the design stage to demolition, from small infrastructures to huge neighbourhood projects have an impact on the environment, society, and economy. The second observation of the COST Action C8 which constituted a starting point of the PETUS project is the need to evaluate the extent of impacts due to urban infrastructures. Many tools have been developed in the past to assess the sustainability of urban projects (Life Cycle Analysis, Social Impact Assessment, set of criteria, etc.) but few of these tools are used in practice. This lack of proper evaluation led to uncertainty over the success of projects and difficulty in assessing their transferability for application elsewhere.

The PETUS project aims at contributing to these two observations and related problems.

The PETUS framework

Studies and explored practices during the project showed that, very often, evaluation tools used in practice are very different to tools developed from theories. Often in practice little time is available to research the most suitable evaluation tool for a project, and the selection of tools is *ad hoc*, not comprehensive in approach, and often based on local experience and personal preference. In many cases this may yield satisfactory outcomes, and the experiences of skilled operators are invaluable to informing the debate on future developments. However, this experience needs to be channelled into a constructive framework of tools and procedures that is replicable and comprehensive.

The project team identified that a framework of Practical Evaluation Tools for Urban Sustainability (PETUS) (2005) would enable a better sustainable management of projects for those working on urban infrastructure projects, in the public and private sectors. The framework would allow the benefits gained and lessons learnt from one project to be of real benefit to others working in similar areas through provision of case studies showing tools for urban sustainability in practice. The PETUS project therefore was formed to develop a framework of practical evaluation tools for urban sustainability, which will enable a consistent approach across a range of urban infrastructure projects, including water/sewage, waste, energy, green/blue space, transport and buildings & landscape and reviewed in a holistic manner.

The PETUS project

The PETUS project was a 3-year research project (2002-2005) funded by the European Commission as part of the 5th Framework Programme Energy, Environment and Sustainable Development, within key action "City of Tomorrow & Cultural Heritage" and includes eight European partners (Welsh School of Architecture and Cardiff University in UK, Institut National des Sciences Appliquées de Lyon in France, Université de Liège in Belgium, Joanneum Research in Austria, Technical University of Denmark, University of Architecture, Civil Engineering and Geodesy in Bulgaria, Etukeno Oy in Finland, Wandsdronk Architektuur in The Netherlands) coordinated by Cardiff University, UK. The eight partners were issued from different areas of expertise in order to share experiences; each partner being 'responsible' for one specific sector of urban infrastructure.

Collaboration with potential end-users such as local authorities, municipalities and industry is essential for PETUS to succeed. It was therefore decided to strongly embed the project development with local partners to provide information, advice and test to ensure that the framework stays practical and to ensure its compatibility with current working practice and local conditions. Each project partner has a group of end-users. A total of 57 organisations participated in the project as end-users. End-users were involved in:

- Providing information about current evaluation tools in use,
- Supplying case studies of projects at different scales and at different life cycle stages,
- Providing information on positive and negative aspects of tools in use and how they assist in the decision making process,
- Advising during development and testing of evaluation tools to ensure that the framework was appropriate, user friendly and compatible with working practices,
- Providing advice on how best to disseminate the project to as wide an audience as possible.

As mentioned above, the main project objectives were:

- Providing clear and concise information on sustainability,
- Analysing and improving the sustainability of urban infrastructure projects,
- Learning about sustainability since, at this time, there is little feedback from projects in terms of sustainability: professionals have to learn sustainability while conducting projects and to conduct projects while learning.

Project methodology

The following methodology was undertaken:

(1) **Literature Review:** tools with a theoretical background were reviewed together with associated benchmark information currently available to evaluate urban sustainability;

(2) **Selection of case studies:** case studies were selected from a range of European projects from different sectors and scales at different life cycle stages; decision-making procedures, tools and benchmark data used in practice have been investigated;

(3) **Development of a Decision Support System** (DSS): this framework was being developed for both sector specific and holistic projects integrating environmental, social and economic factors which will provide performance criteria that will be related to benchmark data; the emphasis has been on consistency, transferability and ease of use in decision making;

(4) **Testing the framework:** it was being tested on previously used and new case studies and has be modified where necessary to suit current working practices;

(5) **Decision making:** this consisted in investigating how the efficiency of decision making and administration of projects can be assisted by the use of evaluation tools.

OVERVIEW OF PETUS FRAMEWORK

The PETUS framework (www.petus.eu.com) has been developed to help people who are involved with, or affected by building and infrastructure to consider impacts on the environment, economy and society. The framework was designed to guide users to sustainability through providing:

- Tools to help incorporate and evaluate sustainability,
- Case studies to exemplify how tools have been used in urban infrastructure projects,
- EU legislation, to provide knowledge of laws that should be attained, particularly within sectors less familiar to the user,
- A monitoring and guidance facility to assist users to make a quick but informed decision on the most suitable approach to incorporate sustainability.

Data are available for different levels of expertise and therefore provide a facility to enable discussion between decision makers and stakeholders from all different levels and help them understand information that is relevant and useful for them. There are two components of PETUS: an information database and a guidance procedure (Figure 1). The two parts are both accessible from the front page of PETUS. The left side of PETUS, the 'Database', contains case studies, tools and legislation whereas the right side, the 'Guidance and monitoring' facility, is designed to assist projects to incorporate sustainability into a project, plan or programme. The PETUS framework was designed with these two components in order to allow it to be used independently by end-users to select information that is interesting or relevant to them from the 'Database', or as part of a 'Guidance' process that provides a data collection system guiding the end-user to appropriate material relevant to their project/plan. The web site map illustrates the layout of PETUS and provides details and access to the two components of the PETUS framework ('Database' and 'Guidance facility'), both being presented below.

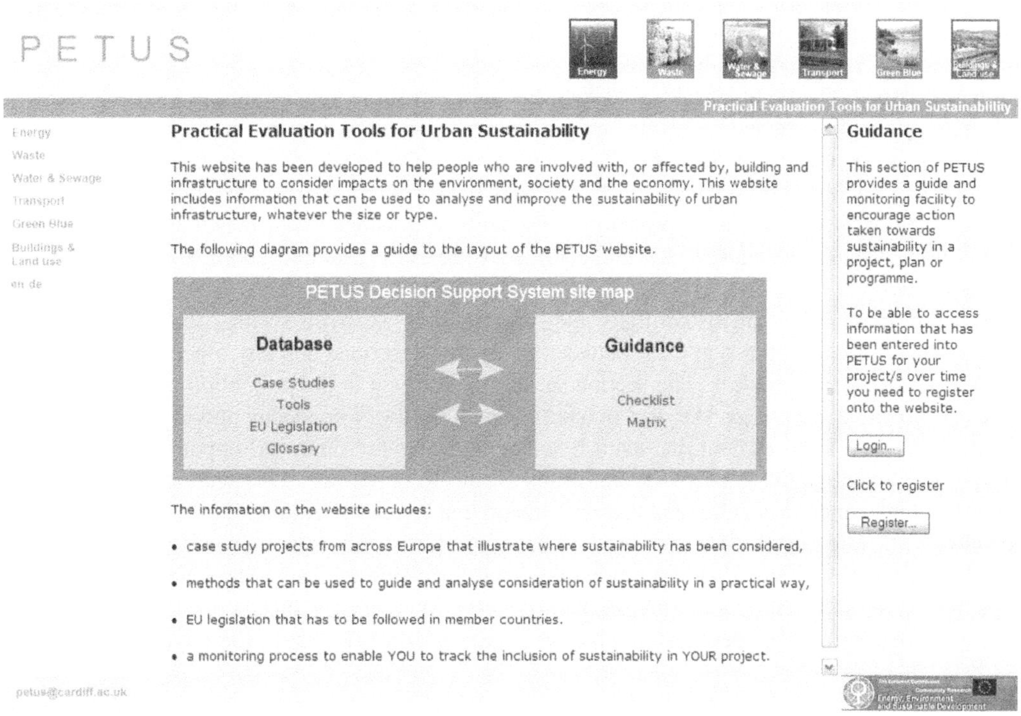

Figure 1: PETUS home page available at www.petus.eu.com.

The 'Database'

The PETUS 'Database' presents information on a sector basis to enable ease of access to data to make PETUS as practical as possible. It is the main part of the framework and could be considered as an informative or "Intelligence" system.

Compared to the structure described by Le Moigne (1990) for the modelling of complex systems decision- making processes, the PETUS 'Database' refers to the first step of the structure: the 'Intelligence' phase. The two other steps being the 'Conception' - the system design intentional actions plans - and the 'Selection' - the system which compares evaluation of action plans. The 'Intelligence' phase is the process by which the decisional problem is built; it is an exercise of problem formulation by the identification of non-satisfactory arguments.

The framework has been designed to allow two routes of entry:

- By sector, *i.e.* energy, waste, water & sewage, transport, green/blue and buildings & land use, under which the relevant sector case studies, tools and EU legislation are listed;
- Through the 'site map': this method provides an explanation of each area and then lists in alphabetical order the case studies, tools, legislation or glossary.

These alternative access routes allow the PETUS framework to be used flexibly by users who wish information only on one sector or alternatively a greater variety, or possibly by selecting some countries only.

Sector description

When access is sought by sector the user is presented with a 'Sector Description'. This part corresponds to the need to:

- Include theoretical knowledge,
- Provide information for end-users who are not specialised in a certain sector with an understanding of the sector,
- Put the emphasis on cross sector links (or conflicts and opportunities between sector information),
- Offer PETUS framework users an easily usable sector entry point.

The structure of the 'Sector Description' is as follow:

(1) **Introduction to sector**. The writing of this part was left more or less to the discretion of the author, depending on the sector described. However some questions were discussed by all partners, such as "What does the sector cover?" and "Which sustainable development and sector (like energy or water) management?" Additional information was included depending on the partner's expertise and any other relevant information that could help the end-user to understand the subject.

(2) **Important issues**. The section seeks to make users aware of some important issues in each sector: it deals with discussions of working space and time scales, presents problems as well as existing and possible solutions related to the issue considered and gives illustrations through links to case studies.

(3) **Benchmarks and indicators.** This section provides examples of benchmarks and indicators which could be useful in terms of sector needs: this part presents both discussions about benchmarks and indicators such as different uses, different categories, etc. and also lists of relevant examples, commonly accepted (if possible) in the sector area.

(4) **Cross sector links.** Also called 'conflicts and opportunities' between sectors, this section presents the main conflicts and/or opportunities that exist between the sector described and any other sectors. The information is based on case study experiences. The section seeks to contribute to the promotion of an enlarged vision of any problems – in an attempt to get the user outside his own field of knowledge and to understand that sectors are interlinked, as well as to explain that even if sector priorities are different, opportunities exist to make urban projects more coherent and to solve several problems at the same time. This section advocates that sustainable development must include an enlarged thematic vision that crosses over sectors.

(5) **References.** The reference list provides the end-user with further information sources.

The 'Sector Description' was intended to provide relevant information to users not specialised in a sector rather than for experts in this sector. The sector descriptions were written looking at subject breadth rather than specifics, the overall aim being to present a sector description which provides information that is attractive and easily understandable for all types of stakeholders.

Case Studies

The following goals were considered as central when reviewing the project description:

- Exploring the consistency of tools used and looking for good examples of tools,
- Exploring how tools were related to decision making,
- Identifying knowledge-gaps and barriers for tools amongst decision-makers,

- Identifying success and practical applicability of tools including their transferability,
- Identifying differences between tools available and tools used in practice.

According to the project description a wide range of case studies should be included:

- Sector specific and holistic projects,
- Different scales from individual building level through to regional scale,
- Geographical areas, selected from a range of European countries,
- Life cycle, different stages of progress from inception to realisation of an operation.

It was agreed that "splendid" and "spectacular" projects on sustainability, e.g. pilot and demonstration projects, were not necessarily good cases in a PETUS context, as many of these projects have almost no impact on the daily operation in the sector and can be difficult to relate to. Instead, it was agreed that case studies should focus on the following criteria:

- Involving new tools and new methods of sustainable decision making; where no tools are used, alternative ways for decision making should be identified;
- Illustrating general trends and activities in the sector (not outstanding or extraordinary examples);
- Both finished and ongoing projects;
- Involving comparability and cost-effectiveness;
- Relevant data and information about decision procedures available;
- Cases distributed to cover all sectors;
- Relating to future or current EU-regulation (e.g. EIA, SEA, Water Framework Directive);
- Relating to other EU-projects.

Box 1: Short description of a case study: "Lyon confluence-holistic study" (Lyon Confluence, 2007), a building and land-use sector within Lyon (France).

"Lyon Confluence" is currently one of Greater Lyon's biggest projects. The project consists of the urban renewal of an old industrial area to create a city centre district containing residential, commercial and leisure facilities. Located between two rivers, the Rhone and the Saone, and long restricted to industry and transport facilities, the peninsula is undergoing a complete and extensive change. This strategic area is now being prepared for the function it should have due to its central and pivotal location South of the main Perrache railway station: a true city-centre location.

For Greater Lyon the Lyon Confluence project is a means of bringing to the city centre jobs, services, institutions and major events that mark a real capital city, and thus establishing the city's future on the international scene. The project aims to:

- Create a new residential neighbourhood that will enhance Greater Lyon's standing,
- Develop an inventive, appealing choice of urban leisure activities,
- Highlight the rivers and the landscape features of the site,
- Reclaim industrial and logistical brown field sites,
- Open up the southern Peninsula, notably via public transport.

In their material, height and density, the blocks resemble the rest of the Peninsula, the same being true of the range of functions: residential, tertiary sector, leisure, retailing, services, public amenities and so on. The river setting, planted areas, riverbanks and inland pools make up a green blue network that distinguishes between the Rhone and Saone Rivers while forming a unified whole and optimising the appeal of public spaces and landscaping.

The project presents step by step land planning:

- A first stage for 12 years (Concerted Development Zone) is intended to set the process of change well under way. It will provide the basis for a central, attractive and lively neighbourhood embodying a dynamic, innovative image matching Greater Lyon's international ambitions;
- A long term project for 30 years defined by "constants", *i.e.* fundamental elements of district development that will not change for 30 years and that will give the area a unity. The long term project aims at being a major urban project well integrated in its environment.

This case was chosen because "Lyon-Confluence" is a large rehabilitation project of an old industrial district, creating a city centre containing residential, commercial and leisure facilities. Due to its central position, the project developed within a particular political context where general stakes were highly important for the influence of Lyon, where political forces have already shown an interest in sustainability.

Sixty case studies were identified and investigated with the selected case studies used at least one tool to assess sustainability. Two approaches were investigated to propose a case study:

- To choose one tool and see how it can be used in practice, highlighting one particular tool and observing its use in different projects. Care-W-Arp (CARE-W, 2005), for example, belongs to this methodology where one particular tool is used in different situations.
- As an example of one project declared as sustainable and to see which tools were used to manage the sustainability aspects of a project. Box 1 presents a short description of a case study which belongs to this approach.

The aim is not to state if the project is sustainable or not, but rather to explore how people have tried to incorporate sustainable elements in the project context, to explore how sustainability has been put into practice. However we have tried to remain critical as regards experiences, in order to allow end-users to get a true idea of the positive and negative points of the project and tools implemented. In each case study, assessment of tool implementation from both the tool user and case study reviewer is available.

Within PETUS, two levels of case study information are available, whether accessing the case studies through the site map or by sector. First, a brief overview of the case study includes the country in which the case is located, the type of activity and a brief summary of the case study. Further information is presented under the headings of:

- 'Project description' providing a description of the case study and the context in which it is set;
- 'Opinion on tools' providing support for and barriers to using the tools within the case study;
- 'Application of tools' providing information about the individual tool/s used in the case study, for example, where a tool is available and who created it;
- 'Decision making process' providing information about the decision making process within the case study; this includes information about how the tool/s used assisted the decision making process, whether the public were involved in the case and how information about the case study was communicated;
- 'Contact details' providing information about websites, references and contact details for people associated with the project; this format links the user directly to reviews of the tools used in the case studies as well as to the general review of the tool outside the same case study.

Tools

On the PETUS website, the tools identified and reviewed are accessible in three ways – through the site map or by sector or through the case studies. Via the site map, background information is provided for the end-user on "What is a tool?", "How can tools help you?", "What are the different characteristics of tools?", "What tools are included on PETUS?" and "Reviewed tools". The headlines of the tool description are: summary, characteristics, application, Source and, Opinions. 126 tools have been reviewed from literature and for each case study, at least one tool has been reviewed. Examples of tools are: evaluation grids, indicator sets, HQE process, BREEAM, cost-benefit analysis, Hydraulic simulation models, etc. In order to simplify the selection of a tool or group of tools that may be useful for a project, tools were grouped into a number of different types: this classification (Box 2) is based both on experience and practice, and made through case study analysis. For each type, significant information was provided: a summary of the type of tool, the stage of a project when the type of tool could be used, the type of output that is produced when using the tool and some comments on experiences from using each type of tool. (Figure 2).

Box 2: PETUS Typology of tools.

1 Process guide
- 1a. Framework tools
- 1b. EIA and SEA procedures
- 1c. Schemes and policies on sustainability (national and local).
- 1d. Checklists and guidelines

2 Calculation
- 2a. LCA-tools
- 2b. Other environmental calculation methods
- 2c. Economic and Social Evaluation Tools;
- 2d. System simulation tools

3. Assessment methods
- 3a Multi-Criteria Assessment tools.
- 3b. Evaluation procedures
- 3c. Surveys.

4 Monitoring
- 4a. Indicators and benchmarks.
- 4b. Accounting tools.

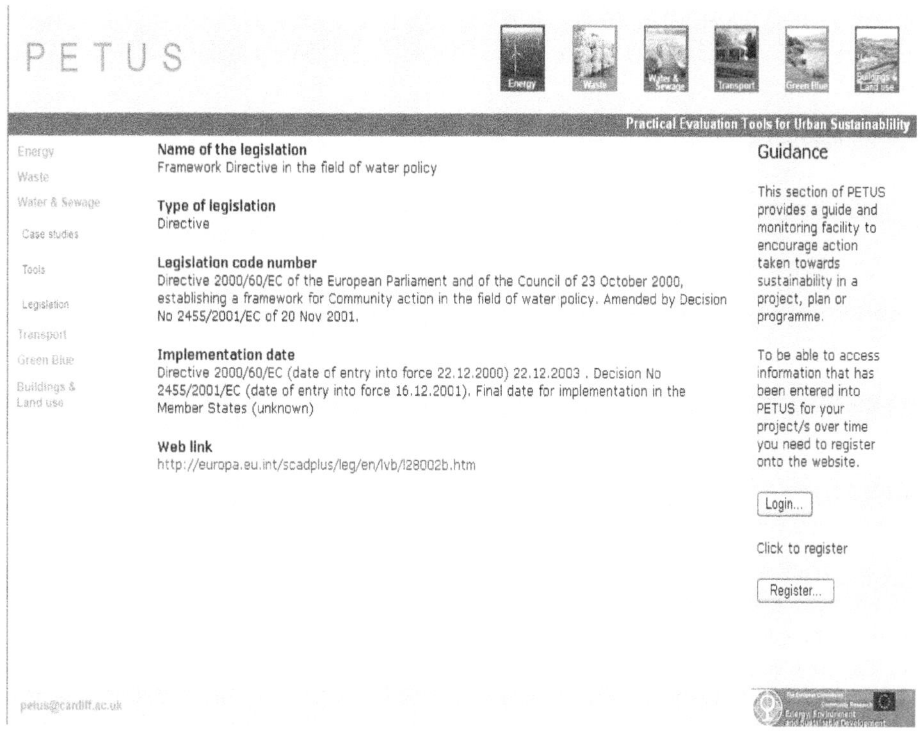

Figure 2: View of legislation part of PETUS framework.

Legislation

People are usually aware of the EU legislation that exists in their own field of expertise or interest but are uncertain of EU legislation in other areas that may be relevant to them. The PETUS framework therefore provides a quick and easy way of accessing legislation in other areas that may be of interest or relevance. This section proposes relevant but basic information - Name of legislation, Legislation Code Number and Implementation Date - and includes a web link (Figure 2). One hundred and nine pieces of legislation, Directives, Green papers or White papers, have been included, all associated with sustainability. The legislation can be accessed through two entry points to PETUS, either through the site map which leads to an explanation of legislation and a full list of EU legislation included in PETUS, or by sector to allow an overview of laws guiding the sector on a European scale.

Guidance part

The PETUS framework provides also a guide and monitoring facility to encourage action to be taken towards sustainability in a project, plan or programme through the provision of the PETUS matrix. This enables end-users to collect information about a project/plan as it progresses. When an end-user has a new project, the PETUS framework provides a matrix related to his own project. Access of the matrix is by registration and logging onto the site. This allows the information to be stored by the user and returned to him during the project, as well as providing access restriction or authorisation to the project (login details can be shared within a project team).

This support tool was designed to support decision makers in their work and to respond to the lack of intelligibility as to what should be done to consider a project in a sustainable way. Decision making is a complex task and requires a comprehensive approach between the stakeholder groups involved. PETUS follows a comprehensive or “holistic” view on the subject of decision making, to help to:

(1) Achieve interaction between stakeholders and understanding for each other’s situation,
(2) Make proposals for new arrangements in negotiation and communication structures,
(3) Gather information for the involved stakeholders.

The PETUS matrix

The concept of a decision making matrix works as an interface for filling in information and structuring information into boxes to provide a guide towards sustainability. The matrix consists of two guiding ‘headlines’ that can be seen in the decision making matrix:

- A vertical headline to the left describing a set of process steps or stages,

- A horizontal headline which relates to issues that should be considered at each step or stage.

The basic structure of the vertical axis of the matrix is shown in Table 1 below whereas the horizontal headlines include sections listed in Table 2.

Table 1: Steps within the PETUS matrix: the matrix vertical axis.

Step	Description
Baseline - gathering information	Describing the background – status quo Giving a problem description, external constraints, etc.
Objectives - identifying alternatives	Sustainability issues should be included and considered.
Key impacts	Definition of key indicators and selection of tools relates strongly to the defined key impacts. A range of indicators and tools is available from the PETUS toolbox.
Evaluating alternatives	Choosing between alternatives
Proposing improvements	Any forms of mitigation – particular measures or follow up activities.
Reporting	Draft reports, milestone reports, information to different stakeholder groups. Dissemination according to the background of the stakeholders.
Monitoring	What should be monitored? How should it be monitored? What means are required for the monitoring?

The boxes of the matrix are composed of a series of questions which aim to guide the user to incorporate sustainability into urban infrastructure projects. The questions relate to the different stages of a project/plan or programme development and also to the different issues that should be considered at the different stages. Not all questions within the matrix need to be answered, but they are included to help or prompt the user to consider sustainability issues, assist with the decision making process and to keep a record of developments of the project over time. It is also possible to include in the boxes any kind of additional information that seems relevant (personal contact, sustainable statements, links with relevant documents, etc.) and need to be shared with other project stakeholders.

The matrix was designed to keep a record of the development of the project and as a potential method of keeping all stakeholders informed: the analysis of PETUS case studies showed that successful application of tools and related procedures requires transparency and easy-to-follow implementation in order to be understood by all involved stakeholders.

Table 2: Description of relevant elements to be considered in decision making: the matrix horizontal axis.

Issue that should be considered	Description
Project details	Describing decisions that are needed in each step, considering reverting and looping back between/to certain steps.
Tools, methods, indicators	Considering the tools that are in use, whether they are helpful. If there is a need for additional tools or specific purposes (e.g. participation, additional indicators etc.) or to improve efficiency?
Public stakeholders, participation	Considerations on stakeholder groups and their role. What kind of process would be appropriate to exchange and co-operate with stakeholders.
Decision making	Considering the existing decision making structure with respect to the defined objectives. Identification of stakeholders that should be integrated in the decision making procedure.
Communication	Selecting the form of participation that would be supportive for the project. Consider different communication pattern for involved stakeholder groups.
Output	Reports and presentation of intermediate results. Considering approved forms of public information, information material and sources.

Different uses of the matrix have been considered:

- The PETUS matrix was designed to be used as a “collaborative platform”, as a way to make people work together, to share information between partners and to keep all the stakeholders involved informed of the evolution of the project. In fact, it is possible to share the login information with different partners and ask them to answer questions or to add points of view related to the sustainability requirements of the project. Thus, the matrix is the place where opinions, points of view can be shared and discussed and therefore contributes to mitigation of potential conflicts.
- The PETUS matrix is a tool which enables the memory and the history of the project to be kept and traced. A project is a sequence of choices. PETUS could help to keep track of it. It is a way of not losing scattered information. Some years after the end of urban project construction, information for technical choice justification is often lost (project team change, etc). The PETUS matrix is a tool enabling project construction to be understood a long time after particular technical choices have been made.

- The PETUS matrix is also a relevant tool to establish post evaluation of a project. In fact, all the required information is already gathered. This enables the project objectives really achieved to be evaluated in terms of sustainability.

The main goal of the decision making matrix is to create inspiration, rather than to create a formulated process that has to be followed exactly. It therefore has to be simple to use. During its design, it was agreed that it would be better to have questions rather than answers within the matrix, together with links to relevant information within or outside the PETUS database. The matrix was designed to be completed directly by decision makers and for use as a guide for the end-users - an incentive rather than a strict scheme.

DISCUSSION OF MAJOR RESULTS OBTAINED DURING THE PROJECT

Literature review

The major results collected during the literature review are presented below; more information and further results are available on the PETUS project website (www.petus.eu.com) where all the deliverables are downloadable.

Tools investigated have been classified as either generic, holistic or sector specific as it is understood that these different types of tool are used for different purposes and provide different types of results. Generic tools are theoretical procedures that were originally developed to investigate sustainability and have been used to create more practical tools over time.

The review of the theoretical/generic tools highlighted how important generic tools have been in promoting sustainability from the highest level of decision making throughout an organisation such as EMS to public education or to procedures such as environmental labelling. Generic procedures tend to focus on one factor only, environmental, economic or social, which is likely to be a significant driver in the development of more practical tools which combine factors enabling a more comprehensive sustainable approach.

The lack of an accepted method for any of the generic procedures assessed, results in a lack of clarity of the concept of sustainability. This makes the application of procedures difficult to use and understand. This also prevents comparison between assessments.

The review of tools demonstrated that there is a limited set of benchmarks available which are realistically achievable. These vary greatly between sectors under investigation and between countries. There is a need for realistic clear legislation to help promoting sustainability. In many cases actors are confused and vary in benchmarks as they are often set with a lack of detailed baseline data to define targets. By using benchmarks set by different organisations or by government, comparisons can be made with other urban infrastructure projects, particularly those in the same sector. Benchmarking requires a base knowledge for continual changes, whether positive or negative, to be measured.

There is a need for clarification of the concept of sustainable development to decision makers. Uncertainties exist as to what sustainable development encompasses: the tendency is to think only in terms of the environment, with economic and social factors not being incorporated in sustainable development assessments. Urban infrastructure projects can often be considered and promoted as more sustainable when all three sectors are incorporated.

Case studies review

The major results collected during the review and analysis of case studies are presented below.

In a few cases the use of only one tool is noticed. In most cases, stakeholders use a combination of pre-existing tools to lead their project in a more sustainable way. These tools can be charters, evaluation grids, agreements, specifications, "logbooks" or indicators. It is the interconnection of all these types of tools that provides a relevant toolkit for the stakeholder that leads a project to sustainability. But it would be wrong to affirm that nothing has been developed and created in the last decade, that the way stakeholders lead their project has not been influenced by the sustainability concept. The combination of the tools used is improved by two new influences: governance and a multidisciplinary approach.

The new tools that have been created generally concern the governance dimension, which means the way that the range of stakeholders, including ordinary inhabitants, plays a role in decision-making. This dimension of sustainable development is essential to put the tools into application; it gives the justification for politicians or in a more global sense for decision makers, to act.

The combination of tools refers also to the multidisciplinary concept, which plays an important role in sustainable development. In fact, sustainability needs to take into consideration the widest range of sectors that can be related to the project (transport, green/blue space, stormwater, drinking and waste water, building, energy…) and also various issues (economic, environmental, social…). To reach this goal, stakeholders use new ways to manage this complex target like coordination cells, workgroups or diagnosis. This is one of the most important key-issues for building an integrated project approach.

The holistic issue of sustainable development, which refers to wide space scales like neighbourhood or city, is definitively more difficult to approach by stakeholders than sector specific issues. The latter, even if not easy to solve, are tackled by the combination of different tools as is mentioned above. However, for the holistic issue, it is noticeable that stakeholders do not know how to tackle the problem because the context (policies, requirements, targets, etc.) in which their project takes place has been radically altered, and the use of governance is the beginning of a solution. To go further in the research area, it could be interesting to cross, to combine the holistic study and the different sector specific case studies that can be explored. This work could underline the lack between the different sectors, improve the multidisciplinary conception, and create new tools for holistic issues.

One of the most important difficulties outlined by case studies is the problem of communication for the stakeholder who carries the project. This communication and/or information problem exists between the project holder and both citizen and politicians (or decision makers). This problem is due to the following features:

- Sustainable development is a new method of project management and its implementation has not yet proved its usefulness; citizens and politicians have not seen the positive results that this new knowledge can contribute to attain: thus they are not absolutely convinced by this new concept.
- Problems of communication and/or information are increased by differing visions of sustainable development by different stakeholders.
- Of course, lack of available clear information about sustainable development increases the communication deficiency.

This problem of communication which restrains project evolution, in terms of time and efficiency, can be reduced (or increased) by the project holder: the importance and the role of this particular actor are essential for its success in each case and maybe even more for the holistic approaches. The permanence and the duration of teams who carry the project improve its coherence.

The question of the "long term approach", when thinking for tomorrow, refers to the right for future generations to take their own decisions: such question is not taken into account by decision makers, and sometimes rejected, which is confusing for the sustainability concept. This long term vision is substituted by speeches about reversibility of decisions (preferring taking decisions that future citizen can later adapt), and by the notion of adaptability (choosing a solution according to its capacity to be adapted to other difficulties). On the one hand, it is an emphasis on the precaution principle but on the other, the decision making process is simplified and it reduces the complex construction of lastingness.

The case studies focused on possible reasons for using or not using tools (Table 3).

Table 3: Reasons for using or not using tools.

Motivations for using tools	Reasons for not using tools
• Legislation and regulation	• Lack of motivation
• Demand by a financial funding stakeholder	• Little or no knowledge of the tool
• Providing a clearer statement of the existing problems and possible alternatives	• Tool is too complicated and requires too many resources
• Part of development of a project	• Tools lack legitimacy, reliability and transparency
• Legitimisation and labelling	• Data are not available or accessible.

The case studies identified a number of different benefits when using tools:

- Demonstrating more sustainable solutions: the use of tools can make decision makers aware of more sustainable solutions available, or take up issues that were not thought of or known about initially;
- New decision procedures involving new actors: using tools can lead to new decision making procedures that involve new actors to implement sustainable solutions;
- Assessment motivating improvements: enabling an assessment or evaluation of a sustainable project, after it has been completed, can motivate the users and owners to make sustainability improvements;
- Legitimisation, documentation and labelling: the desire of end-users and decision-makers for a project to receive authorised certification as 'sustainable' is a central feature of most projects and policies;
- Spin-offs from using tools: there are a number of examples of spin-offs from using these tools in terms of unexpected advantages or benefits; this could include more efficient administrative procedures, more cost-efficient procedures, more careful design, more satisfied customers or a better image amongst clients.

As the concept of sustainability and its concrete applications is quite recent, the feedback that end-users can obtain from projects considered as sustainable is really rare, not to say non-existent. Still, end-users have to operate while learning and learn while operating.

CRITICAL APPRAISAL OF THE PETUS FRAMEWORK

PETUS testing

PETUS has been tested within public and private sector organisations and modifications have been and will be made to the framework in order to ensure compatibility with current working practices and local conditions. Each PETUS partner is developing an extensive network of end-user partners who have tested PETUS and who are helping to guide the project to produce a user-friendly evaluation framework. Some feedback conclusions of PETUS testing with end-users are presented below.

The first general feelings on PETUS DSS are often positive; yet its practical implementation in every day work is not so clear: problems of language, time for its implementation etc. have been raised. Through discussions, PETUS appears as a way to learn about sustainability and not to improve technical competence: as PETUS investigates a large area (several sectors, different space and time scales and economic, social and environmental items being investigated), it is not possible to propose a comprehensive technical database. Nevertheless PETUS provides more approaches, discussions, context specific examples etc. related to sustainability.

PETUS is essentially available in English (only some parts are translated into German and Turkish). It appears that English is a barrier to French users. Professionals will not waste time on translation, if they can find equivalent information in French. The need for translation to promote the framework in everyday work is essential.

PETUS helps to enlarge the vision of problems and projects (for instance between partners of an urban infrastructure project) in terms of impact assessment.

Here are presented some examples of conclusions (not exhaustive). The varied feedback from end-users is displayed and discussed in the 'Testing report' available on the PETUS project website. The feedback from end-users is essential for the development of PETUS. As mentioned, it ensures that the development of PETUS stays close to the end-users' needs and work realities. Additionally, end-users have detachment from the framework development and have a different view of the problems, technical improvements, missing parts or the different potential uses of PETUS. The PETUS team tried to pay special attention to delivering easy to understand information (knowledge popularisation). This point sometimes also presents a risk of superficiality.

PETUS for whom?

We hope that PETUS provides feedback about sustainability experiences through its different parts. This 'indirect' feedback can have four objectives in relation to different type of PETUS end-users:

- Experts who are interested in sustainable projects, examples, ideas in their own field of expertise: they need detailed studies, which present a lot of information and different local contexts; these case studies should for example tackle the questions of transferability, local context…;
- Experts interested in projects, opportunities…in another sector than their own field of expertise: they know what is possibly sustainable in their own sector and would like to improve, enlarge their view, their way of thinking; these case studies should present them with opportunities, conflicts, benchmarks, propositions of new possibilities of collaboration between new actors that have to learn how to work together (multidisciplinary approach), propositions for new kinds of work organisation and management…;
- Decision makers: case studies should here help end-users to see what is possible in relation with decision making, what could be enhanced…; it deals with policy, plans and programmes;
- A fourth category could be young technicians or engineers that need to learn easily and quickly elements of sustainable development (examples of sustainable urban projects): case studies appear here as (professional) training.

CONCLUSION

Is the PETUS framework a Decision Support System (DSS) or an Informative Support System (ISS)? Indeed the question is: "Does PETUS really support the decision making process?" We think that in the current situation, PETUS is more an ISS than a real DSS. To know if the online PETUS framework really supports decisions, it should be tested and used in real urban infrastructure projects, which typically have a time scale of 2, 5 or 10 years. Such testing time scale obviously does not correspond to the time scale of the PETUS research project (3 years) and of the testing period (about 8 months).

The success of PETUS to really support decisions made during an urban project depends on the framework appropriation and use by end-users. Now the questions are: "How will the PETUS matrix be used? How can the framework improve project management? Are there enough case studies to create inspiration and to present transferable information from relevant projects? How to maintain information on PETUS? How will new tool descriptions and new relevant case studies be inserted in the future when the PETUS project is finished? Etc."

To conclude, we think that firstly, PETUS should be tested in several urban infrastructure projects throughout Europe where stakeholders have an interest in sustainability to check if it is really supporting (and improving) decision making through time. Then the PETUS framework could be improved with some slight modifications (specially on the guidance part), and finally, if time is spent to explore how the matrix can be controlled and used, PETUS could become a real help and provide benefits for stakeholders involved in urban infrastructure projects related to sustainable development and become a real DSS.

REFERENCES

CARE-W (2005) Computer Aided Rehabilitation of Water Networks, http://care-w.unife.it (verified August 2007).

COST UCE C8 (2004) Best Practice in Sustainable Urban Infrastructure, Sustainable urban infrastructure: approaches-solutions-methods, B. Zanon ed., TEMI EDITRICE Italy, ISBN 88-85114-8066, 426 p.

Le Moigne, J.–L. (1990) La modélisation des systèmes complexes [Modelling complex systems],. Bordas Editions, Paris, in French.

Lyon Confluence (2007) http://www.lyon-confluence.fr (verified August 2007).

PETUS (2005) Practical Evaluation Tools for Urban Sustainability, http://www.petus.eu.com (verified August 2007).

Chapter 24

Risk analysis and impact assessment of urban stormwater – with emphasis on the EU – Water Framework Directive

Carolina Engelhard and Wolfgang Rauch

INTRODUCTION

Urban stormwater can create a risk to both environment and human health along different exposure routes. On the one hand, the pollutants contained in surface runoff are discharged in separate systems directly to receiving waters or introduced via infiltration to soil and groundwater. On the other hand, stormwater can lead to overflows in combined sewer systems and so to the discharge of both runoff and wastewater pollutants. Source control measures can help to reduce the amount of stormwater discharged into surface waters. By helping to identify the most effective source control measure the ADSS developed by the project DayWater facilitates the achievement of good ecological status in surface waters impacted by urban drainage (Chapter 1).

The main points of the European Water Framework Directive (WFD) will be explained and it will be shown how this regulation will determine various aspects of future urban stormwater handling in Europe, as e.g. the combined approach, the list of priority substances, the good ecological status of surface waters, or the good quantitative and chemical status of groundwater bodies. Then an approach is presented for the evaluation of measures in the wastewater system which tries to integrate the good ecological status and the list of priority substances. This approach is based on a classification scheme for receiving waters, which has been developed on the basis of the main impacts of urban stormwater on running waters. Finally the problems encountered at assessing impacts on the ecological quality by applying ambient water quality limits are described.

THE EUROPEAN WATER FRAMEWORK DIRECTIVE (EU – WFD)

The European WFD (2000/60/EC) aims at integrated water policy. It came into force in the year 2000 and is supposed to ensure consistent and clear water policy in the European member states by combining various earlier directives in the field of water (like for example the directive concerning the quality required of surface water intended for the abstraction of drinking water in the Member States (75/440/EEC) or the directive of on the quality required of shellfish waters (79/923/EEC), etc.).

The WFD features the combined approach, meaning that the directive contains both emission based regulations and ambient water quality approaches. Therefore it demands the setting of emission limit values and of environmental quality standards. Emission limit values are also called end-of-pipe limits. They are limits for loads discharged over a

 DayWater: an Adaptive Decision Support System for Urban Stormwater Management. Edited by Daniel R. Thévenot. ISBN: 9781843391609. Published by IWA Publishing, London, UK.

certain time. The limits are applied to facilities' discharges, without concerning the receiving water. Emission based limits are widely used in regulations because they are easy to apply. Environmental quality standards - on the other hand - are concentrations of pollutants which should not be exceeded in order to protect human health and/or the environment, and are to be defined for water, sediment and biota. The combined approach in the WFD requires the member states to reduce pollution from point and diffuse sources, by applying emission limits on the basis of best available techniques or best available controls. If the pollution reduction achieved by the emission limits is not sufficient to protect specific receiving waters (i.e. if the quality objective or quality standard set for the specific receiving water is exceeded) more stringent emission controls have to be defined. The combined approach of the WFD does not permit to reduce the requirements for wastewater treatment below the emission limits, even if the receiving water could cope with the pollution. However, it has to be noted that the EU WFD is not European law but instead a guideline to be adapted by the member-states: therefore, national water law may occasionally divert from the above.

The WFD aims at a good status for all European water bodies (Figure 1). Further important traits are cross national administration of river basins on the basis of management plans, a reduction of the emission of priority substances, the no-deterioration clause for water bodies, a widespread monitoring of the waters and recovery of costs approach for all water services.

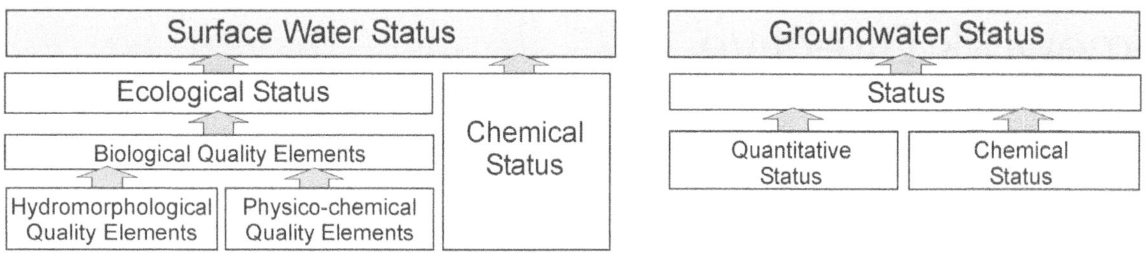

Figure 1: Components of the status of surface water bodies (left) and groundwater bodies (right).

"Surface water status" is determined by the poorer of its ecological status and its chemical status (Figure 1, left side). Good chemical status means that pollutant concentrations in the surface water do not exceed limits set in relevant Community legislation. The good ecological status for surface waters is in the WFD only described normatively on the basis of biological elements which should show only small deviations in composition and abundance from the natural status. The implementation of this normative description is to be made by the member states. Supporting for the biological elements, hydromorphological elements as well as chemical and physico-chemical elements are to be applied (Figure 1). Hydromorphological elements are hydrology, river continuity and morphology. Physico-chemical and chemical elements are, on the one hand, general parameters, like nutrients or temperature. On the other hand, they include specific pollutants, e.g. priority substances. All surface water bodies have to achieve a good ecological status. Only waters classified as heavily modified or artificial water bodies do not have to comply with the good ecological status but with the reduced requirements of the good ecological potential and a good surface water chemical status.

For groundwater bodies also the achievement of a good status is required in the WFD (Figure 1, right side). The groundwater status is composed of the quantitative status and the chemical status of the water body. A good quantitative status means that withdrawal and groundwater recharge have to be balanced. The good chemical status is defined in the WFD only normatively for conductivity and pollutant concentrations. Therefore currently only limits for nitrate (91/676/EEC) and pesticides ((91/414/EEC) and (98/8/EC)) exist on a European level. Thresholds for other pollutants are not specified yet but are to be defined later in a daughter directive of the WFD. There is a proposal for this daughter directive: the proposal for a directive of the European parliament and of the council on the protection of groundwater against pollution (COM(2003)550), which still postpones the definition of pollutant thresholds but demands that the European member states report propositions until the end of 2005. These propositions shall then be used as basis for the prescription of limits for groundwater pollution. The proposal gives a minimum list of substances for which the member states have to give propositions for thresholds, including for example cadmium and lead. In the proposal for the daughter directive, it is defined that indirect discharges to the groundwater (i.e. discharge of pollutants into groundwater after percolation through the ground or subsoil) have to be reduced progressively.

A list of priority substances is to be developed and regularly updated as prescribed in the WFD. The list of priority substances (see 2000/60/EC, annex X) is the basis to establish community-wide harmonised quality standards and emission controls for substances which pose a significant risk to or via the aquatic environment (Table 1). The priority substances have been identified by using the COMMPS (combined monitoring-based and modelling based priority setting) procedure (Klein *et al.*, 1999). Substances are classified as hazardous, when they are toxic, persistent and liable to bioaccumulate. From the priority substances (PS) so called priority hazardous substances have been identified by using information from different hazard and risk assessments and other relevant information like international treaties.

According to their "level of concern" (derived by toxicological characteristics and distribution) the substances from the list were classified as priority substances (currently eight substances), priority hazardous substances under revision (14 substances) and priority hazardous substances (11 substances) (Table 1). The WFD aims at the progressive reduction of the emission of priority substances and at the cessation or phasing- out of discharges, emissions and losses for priority hazardous substances during the next 20 years.

The European WFD contains in annex VIII a list of main pollutants. But the decision of the substances, which should be monitored and thus for which environmental quality standards are needed, has to be made by the member states. For the priority substances Europe-wide environmental quality standards will be defined after an intercalibration process, but for the other specific pollutants, defined by each member state, the limits will only be valid on a national basis.

Table 1: List of Priority Substances in the Field of Water Policy (2000/60/EC annex X, as identified in (2455/2001/EC)): where groups of substances have been selected, typical individual representatives are listed as indicative parameters (in brackets and without number).

	PHS	Name of priority substance
1		Alachlor
2	(X)	Anthracene
3	(X)	Atrazine
4		Benzene
5	X [2]	Brominated diphenylethers [1]
6	X	Cadmium and its compounds
7	X	C10-13-chloroalkanes [1]
8		Chlorfenvinphos
9	(X)	Chlorpyrifos
10		1,2-Dichloroethane
11		Dichloromethane
12	(X)	Di(2-ethylhexyl)phthalate (DEHP)
13	(X)	Diuron
14	(X)	Endosulfan (alpha-endosulfan)
15		Fluoranthene [3]
16	X	Hexachlorobenzene
17	X	Hexachlorobutadiene
18	X	Hexachlorocyclohexane (gamma-isomer, Lindane)
19	(X)	Isoproturon
20	(X)	Lead and its compounds
21	X	Mercury and its compounds
22	(X)	Naphthalene
23		Nickel and its compounds
24	X	Nonylphenols (4-(para)-nonylphenol)
25	(X)	Octylphenols (para-tert-octylphenol)
26	X	Pentachlorobenzene
27	(X)	Pentachlorophenol
28	X	Polyaromatic hydrocarbons (Benzo(a)pyrene), (Benzo(b)fluoranthene), (Benzo(g,h,i)perylene), (Benzo(k)fluoranthene), (Indeno(1,2,3-cd)pyrene)
29	(X)	Simazine
30	X	Tributyltin compounds (Tributyltin-cation)
31	(X)	Trichlorobenzenes (1,2,4-Trichlorobenzene)
32		Trichloromethane (Chloroform)
33	(X)	Trifluralin

PHS Priority Hazardous Substance.
X Identified as priority hazardous substance.
(X) This priority substance is subject to a review for identification as possible "priority hazardous substance". The Commission will make a proposal to the European Parliament and Council for its final classification not later than 12 months after adoption of this list. The timetable laid down in Article 16 of Directive 2000/60/EC for the Commission's proposals of controls is not affected by this review.
(1) These groups of substances normally include a considerable number of individual compounds. At present, appropriate indicative parameters cannot be given.
(2) Only Pentabromobiphenylether.
(3) Fluoranthene is on the list as an indicator of other, more dangerous Polyaromatic Hydrocarbons (PAHs).

Implementation of the WFD - case study Austria

In Austria it was proposed to describe limits for general physico-chemical elements only for the class "very good ecological status" whereas good ecological status is only described by biological elements. For the specific pollutants, substances were chosen from relevant European regulations as well as substances which are discharged in significant amounts to the Austrian waterways. From this list of 320 substances, those chosen as relevant were identified either for local pressures or for water quality impacts. For the 33 priority substances identified on a European level and for 43 other relevant substances, identified for the Austria situation, environmental quality standards (EQS) were determined. In Austria, the procedure for the determination of the EQS is based on a study for the European commission (Lepper, 2002). The WFD requires EQS for water, sediment and biota. But only for water there is currently a good scientific basis and harmonised instructions for sampling and analysis. Therefore, in Austria currently only the definition of environmental quality standards for water has been made (Wimmer *et al.*, 2003). For non-metallic pollutants, the environmental quality standards are predicted no-effect concentrations (PNEC) which are calculated from toxicological data, like lethal concentrations (LC) or no-observed-effect-concentration (NOEC), and a corresponding safety factor according to the WFD (2000/60/EC, annex V). For metals the situation is more complicated. The bioavailability of metals depends on the water chemistry (e.g. pH, alkalinity, water hardness, presence of chelating agents, etc.). Further metals have geogenic sources and thus, depending on geochemical and hydrological conditions, the natural concentrations (also called background concentrations) in running waters are highly variable (Wimmer *et al.*, 2003). Therefore the environmental quality standards were defined on the basis of an added risk approach: environmental quality standard = background concentration + maximal tolerable additional concentration. The added risk approach is based on the assumption that the organisms are adapted to the natural metal concentration and that the same additional exposure leads to the same effect, independent of the background concentration. The maximal tolerable additional concentration was derived by a statistical interpolation method as predicted NOEC from the 5-percentil of a logarithmic transformed distribution of NOEC values from different species. The added safety factors vary between one and five. Based on the previously done work, a regulation concerning the environmental quality standards in surface water came into force in the year 2006 (BGBl. Nr. 96, 2006).

Every member state had to assess which of its water bodies might fail to achieve the good status and report the outcomes to the European commission. For surface water bodies, Austria defined a set of criteria to assess significant pollutant and hydromorphological impacts. For specific pollutants, the environmental quality standards were applied. The general chemical status was investigated using the limits defined in a draft for an ambient water quality directive (AImVF, 1995) and the saprobic index (a biological index which covers the impact of organic pollution). As hydromorphological impacts, the following instances were defined: decreased flow by hydropower plant withdrawal, high flow variations by hydropower plant operation, backwater, migration obstacles and changes in watercourse structure. The assessment found that only about 20% of the Austrian running surface water bodies will achieve the good status. About 40% will surely fail to achieve the good status and for the rest, the data (either the information on the water body or the data used to define the criteria for the assessment) is not sufficient to clearly define their status. Surface water bodies fail to achieve the good status mostly because of hydromorphological problems. Austria plans to declare between 35% and 81% of its watercourses as heavily modified or artificial water bodies.

For groundwater bodies, the assessment of the status was made for qualitative and quantitative criteria (Figure 1). The qualitative criteria used are mainly nitrate and pesticides, because these are the only parameters for which currently thresholds are defined at European level. The assessment included a trend analysis: a water body was seen to possibly fail the good status, if a pollutant concentration exceeded 75% of the threshold value in the last six years. Because of the qualitative assessment eight areas were identified which fail to achieve the good chemical status. A groundwater body was determined to fail the good quantitative status, if the trend analysis found that at more than 25% of the measurement points the level falls below the lowest groundwater level ever measured, or when the withdrawal exceeded 75% of the groundwater resource. On the basis of the data which was available, no groundwater body failed to achieve the good quantitative status but data was scarce.

POLLUTION OF URBAN RUNOFF

Surface runoff contains various pollutants from different sources (Chapter 11). Atmospheric pollution can affect the quality of rain water even before it reaches the ground with inorganic substances (mainly: chloride, sulphate, ammonia, nitrate, nitrite, phosphate), trace metals (e.g. lead, cadmium, iron, copper, zinc or mercury) but also with various toxic organic substances like PAH from combustion processes, organic chlorinated compounds and chlorinated carbons (from chemical industries and herbicides), small amounts of nitro phenols and volatile chlorinated hydrocarbons, as well as aliphatic and aromatic hydrocarbons. Moreover the pollutants of the urban surface contribute to the runoff contamination. Roof runoff can contain elevated concentrations of heavy metals (e.g. lead, copper, zinc) from metallic down pipes, eaves gutters and especially full metal roof covers. Street runoff is polluted with grit from tyres (e.g. rubber, grime, zinc, chromium, copper, nickel), grit from brakes (e.g. nickel, chromium, copper lead and iron), drip losses (fuel, motor- and gear oil, etc), emissions from burning fuels, products from corrosion as well as grit from road surface (Dierkes, 2000). A literature review (Brombach *et al.*, 2005), in which worldwide measurement data on urban runoff pollution were collected, shows that pollutant concentrations in wastewater as well as in stormwater vary widely between the minimum and the maximum concentrations found for the different parameters (Table 2). Furthermore it shows that the heavy metal pollution in surface runoff can be very high, the mean values being in the same range as in wastewater and the maximum concentration of heavy metals measured in stormwater being even higher than those in wastewater (Table 2).

Table 2: Pollutant concentrations found in dry weather flow and stormwater, data taken from a worldwide literature review (Brombach *et al.*, 2005).

	BOD_5 (mg O_2/l)	COD (mg O_2/l)	P_{total} (mg P/l)	N_{total} (mg N/l)	Cadmium (µg/l)	Copper (µg/l)	Zinc (µg/l)
Dry weather flow							
Minimum	17	45	0.12	13.9	0.8	18	45
Median	178	403	4.50	34.2	2.0	58	232
Maximum	503	1070	27.00	93.8	10.0	181	600
Stormwater							
Minimum	2	6	0.03	0.7	0.3	3	1
Median	13	81	0.42	2.4	2.3	48	275
Maximum	162	551	11.58	8.8	37.0	1800	3563

BOD_5: five days biological oxygen demand; COD: chemical oxygen demand.

Urban wastewater can contain hundreds of different chemicals at small concentrations, so called micropollutants. Most of these micropollutants are xenobiotic organic compounds. For a comprehensive assessment of the consequences of these micropollutants on the environment, there is today not enough data available. Especially in the dry weather flow numerous pollutants can be found: in a study made in Denmark all substances that are present in household chemical products were collected and it was found that 900 different xenobiotic organic compounds are potentially contained in grey water. From these 900 substances only for 211 substances data on toxicity, bioaccumulation and degradation was available (Eriksson *et al.*, 2002). Therefore the consequences of micropollutants on the environment cannot be assessed accurately today. To identify the most important pollutants and so set priorities in urban water management, the DayWater project developed the tool CHIAT (Chemical Hazard Identification and Assessment Tool) (Chapter 11).

IMPACT ON THE ENVIRONMENT

The good ecological status is defined on the basis of biological indicators. Unfortunately there is only little known about the cause-effect relationship between pollution and biological changes. But some impacts of stormwater and combined sewer overflows are known to affect good ecological status (Figure 2). The time scales and the types of impacts that are generated by the different parts of the urban wastewater system depend on the composition of the wastewater as well as on the modus of discharge. The modus of discharge varies: while the wastewater treatment plant discharges continuously into the receiving water, stormwater induced discharges (combined sewer overflows, storm sewer outlets and discharges to infiltration facilities) occur intermittently. Therefore the impacts that they cause differ from each other. The discharge of polluted water to stagnant waters should be avoided because pollution, especially nutrients and persistent substances like heavy metals, accumulate there.

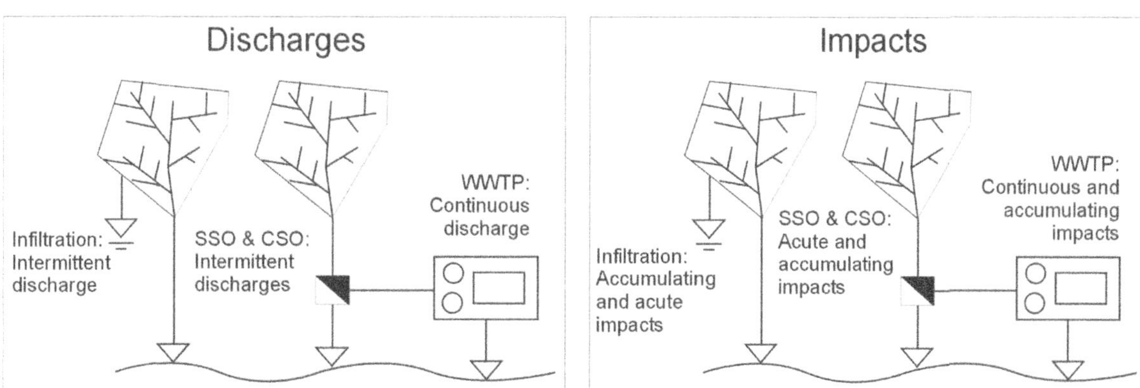

Figure 2: Modus of discharge and types of impacts caused by different parts of the wastewater system: (SSO) separate and (CSO) combined sewer overflow, (WWTP) waste water treatment plant.

If the surface runoff is infiltrated, its pollution is kept away from the surface water but the infiltration can lead to accumulation of persistent substances in the soil of the infiltration device. The pollution of soil caused by infiltration of urban stormwater is difficult to predict. In many infiltration guidelines, pollution of runoff is described according to the origin of the runoff. But although it was often searched, for example, for relations between traffic density and pollutant concentrations in runoff, such relations could not be identified. For example Welker and Dittmer (2005) found, although providing a comprehensive review of measurements, no significant difference between highways and other roads. A possible reason for this lack of relation between traffic load and contaminant level could be the drift of pollutants by air. Boller and Steiner (2002) quantify the load of copper in wet weather surface flow from a motorway to be only about 40 % of the total diffuse copper emission generated, whereas remaining parts are deposited directly on the motorway banks by dust and aerosol drift. Moreover the metal concentrations in the soil of infiltration devices are affected very much by the natural background concentrations. The retention capacity and the bioavailability of heavy metals in soil depend on various factors, e.g. soil composition, redox potential or pH (DVWK M212, 1988). This makes it difficult to predict the impact of stormwater infiltration on soil and groundwater.

Because they are important receiving waters for stormwater, in the following sections the impacts on running waters are described in detail. The wastewater treatment plant creates continuously elevated levels of pollutants. The receiving water's biocoenosis can adapt to a certain degree to these permanent changes. Overflow events caused by stormwater, irrespective whether they come from a combined or a separate sewer system, occur intermittently and thus the biocoenosis cannot adapt to them.

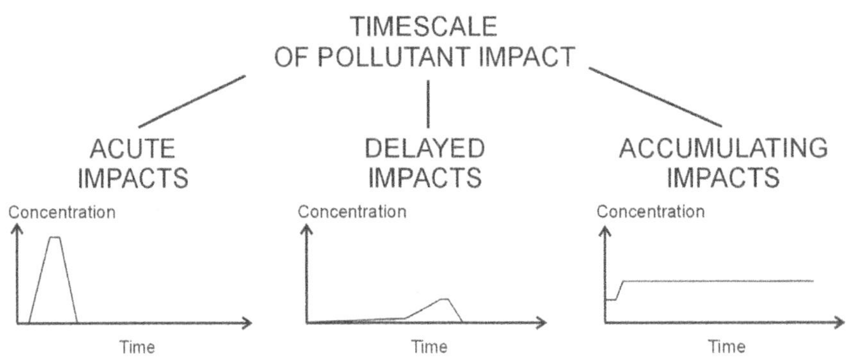

Figure 3: Timescales of pollutant impacts of urban stormwater discharges.

The effects of the discharges in the receiving water can be classified as acute (duration within hours), delayed (within days) or accumulating (weeks/months/years) (Schilling *et al.*, 1997) (Figure 3 and Table 3). While some impacts are mainly caused by combined sewer overflows (e.g. oxygen depletion) others can also derive from surface runoff (e.g. hydraulic impacts or the accumulation of heavy metals). The impacts of urban discharges can cause lasting changes in the abundance and occurrence of the aquatic organisms and therefore also on the ecological status. The appropriate biological indicators to distinguish different kinds of impacts are still subject of research (Hering *et al.*, 2004). Besides the largely unknown cause-effect relationship between pollution and biological changes, another problem is the immense number of pollutants which makes it impossible to measure them all. Toxicity measurements – as they also have been applied in the frame of the DayWater project – give at least an idea of whether a particular discharge could be a threat for the ecological quality of its receiving water (Chapter 12).

Table 3: Impacts of urban discharges on receiving waters (Schilling *et al.*, 1997).

Time scale	Characterization	Indicator variable
Acute	Hydraulic	Flow, shear stress, bed erosion
(hours)	Chemical	Toxic substances (NH_3)
	Physical	Suspended solids
	Bio-chemical	Oxygen depletion in the water body
	Hygienic	Bacteria, virus
	Aesthetic	Floating material, odour
Delayed	Hydraulic	Sediment carrying capacity
(days)	Chemical	Toxic substances (NH_3, NO_2)
	Bio-chemical	Oxygen depletion in the sediments
	Hygienic	Bacteria, virus
	Aesthetic	Floatables, debris, oil
Accumulating	Hydrologic	Flow regime, morphology
(weeks, years)	Chemical	Heavy metals
		Persistent organic substances
		Inorganic and organic sediments
	Bio-chemical	Oxygen depletion (eutrophication)

RECEIVING WATER CLASSIFICATION

For integrated wastewater management in the light of the WFD it is important to develop tailored measures that take the receiving water into account. For the development of such measures a thorough knowledge of the consequences in the integrated wastewater system is prerequisite. Measures to improve one parameter can lead to significant deteriorations regarding other parameters. For example the extension of combined sewer overflow (CSO) storage capacity can be a good measure at a receiving water with high morphological sensitivity, but the stored stormwater may deteriorate the sewage plant performance (Krebs and Rauch, 2002): thus it can be a bad solution for a receiving water that is impacted by plant nutrients. Especially in the implementation of integrated real time control in the wastewater system, the consequences in all parts have to be considered. An example of how different the consequences of real time control are depending on the implemented strategy can be found in Rauch *et al.* (2001).

In the light of the WFD it is useful to develop a classification of receiving waters with regard to their sensitivity to be impacted by different pollutants (Table 4). Such an approach has been developed to evaluate measures in integrated wastewater management of the European project "Cost- effective development of urban wastewater systems for WFD compliance" (www.cd4wc.org). Different classes of river types have been determined which reflect important basic characteristics of rivers regarding their response to impacts from the wastewater system. These classes highlight parameters of special importance, for which a compliance with limits is vital. The river types reflect the specifics of the receiving water, which basically determine the reaction to impacts caused by the urban wastewater system and can thus describe the performance of the system "receiving water".

Table 4: Different river types and corresponding parameters important in wastewater treatment.

River type	Parameter important in wastewater treatment
Sensitivity to increase of biodegradable matter	e.g. biochemical oxygen demand (BOD), dissolved oxygen (DO), suspended solids (SS)
Sensitivity to plant nutrients	e.g. total nitrogen (TN), total phosphorus (TP), dissolved oxygen (DO)
Sensitivity to acute toxic effects	e.g. ammonia (NH3 and NH4), nitrite (NO2)
Morphological sensitivity	e.g. combined/ separate sewer overflow volume and frequency
Sensitivity to accumulation of persistent toxics	e.g. heavy metals, priority hazardous substances

There are several river characteristics known to increase the risks resulting from the impacts (like low flow velocity increases the potential impact of oxygen depleting substances). These characteristics were used to define the five different categories of "sensitivities" (Table 4). Each is associated with parameters which are the main risks for receiving water of high sensitivity (e.g. at a highly plant nutrient sensitive river, the discharged amount of phosphorus and nitrogen is of special importance). So this scheme reveals on which parameters the wastewater treatment has to focus. The more endangered the receiving water is by an impact, the higher is the probability to fail the "good ecological status". For this reason, the consideration of the sensitivities of the receiving water in wastewater management should secure a "good ecological status". Figure 4 shows examples how the receiving water classification can be used to derive adequate measures in the urban drainage system. For example if the receiving water is classified as sensitive to plant nutrients, those measures in the urban drainage system, which reduce the input of nitrogen and phosphorus, should be given special attention. This method can be applied for a decision support system (DSS) intended to support the choice of the best measures for urban drainage system improvement. Here the classification system can be applied to assign weights in the ranking procedure.

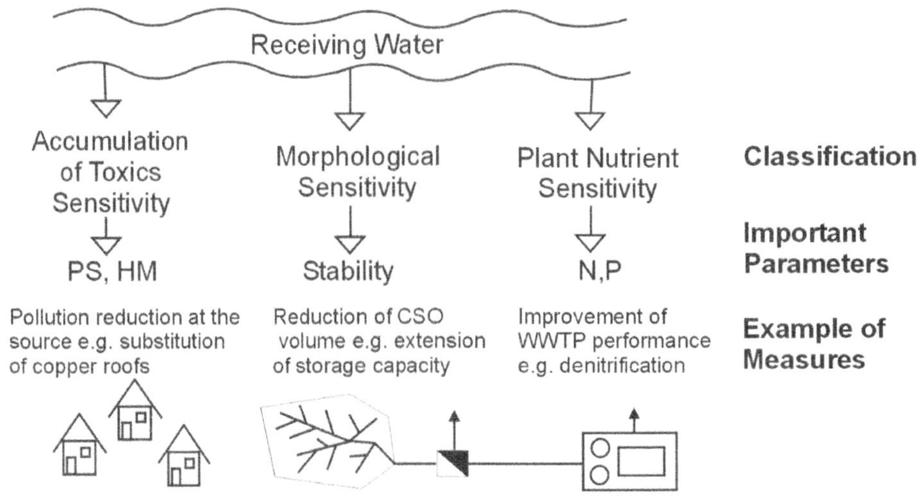

Figure 4: Example for the application of the receiving water classification (PS: priority substances; HM: heavy metals; CSO: combined sewer overflow; WWTP: wastewater treatment plant, N: nitrogen, P: phosphorus).

Sensitivity to increase of biodegradable matter

Oxygen depletion can occur if organic matter or ammonia is discharged to the receiving water. Oxygen depleting substances are measured in terms of chemical oxygen demand (COD) or biochemical oxygen demand (BOD). BOD is the biologically degradable fraction and contains of different groups of substances. The degradation of these substances leads to oxygen depletion in the watercourse (Burton and Pitt, 2002). Usually impacts on the oxygen during wet-weather will be produced by CSO and less by storm sewer outlets because the CSO discharges contain higher concentrations of dissolved BOD than surface runoff. The BOD can consist of readily degradable dissolved material which can lead to considerable acute oxygen deficits in the receiving water. Higher organisms need sufficient oxygen concentrations to survive - fish are among the most sensitive species. Therefore the effects on fish are used to describe the impact of oxygen depletion. The BOD often also consists of particulate material which is degraded slower and therefore can lead to delayed impacts. The particulate matter can be deposited on the river bed. The deposit can derive both from input of particles and from dead biomass built by introduced dissolved BOD, and can lead there to clogging of the interstitial and to an anoxic status in the river bed. This could cause die-off of fish eggs, e.g. of salmonids, which are deposited in the river bed and need sufficient oxygen concentrations (Greig *et al.*, 2005). Therefore regular sedimentation of fine material on the ground could lead to a lasting damage the fish population.

The class “sensitivity to increase of biodegradable matter” was defined to assess the impacts of oxygen depleting substances (Table 4). There are several characteristics of a river that magnify the impact of discharged biodegradable matter. One of these characteristics is a low reaeration rate which leads to a low oxygen input in the receiving water. The reaeration rate increases with the roughness of the river bed and the flow velocity, and decreases with the river depth (BWK, 2001). The flow velocity is also important because of the deposition of sediment, given that a river with low flow velocity also has a low sediment transport capacity. Sensitivity to plant nutrients enhances the problems since increased primary production leads to higher content of BOD. The parameters which are important for wastewater treatment, if the receiving water is sensitive to an increase of biodegradable matter, are for example BOD, dissolved oxygen and suspended solids. Special precautions have to be taken in cases of salmonid spawning grounds, as the eggs and the fry are extremely sensitive to oxygen depletion (Orth *et al.*, 2003). The “good ecological status” is defined as a status with only slight deviation in abundance and occurrence of the indicator organisms. An input of nutrients can lead to deviations in abundance and occurrence of species (Schwoerbel, 1999) especially in waters with a naturally low nutrient content as their biocoenosis is adapted to that. Such acceptable deviations have not been defined yet.

Sensitivity to plant nutrients

Trophy is the intensity of the photoautotroph production in the river (Schwoerbel, 1999). So eutrophication can be seen as an increase of the trophy. The typical limiting factors of primary production in aquatic environments are nitrogen and phosphorus. Eutrophication related with urban stormwater pollution is of minor relevance. In urban surface runoff the concentration of nitrogen and phosphorus are usually small and also the input of these substances by CSO is small compared to diffuse pollution. A study estimated that the input of nitrogen by CSO is only 1-2% and below 2% of phosphorus total emission into surface waters in Austria (Fenz, 2002).

The class "sensitivity to plant nutrients" was defined to assess the susceptibility to eutrophication (Table 4). There are several characteristics of a river that increase the effect of elevated trophy. For example a low flow velocity increases the risk, contrary to a high flow velocity which controls plant growth mechanically (Welch and Jacoby, 2004). Long backwater stretches in rivers react, in the worst case, like a lake so that plant nutrients accumulate there. High water temperatures lead to increased reaction rates, as well as decreased oxygen solubility. A low reaeration rate means little oxygen input to compensate plant respiration in the night, possibly leading to anaerobic conditions. In receiving waters with low buffer capacity, the carbon dioxide consumption of photosynthesis reduces the carbonate-bicarbonate buffer. This can result in un-buffered water and thus lead to high pH variations. High pH variations limit, on the one hand, plant growth but, on the other hand, strongly impact the aquatic organisms. The parameters which are important for wastewater treatment, if the receiving water is sensitive to eutrophication, are total nitrogen and total phosphorus because these two are the main plant nutrients.

Sensitivity to acute toxic effects

The acute toxic effect mostly considered regarding urban wet-weather pollution is un-ionised ammonia. It was realised quite early that CSO contain high concentrations of ammonia which endanger the aquatic organisms. Un-ionised ammonia (NH_3) is highly toxic and easily soluble in water. The toxicity of un-ionised ammonia results partly from an increase of pH in the body, which leads to changes in the tertiary structure of proteins, further it impacts some ion transport mechanisms (Eckert *et al.*, 1993). Impacts by un-ionised ammonia are unlikely to be caused by surface runoff. Ammonia in wastewater mainly originates from urine and is only present in small amounts in surface runoff. As for oxygen, fish are among the most sensitive species and are therefore generally used to express the effect of CSO on the receiving water. Ammonia (NH_4^+) and un-ionised ammonia (NH_3) are in equilibrium. The status of this equilibrium is controlled mainly by the pH value, the higher the pH the more un-ionised ammonia is present. High concentrations of un-ionised ammonia, for a short time, are lethal. But also lower concentrations of un-ionised ammonia with several hours duration damage the organisms as they lead to stress and weaken them (FWR, 1998).

The class "sensitivity to acute toxic effects" was defined to describe the impact caused by the short discharge of high concentrations of toxic substances (Table 4). Characteristics of a river which increase the risk, that the receiving water is impaired, are small receiving waters relative to the size of connected catchment, because this results in low dilution of the discharged overflow water. Further receiving waters with a low recolonisation potential are endangered as there is little compensation for killed organisms. The recolonisation potential depends on how natural the river's and its confluents' morphology are - especially in terms of linear continuity and availability of hideaways (BWK, 2001). The first two characteristics are generally valid for all substances which can cause acute toxic impacts. With regard to un-ionised ammonia, salmonid fishes are more sensitive than cyprinids. If the river's water has an alkaline pH and a high water temperature, the proportion of un-ionized ammonia is increased, because the ammonia/un-ionised ammonia equilibrium is shifted on the side of the un-ionised ammonia. The parameters which are important for wastewater treatment, if the receiving water is sensitive to acute toxic effects, are all acute toxic substances.

Morphological sensitivity

High discharges drift organisms downstream: this is a naturally occurring process. The urban discharges however increase the number of times per year when drift occurs. Often the discharges from the urban catchments produce flows as high as floods which lead to catastrophic drifts (Borchardt, 1992) at which high percentages of the benthic population can be lost at the point of discharge. The population has to recover on the basis of individuals that withstood the flood in suitable hiding places and from unimpaired stretches, like small tributaries. But the population can only recover if the recolonisation potential is good enough and frequency events not too high (Frutiger and Gammeter, 1996).

The class "morphological sensitivity" was defined to assess the impacts of increased discharge volume (Table 4). Several characteristics of a river determine its morphological sensitivity like small base flow compared to discharge volume, a naturally straight river plan, fine bed material, low bank stability, little floodplain (no possibility of flood retention or river widening) or a low recolonisation potential. Wastewater treatment should focus on stormwater treatment (e.g. the reduction of overflow frequencies or discharge volumes) or the application in-stream measures if the receiving water is morphological sensitive.

Sensitivity to accumulation of persistent toxics

Both stormwater and wastewater contain persistent pollutants which can accumulate in the receiving water. Especially pollutants that are identified as priority or priority hazardous substances pose a special risk to humans and to the environment by being toxic, persistent and liable to bio-accumulate, or otherwise giving rise to an equivalent level of concern. To this group of pollutants belong for example many heavy metals and aromatic organic compounds (e.g. pesticides, fungicides). Heavy metals (HM) are present in rather high concentrations in wastewater and surface runoff.

The class "sensitivity to accumulation of persistent toxics" was created with regard to discharge of priority and especially priority hazardous substances (2455/2001/EC) (Table 4). All receiving waters should be treated as if they were susceptible to accumulation of persistent toxics, in order to reduce the risk from emission of such substances.

LIMITS

Ambient water quality limits

The European WFD emphasises the biological elements for the determination of the ecological status. But, on the one hand, it is not defined today what a good biological status is. On the other hand, there are no quantitative cause-effect relations established today between concentrations of pollutants and the aquatic biocoenosis. Still for planning purposes in wastewater management, it is necessary to have figures to work with. Therefore a study was conducted which aimed at comparison of various existing limits, regarding ambient water quality, to identify reliable limit values. It was assumed that compliance with such limits should secure the good ecological status.

It was found that regulations show great differences. Figure 5 shows as an example a comparison of two different regulations for copper standards). Probably this is due to the fact of the different philosophies which can be the basis of ambient water quality standards. One approach is the technology-based approach, based on the pollution prevention principle. Here it is assumed that all pollution is potentially harmful and should thus be prevented, as far as it is technological and socioeconomic feasible. The other approach is the environmental quality objective approach, which is based on the carrying capacity principle. It is assumed that the receiving water can cope with a certain level of pollution (Ragas *et al.*, 2005). Furthermore the limits are often not comparable because of different basic conditions: some directives distinguish between salmonid and cyprinid waters, while others do not. Especially the limits for heavy metals show high discrepancies because of the various applied dependencies of limits on water hardness.

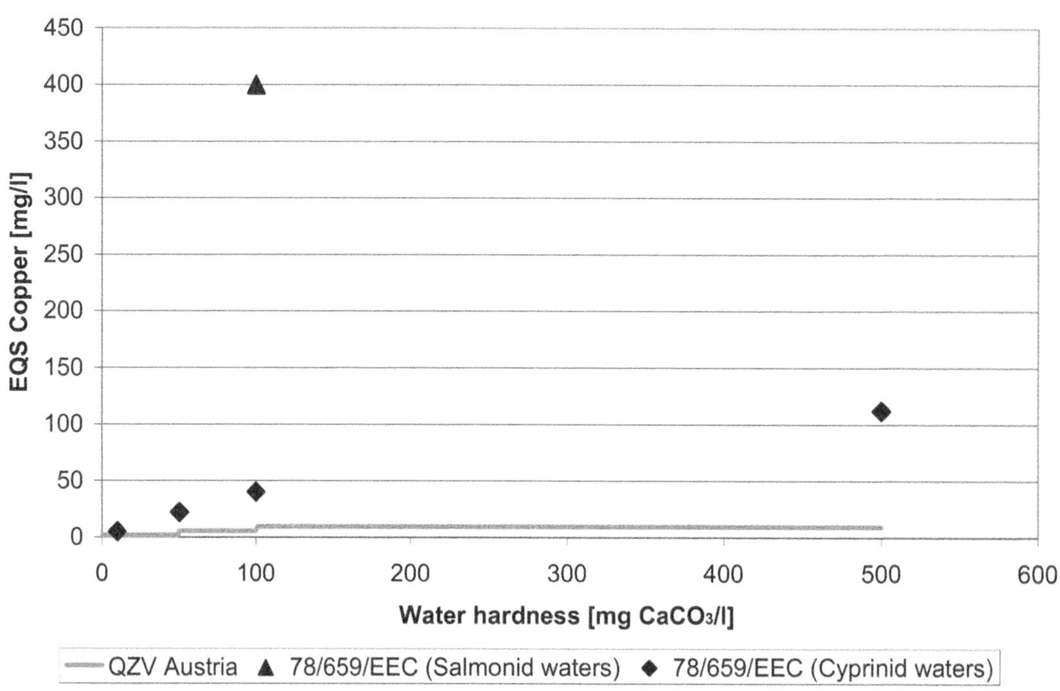

Figure 5: Comparison of the EQS for copper of the directive on the quality of fresh waters needing protection or improvement in order to support fish life (78/659/EEC) and the new Austrian regulation for surface water standards (QZV) (BGBl. Nr. 96, 2006). EQS depend on the hardness of the water.

Figure 5 shows the standards for copper of the European directive on the quality of fresh waters needing protection or improvement in order to support fish life (78/659/EEC) and the new Austrian QZV- the regulation of quality standards for chemistry of surface waters (BGBl. Nr. 96, 2006). These two directives are a good example for the outcome of the literature review on different EQS: on the one hand it is difficult to compare them, because of the different boundary conditions on which they are based. On the other hand, the concentrations specified in the two regulations differ in the order of magnitudes (e.g. for water hardness of 100 mg $CaCO_3$/l, the EQS of the 78/659/EEC are 40 and 400 µg Cu/l, the one of QZV is 4.8 µg Cu/l + background concentration). The finding for other substances where mostly similar, therefore it seems currently not possible to deduce reliable concentrations for pollutants, which could be used to represent the good ecological status, from literature.

Infiltration

For infiltration the legal situation is often unclear. In earlier days stormwater was seen as wastewater, for example in Germany stormwater was treated in regulations as wastewater until the early 1990s. Today there are many guidelines for infiltration of stormwater in Europe. Usually surface runoff pollution is classified according to its origin (e.g. metal roofs or non-metal roofs; streets with high traffic density – housing areas) and, based on this classification, different kinds of treatment (like infiltration shafts or grassed swales) are proposed (ATV-DVWK-A 138, 2002;OEWAV R35, 2003). But there are no ambient quality limits for water or soil specified for infiltration devices. For groundwater there are often limits for pollutants in the water which can be used to investigate the impact of infiltration devices, but for the soil the situation is unclear in many European countries. Although there are limits when soil is hazardous and has to be disposed, no limits for pollutants in soils of infiltration devices exist.

CONCLUSIONS

The WFD signifies for European wastewater management that both emission-based limits and the receiving water have to be taken into account. It is an instrument for administration but not for detailed impact assessment of stormwater management. Emission-based approaches are relatively easy to handle. The good ecological status on the other hand is problematic. First of all the cause-effect relations between pollutant concentrations and biological quality are insufficiently known which makes it impossible to predict reliably the consequences of pollution on receiving waters. Furthermore taking the receiving water quality into account is a complex task which requires high data knowledge. As the good ecological status can currently not be applied for wastewater management, for planning and design, still ambient water quality standards have to be used. Today simulations are widely applied in integrated wastewater management. Many of these tools also allow the consideration of the river water quality. But the evaluation of such results, based on ambient water quality limits, contains high uncertainties as these limits have been found to differ highly from each other. This also results in large uncertainties in the planning process. Especially if measures are planned on the basis of such limits, also large uncertainties regarding investment costs occur. The classification of receiving waters described above gives the possibility to choose the most important parameters, for a specific receiving water, as well as for a prioritisation of measures. But still knowledge on the specifics of the particular receiving water will be prerequisite for future wastewater management.

ACKNOWLEDGEMENT

The results presented in this publication have been elaborated in the frame of the EU project CD4WC - as part of the CityNet framework - (www.cd4wc.org), contract no EVK1-CT-2002-00118, coordinated by TU Dresden (D) and including RUG / BIOMATH (B), IUT Innsbruck (A), Panteion University of Athens (GR), Ruhrverband (D), Tiroler Wasserkraft AG (A), AQUAFIN NV (B) and PVK Prague (CZ). The programme is organised within the Energy, Environment and Sustainable Development Programme in the 5th Framework Programme for Science Research and Technological Development of the European Commission.

REFERENCES

75/440/EEC. *Council Directive of 16 June 1975 concerning the quality required of surface water intended for the abstraction of drinking water in the Member States.*

78/659/EEC. *Council Directive of 18 July 1978 on the quality of fresh waters needing protection or improvement in order to support fish life*

79/923/EEC. *Council directive of 30 October 1979 on the quality required of shellfish waters.*

91/414/EEC. *Council Directive of 15 July 1991 concerning the placing of plant protection products on the market.*

91/676/EEC. *Council Directive of 12 December 1991concerning the protection of waters against pollution caused by nitrates from agricultural sources.*

98/8/EC. *Directive 98/8/EC of the European Parliament and of the Council of 16 February 1998 concerning the placing of biocidal products on the market.*

2000/60/EC. *Directive of the European Parliament and of the Council establishing a framework for Community action in the field of water policy.*

2455/2001/EC. *Decision No 2455/2001/EC of the European parliament and the council of 20 November 2001 establishing the list of priority substances in the field of water policy and amending Directive 2000/60/EC.*

AImVF (1995) *Entwurf für eine Verordnung des Bundesministers für Land- und Forstwirtschaft betreffend die allgemeine Beschränkung von Immissionen in Fließgewässern [Draft of a regulation of the federal minister for agriculture and forestry regarding the general restriction of immission to running waters].* Vienna, Austria.

ATV-DVWK-A 138 (2002) *Arbeitsblatt A 138: Planung, Bau und Betrieb von Anlagen zur Versickerung von Niederschlagswasser [Fact sheet A138: Planning, construction and operation of facilities for infiltration of stormwater].* ATV-DVWK, Hennef, Germany.

BGBl. Nr. 96 (2006) *Qualitätszielverordnung Chemie Oberflächengewässer [Quality standard regulation chemistry surface waters].* Vienna, Austria.

Boller M. A. and Steiner M. (2002) Diffuse emission and control of copper in urban surface runoff. *Wat. Sci. Technol.*, **46**(6-7), 173-181.

Borchardt D. (1992) *Wirkungen stoßartiger Belastungen auf ausgewählte Fliessgewässerorganismen [Effects of intermittent impacts on selected running water organisms].* Universität- Gesamthochschule Kassel, Germany.

Brombach H., Weiss G. and Fuchs S. (2005) A new database on urban runoff pollution: comparison of separate and combined sewer systems *Wat. Sci. Technol.*, **51**(02), 119-128.

Burton G. A. and Pitt R. (2002) *Stormwater effects handbook: a toolbox for watershed managers, scientists, and engineers.* Lewis Publisher, USA.

BWK (2001) *M3: Ableitung von immissionsorientierten Anforderungen an Misch- und Niederschlagswassereinleitungen unter Beruecksichtigung oertlicher Verhaeltnisse - Merkblatt [M3: Derivation of immission based requirements to combined and stormwater discharges with consideration of local circumstances].* Bund der Ingenieure für Wasserwirtschaft, Abfallwirtschaft und Kulturbau (BWK), Duesseldorf, Germany.

COM(2003)550. *Proposal for a directive of the European parliament and of the council on the protection of groundwater against pollution.*

Dierkes C. (2000) Anlagerung von Schadstoffen in wasserdurchlässigen Flächenbelägen und Böden [Adsoprtion of pollutants in water permeable pavements and soils]. In: *Entwässerungskonzepte - ATV-Fortbildungskurs für Wassergütewirtschaft und Abwassertechnik*, pp. 9-1, ATV-DVWK, Kassel, Germany.

DVWK M212 (1988) *DVWK-Merkblatt 212: Filtereigenschaften des Bodens gegenüber Schadstoffen [Fact sheet 212: filter characteristics of the soil regarding pollutants].* DVWK- Deutscher Verband für Wasserwirtschaft und Kulturbau, Bonn/Hennef, Germany.

Eckert R., Randall D. and Augustine G. (1993) *Tierphysiologie [animal physiology].* Georg Thieme Verlag, Stuttgart; New York.

Eriksson E., Auffarth K., Henze M. and Ledin A. (2002) Characteristics of grey wastewater. *Urban Water*, **4**(1), 85-104.

Fenz R. (2002) *Gewässerschutz bei Entlastungsbauwerken der Mischkanalisation [Watercourse protection at overflow structures of combined sewer systems].* Wiener Mitteilungen, Band 174, Institut für Wassergüte und Abfallwirtschaft, Technical University Vienna, Austria.

Frutiger A. and Gammeter S. (1996) *Zustandsbeurteilung Fliessgewässer - ein Methodenvorschlag [Assessment of status of running waters - a method proposal].* VSA Verbandsbericht No. 503. Verband Schweizer Abwasser- und Gewässerschutzfachleute, Zurich, Switzerland.

FWR (1998) *Urban Pollution Management Manual - A planning guide for the management of urban wastewater discharges during wet weather.* Foundation for Water Research (FWR), Buckinghamshire, Great Britain.

Greig S. M., Sear D. A. and Carling P. A. (2005) The impact of fine sediment accumulation on the survival of incubating salmon progeny: Implications for sediment management. *Science of The Total Environment*, **344**(1-3), 241.

Hering D., Meier C., Rawer-Jost C., Feld C. K., Biss R., Zenker A., Sundermann A., Lohse S. and Böhmer J. (2004) Assessing streams in Germany with benthic invertebrates: selection of candidate metrics. *Limnologica*, **34,** 398-415.

Klein W., Denzer S., Herrchen M., Lepper P., Müller M., Sehrt R., Storm A. and Volmer J. (1999) *Revised proposal for a list of priority substances in the context of the water framework directive (COMMPS procedure).* Fraunhofer-Institut Umweltchemie und Ökotoxikologie, Schmallenberg, Germany.

Krebs P. and Rauch W. (2002) Von der Abflusssteuerung zur immissionorientierten Steuerung [From the control of discharge to immission oriented control]. *Dresdner Berichte, Wasserbauliche Mitteilungen*, **19**(21), 87 - 109.

Lepper P. (2002) *Towards the derivation of quality standards for priority substances in the field of water policy (Excerpt).* Fraunhofer-Institute of Molecular Biology and Applied Ecology, Vienna, Austria.

OEWAV R35 (2003) *OEWAV- Regelblatt 35: Behandlung von Niederschlagswässern [Fact sheet 35: treatment of stormwaters].* Oesterreichischer Wasser- und Abfallwirtschaftsverband, Vienna, Austria.

Orth H., Londong J., Podraza P. and Halle M. (2003) *Ein integriertes Gesamtkonzept für Entwässerungsgebiet, Kanalnetz, Kläranlage und Gewässer mit dem Ziel eines nachhaltigen Gewässerschutzes und einer Reduktion der Kosten [An integrated overall concept for catchement, sewer, wastewater treatment plant and receiving water with the aim of a sustainable watercourse protection and a reduction of costs].* Ruhr-Universität Bochum, Wupperverband, Germany.

Ragas A. M. J., Scheren P. G. A. M., Konterman H. I., Leuven R. S. E. W., Vugteveen P., Lubberding H. J., Niebeek G. and Stortelder P. B. M. (2005) Effluent standards for developing countries: combining the technology- and water quality-based approach. *Wat. Sci. Technol.*, **52**(9), 133-144.

Rauch W., Bertrand-Krajewski J. L., Krebs P., Mark O., Schilling W., Schütze M. and Vanrolleghem P. A. (2001) *Mathematical Modelling of integrated urban drainage systems.* INTERURBA II, Lisbon, Portugal.

Schilling W., Bauwens W., Borchardt D., Krebs P., Rauch W. and Vanrolleghem P. (1997) *On the Relation Between Urban Wastewater Management Needs and Receiving Water Objectives*. XXVII IAHR Congress "Water for a Changing Community", 10 - 15 August 1997. San Francisco, USA.

Schwoerbel J. (1999) *Einführung in die Limnologie [Introduction to limnology]*. Gustav Fischer Verlag, Stuttgart, Jena, Lübeck, Ulm.

Welch E. B. and Jacoby J. M. (2004) *Pollutant effects in freshwater: applied limnology*. Spoon Press, London, New York.

Welker A. and Dittmer U. (2005) Belastung von Verkehrsflächenabflüssen mit Schwermetallen und PAK [Loading of traffic area runoff with heavy metals and PAH]. *GWF- Wasser, Abwasser*, **146**(4), 320-332.

Wimmer M., Meisriemler P., Enzelberger-Heis E., Friesenecker R., Deutsch K., Unterweger A., Rauchbüchl A., Friehs B., Krämer D., Lesky U., Pattermann M., Sossau C., Szinovatz H., Buhmann D., Polzer E., Nagel G., Traer K., Philipp A. and Konheisner G. (2003) *Wasserrahmenrichtlinie: Qualitätsziele für chemische Stoffe in Oberflächengewässern [Water Framework Directive: Quality standards for chemical substances in surface waters]*. Arbeitskreis Chemie/Überwachung und Ziele, Bundesministerium für Land- und Forstwirtschaft, Umwelt und Wasserwirtschaft (BUWAL), Vienna, Austria.

Chapter 25

Criteria and indicators in delivering sustainable water systems: from Sustainable Water Asset Resource Decisions (SWARD) to Water cycle management for New Developments (WaND)

Richard Ashley, David Butler, Louise Hurley and Fayyaz Ali Memon

INTRODUCTION

The European Parliament and Council 'EU Water Framework Directive' (WFD, 2000) has taken issues of sustainable development into account and sets out an integrated water resources management (IWRM) approach, combining the measures taken to control pollution at source with an inventory of their effects and appropriate pricing that reflects true costs (Chapter 24). Crucially, stakeholder engagement is also explicit. The concept of sustainability is different for different individuals depending on their standpoint (Sahota and Jeffrey, 2005), however, it is widely accepted that sustainability should encapsulate the three pillars of 'society', 'the environment' and 'the economy'. Taking into account varied points of view when deciding how to act is a difficult process; taking into account varied and often conflicting ideas of the future further compounds this complexity. Support for decision making processes is obviously necessary given their complexity, but practitioners have in the past been reluctant to use integrated decision aid methods that go beyond cost-based techniques (Brunner and Starkl, 2004; Kapelan *et al.*, 2005). Implementation of the WFD requires a change in technological resource management regimes, many of which are inflexible and are unable to adapt to changes in environmental, economic or social circumstances (Pahl-Wostl, 2002; Whitelegg and Weber, 2002). Decision support must therefore encourage flexibility, adaptability and enable innovation, with easy inclusion of considerations other than finances.

The challenge of the 'measurement' of sustainability as a decision-making tool has traditionally been met by breaking down the broad concept principles into constituent criteria that must be fulfilled in some equitable way in order to provide a balanced decision (Figure 1). To know whether the criteria have been addressed, some form of measurement is ultimately required that takes the form of indicators which account for changes in a 'driver-pressure-state-impact-response' (DPSIR) system and are appropriate to criteria fulfilment. The use of criteria and indicators gives a framework for the measurement of relative sustainability (e.g. Brunner and Starkl, 2004 and Chapter 7). Depending on the criteria and indicators used, this framework may be able to compare between options or track improvements of a scheme, often using a codified multi-criteria decision support methodology (DSM) (Kapelan *et al.*, 2005; Chapter 9). However, there are fundamental problems of sustainability measurement: the scale of incorporation of wide-ranging points of view, conflict resolution, risks (and their perceptions), uncertainty, the scale of the problem in time and space, how criteria should be chosen and who should choose them, the relative importance given to sets of indicators, the constancy or otherwise of indicators and indeed whether this is an appropriate way in which to tackle the problem. Within urban water management the development of criteria and indicators and use of a DSM has been criticised by practitioners for lack of feasibility in implementation in that prefabricated software solutions may not always fit their needs and the data required can be prodigious (Brunner and Starkl, 2004).

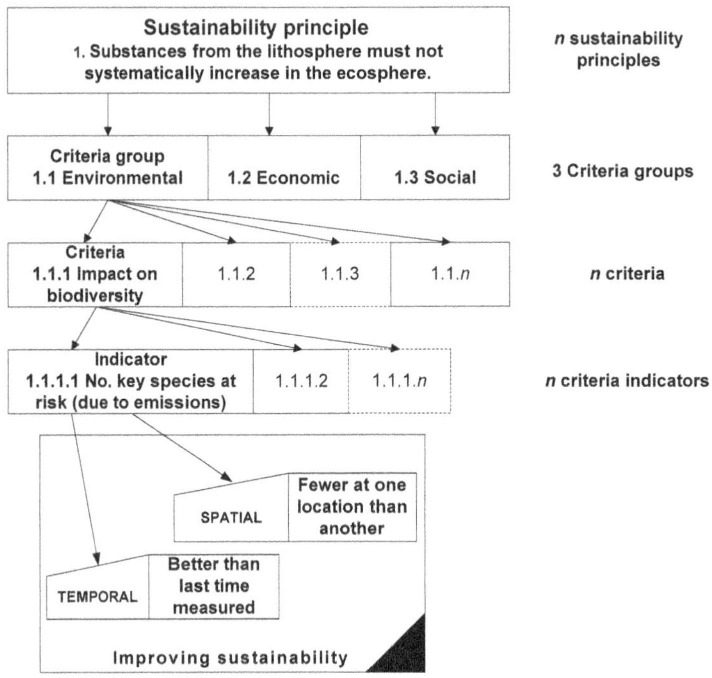

Figure 1: The 'principle – criteria - indicator' cascade for sustainability assessment (showing illustrative examples).

Decision making practices within organisations are often 'top down' and mechanistic following a pattern of issue definition, criteria setting, option generation, assessment and final solution. The inclusion of a wider stakeholder group from outside the organisation is problematic. Pre-agreed preferences within the organisation (or organisational vetoes) can mean that weights or scores are allocated to criteria which greatly influence any trade-offs made. The final decision satisfies preferences within the organisation but the trade-offs are only transparent and inclusive to those in agreement with the organisational vetoes. Reasons for arrival at a decision may be incomprehensible to stakeholders without 'inside information' and may be unacceptable to those with different preferences.

Primary criteria may be set that are encompassed within sustainability principles and are generic to decision making processes to give a framework by which to encourage broad consideration of sustainability and ensure transparency to all stakeholders. There are two main approaches to setting criteria (Wrisberg *et al.,* 2002). Pre-definition, which provides *a priori* goals to be aimed at in option selection, or alternatively 'undefined' *a posteriori* criteria that are considered informally once the data and information about a development have been assembled with a consensus view taken on option selection. Here only the former approach is considered in detail.

The use of indicators tends to focus on predominately quantitative data. This is problematic for the representation of social criteria, indicators of which may be qualitative in nature. Where quantitative data can represent indicators of social criteria, the means of collection is as important as the data that are eventually utilised. There is a need for decision makers to understand the level of involvement required for a project; the level of participation should increase with increasing complexity of a problem but hence so does the number of conflicting viewpoints and the level of co-operation required and capacity within the actors to engage (Brunner and Starkl, 2004). The WFD explicitly mentions information, consultation and involvement, delineations of which may not be immediately clear to decision makers, much less the appropriate processes for their inclusion.

The less prescriptive *a posteriori* decision making process does not pre-define either preferences or criteria. It may be that criteria have not been defined, or that there exist sets of conflicting criteria. In either case, in order to arrive at a decision the criteria for making the decision need to be agreed on by the stakeholders at some point in the process. Decision making in this context is therefore 'bottom up'. Firstly, the legitimate stakeholders must be defined. Deliberation with stakeholders is then necessary to achieve consensus on the criteria used (Chapter 4). In order to delineate criteria it may be necessary to begin with overarching principles such as the four sustainability groups used in the Sustainable Water industry Asset Resource Decision (SWARD) project (Ashley *et al.*, 2004): economic, environmental, technical and social or alternative sets (Malmqvist *et al.*, 2006). These may then be used for criteria setting under which equal consideration must be given to each.

In this chapter the evolution of a suite of studies carried out over more than a decade to devise sustainability assessment support tools in the water sector are outlined. Case examples are provided to illustrate the development of criteria and their utilisation in sustainability assessment.

THE DEVELOPMENT OF CRITERIA AND INDICATORS FOR WATER MANAGEMENT DECISION MAKING

Research into the choice of least unsustainable options has led to the development of frameworks for decision making in conjunction with decision makers and those affected by the decisions made (e.g. Malmqvist *et al.*, 2006). The following section describes the development of criteria and indicators formulated to aid in decision making processes within different areas of water management.

The 'Think Before You Flush' (TBYF) campaign

The toilet is seen habitually in many developed countries as a waste disposal unit for the disposal of domestic and sanitary solids. This practice stems from the historical link associating human waste with risks to health (Ashley *et al.*, 2005). The large quantity of (treated) water used to flush the toilet and the downstream costs of dealing with the waste have raised the question of whether this is a sustainable practice and whether disposal of these items, particularly female sanitary items which include non-degradable plastics, through the solid waste stream might be more sustainable.

Two options for the disposal of sanitary solids, via the toilet or via the solid waste stream, were examined for their relative sustainability. A detailed multi-disciplinary study was carried out to look at the problem holistically regarding economic, social and environmental aspects and is detailed by Ashley *et al.*, (2005). Defining the more sustainable option necessitated the use of indicators (Table 1), despite their known limitations when considered in isolation from the end-users (Ashley *et al.*, 1998).

The economic indicators deal with direct costs (Chapter 10). The environmental indicators have been represented by equivalent value-based economic indicators (despite the limitations of this approach given in the Foundation for Water Research manual, 1996) and the social indicators were based on the results of surveys of the public. The single indicator, C1 was found to represent an aggregation all the other indicators, as when it was found to be decreasing it represented (Table 1) increasingly responsible public behaviour (S1 tends to unity, S2 tends to zero); less aesthetic pollution (EN1 and EN2 can only be assessed in conjunction and whilst EN1 declines it is assumed willingness to pay will also decline) and economic cost savings (when E1-E2 is positive). The indicators were shown to be robust and largely insensitive to changes in assumptions made in their determination.

Focus groups demonstrated the awareness of problems of flushed sanitary waste but elicited admissions of flushing a range of sanitary and other items. Qualitative data were used to plan quantitative studies undertaken before and after a campaign to promote habit change, the TBYF campaign which included the distribution of free disposal bags. The campaign resulted in a significant reduction in numbers of items flushed, particularly in areas of high community identity. The study concluded that it is more sustainable from each of the environmental, technical and social aspects to dispose of domestic sanitary wastes via the solid waste route and that a single criterion (C1) incorporating a number of sub-indicators that are sufficiently robust to allow for uncertainties in their assessment could be used for sustainability assessment. In addition, it was clear that it was feasible to deliver a strong message about unsustainable behaviour in such a way that individuals and communities would engage and be willing to change habits.

Table 1: Indicators of the relative sustainability of sanitary waste disposal routes (Ashley *et al.*, 2005)

C1. Number of sanitary waste items flushed		
Economic	Environmental	Social
E1. Cost-benefit: waterborne disposal route (present value per person, £PV/cap) E2. Cost benefit: solid disposal route (£PV/cap)	EN1. Aesthetic: numbers of sanitary waste items on beaches/riverbanks (items) EN2. Environmental evaluation (public willingness to pay for sanitary waste free beaches/watercourses) (£/cap)	S1. Public willingness to change from toilet to solid waste disposal (number willing to change ÷ population currently flushing) S2. Actual change in disposal habits (number flushing ÷ original population flushing)
Beneficial results if E1-E2 is cost effective for society	Beneficial results if a reduction in numbers and a willingness to pay	Beneficial results as S1 tends to unity (all willing to change) and S2 tends to zero (no flushers)

The surveys and intensive public awareness campaigns in four test areas in Scotland established that a significant number of people would be willing to change their behaviour and place sanitary waste into the solid waste route rather than into the toilet and that a rapid change of at least 50% of those currently flushing this waste could be achieved by the mounting of awareness campaigns. Subsequently the process was 'rolled-out' across a number of the smaller communities across Scotland and additional detailed studies of what sanitary items were actually being flushed undertaken. The unfortunate abandoning of the project due to reorganisation and notably the lack of a champion within Scottish Water illustrates the need for such a person or persons with the will and resources to: (i) engage colleagues within organisations such as Water Service Providers (WSPs), (ii) not be averse to mounting public awareness campaigns with communities to build capacity and change behaviours (Brown *et al.*, 2005).

SWARD – development of core criteria

The SWARD project (Ashley et al., 2004) was undertaken by a consortium of UK academics in collaboration with water service providers in Scotland, England and Romania and in consultation with a wide range of stakeholders concerned with water services. It was recognised from the beginning of the project that current decision-making processes needed to be understood to enable support for the processes to align with existing procedures (Foxon *et al.*, 2002) and this was analysed by means of focus groups and decision-mapping exercises. The way in which decisions were made was also examined in a more formal sense and flow-diagrams of the processes produced. Despite differences in operation by the various WSPs, areas of commonality were identified including: the priority to address statutory requirements; costing; and customer service provision. The decision methodology allowed for collaborative research in the production of criteria rather than their imposition by external expertise. A key finding of the research was that the regulatory framework imposed no sustainability duty at that time on WSPs and hence this was poorly reflected in the way in which performance targets were set (Ashley and Hopkinson, 2002). However, the environmental regulators in England and Scotland were incumbent to consider sustainability in consideration of developments. Unfortunately their procedures did not include the requisite breadth of perspective needed to ensure this.

A guidebook was developed (Ashley *et al.*, 2004) as a practical tool to facilitate the explicit inclusion of 'sustainability' in the decision-making process. A 'generic' set of sustainability criteria informed by the then UK government and Water UK principles and indicators (Water UK, 2000) and other key publications were developed. The two key factors addressed in developing the criteria were: the use that would be made of them and to what extent any set of criteria can encompass the range of issues to be considered under the heading of 'sustainability'. A common approach and the strength of these were found to be that of the production of a structured classification of criteria relevant to the decision-making process in question. This led to a hierarchy of a small number of high level criteria cascading down to a larger number of more specific secondary criteria, *i.e.* Figure 1 (Foxon *et al.*, 2002). Primary criteria were classified in SWARD under four categories: economic, environmental, social and technical as shown in Table 2.

Table 2: The categorisation of SWARD primary criteria.

Category	Primary criterion
Economic	Life cycle costs
	Willingness to pay
	Affordability
	Financial risk exposure
Environmental	Resource utilisation
	Service provision
	Environmental impact
Social	Impact on risks to human health
	Acceptability to stakeholders
	Participation and responsibility
	Public awareness and understanding
	Social inclusion
Technical	Performance of the system
	Reliability
	Durability
	Flexibility and adaptability

The secondary criteria were then derived to support the primary criteria in Table 2 (Figure 1) for the types of case study being investigated which related to asset management in the water industry:

- Economic criteria: life cycle costs cover all stages of the life cycle; willingness to pay is the willingness of the operator to pay for the service provision; affordability relates to the ability of householders to pay for services delivered and financial risk exposure relates to the risk of loss to the company associated with particular kinds of investment;
- Environmental criteria: resource utilisation covers use of water resources, land, energy, materials and chemicals; service provision relates to the provision of water and wastewater services by the WSP; environmental impact covers a range of impacts on water, land, air and biodiversity;
- Social criteria: impact on risk to human health includes availability of clean water, infection risk and exposure to toxic compounds; acceptability to stakeholders covers acceptability and perceptions of impacts on health and environment; participation and responsibility includes individual actions, participation in decision making and willingness to change behaviour; public awareness and understanding covers awareness of implications of behaviour and consumer information; social inclusion covers poverty alleviation, voluntary activity and access to water courses;
- Technical criteria: performance of the system includes the quality of supplied water and treated wastewater; reliability is of the system; durability of the system is its expected lifetime; flexibility and adaptability covers the ability to make future changes to the system.

Validation of the above criteria (acceptability and usefulness) was conducted during workshops held with WSP project partners in the UK and Romania. These exercises showed broad support for the criteria chosen, although it was recognised that many of the primary criteria could occupy more than one category. The choice of criteria chosen would clearly have an influence on the outcome of the decision being made as would aggregation and methods of comparison, but the SWARD processes are intended to provide guidance, with users reviewing use and learning appropriate applicability across a range of different decisions.

The Guidebook outlines seven component phases in the decision-making processes that may be used by a WSP for evaluating the contribution to sustainability of a range of options as shown in Figure 2.

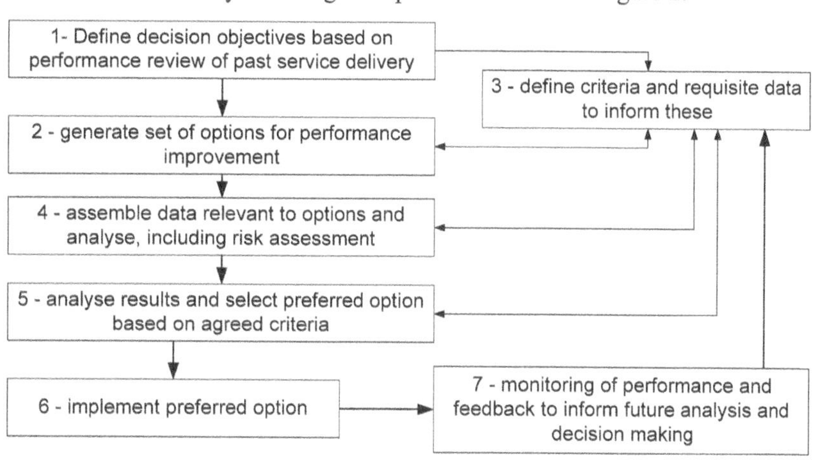

Figure 2: SWARD decision support process phases.

These phases are not sequential as is illustrated in Figure 2 and comprise a number of loops and feedback processes. Guidance is given in the Guidebook for the selection of criteria (phase 3) to ensure comprehensive coverage of the four categories of criteria and the applicability, tractability, transparency and practicability of the criteria chosen. Appropriate modelling tools are required in phase 4 to assess future performance based on the interactions of criteria chosen over the life cycle of a project (Gouda *et al.*, 2001). Indicators for assessing the performance of the options may be developed for the chosen criteria by stakeholder negotiation (examples of which are given in Ashley *et al.*, 2004) and option selection may entail information from a formalized multi-criteria analysis in which weightings or choices used are open to criticism by other stakeholders. The final stages of implementation and monitoring of the preferred option will help to inform future decision making processes as well as giving some confidence in the relative sustainability performance with time.

The criteria selection stages have subsequently been tested in workshops in India, with Indian professionals in the UK and under the auspices of the US-EPA in the USA. In each case the criteria and the approach recommended in SWARD were found to be appropriate. Hence the robustness is clearly evident and has informed subsequent activities.

Criteria development for stormwater drainage

The use of Sustainable Drainage Systems (SUDS) or Best Management Practices (BMPs) (Chapter 8) has been encouraged in the UK by environmental regulators and local authorities due to their perceived environmental benefits (National SUDS working group, 2003). The epithet 'sustainable' for these systems is largely based on qualitative and subjective assessments with a bias towards environmental considerations (Ashley *et al.*, 2001). Unfortunately the use of the term 'sustainable' in this context is highly misleading and can be taken by the uninformed to mean that if such systems are used then 'self-evidently' they will deliver a sustainable solution. Sustainability is primarily about context, hence to say that SUDS are more or less sustainable than traditional systems without considering where, when and how they are being implemented is not only naïve but likely to lead ultimately to a loss of faith in so-called 'sustainable' systems. Those responsible for the terminology should be ashamed that they may have blighted in the short to medium term any real progress to delivering systems that are more sustainable. It is fortunate that the UK terminology is not used elsewhere in the world and that DayWater project has preferred the BMP term (Chapter 1), just as done below in this chapter. At some point in the future the terminology will have to be changed if the word 'sustainable' is not to become discredited.

In order to provide a more objective view of the relative sustainability of BMPs in the SWARD project, a whole systems approach was taken to their assessment in a case study involving a Municipality and a leading civil engineering company. The case study illustrated use of the SWARD methodology in a scenario in which the water utility is not the decision maker.

The Municipality involved in the case study had a positive bias towards BMPs and had recommended their implementation on numerous occasions, but was met with apathy from developers. The reasons for this appeared to include cost and risk worries, technical considerations, lack of long-term performance data, lack of information on design and best practice, and issues of maintenance and ownership uncertainty (Ellis, 2000; Everard, 2001). As a consequence, developers were unwilling to pioneer the systems due to the potential risks involved.

The case study was selected as a result of previous studies into the retrofitting of BMP systems in the study area to alleviate a number of unsatisfactory Combined Sewer Overflows (CSOs). The study site was the main refuse vehicle depot from which surface runoff drained into a pond that overflowed in heavy rain. The pond was originally a natural pond but had been enlarged and landscaped and also comprised two grassed swales and a filter drain with an oil separator which fed into the pond.

The case study involved the hypothetical scenario that SWARD had been used in the decision making process before the drainage plans had been put into place; the decision objective (phase 1, Figure 2) was to determine the management of runoff in the most sustainable manner. For a holistic and comparative evaluation it is essential that all criteria used can be applied to all options generated. The three options in this case were: (i) traditional piped drainage for the whole site with all foul and stormwater draining into the local sewer system, (ii) part BMPs and part traditional drainage and (iii) BMPs for stormwater management of the whole site (phase 2). The baseline option of no drainage installed, whilst assessable by SWARD was not considered relevant in this case. The primary and secondary criteria and indicator sets chosen for the case study are shown in Table 3. This case study is used here only to illustrate phases 1-3 of the SWARD processes.

Some of the criteria chosen for this case study will be relevant for the assessment of the relative sustainability of generic BMP systems, but it is recognised that certain criteria will be more relevant in alternative locations. For example, decision making for potential installation of BMPs in residential rather than industrial areas would employ greater emphasis on many of the social criteria and may include additional secondary criteria such as 'voluntary activity' where wetlands or ponds may elicit action from local community groups. The importance of local values to criteria selection cannot be overstressed; questions such as "what are the economic opportunities?" and "what is the relationship to natural resources and nature?" will elicit a unique set of concerns and values not universally applicable (Lawrence, 1998). Social research methodologies are well developed in other areas of water management (Ashley *et al.*, 2001) and may be used for assessment of many of the criteria in Table 3.

The case study confirmed that the way in which the evaluation was approached by SWARD at the decision stage was realistic and provided a means to refine generic and specific criteria, the database and indicators and to feed back any improvements or enhancements to the knowledge required for the decision making process.

Table 3: Categories, primary and secondary criteria and indicators chosen for assessment of a drainage decision case study.

Category	Primary criteria	Secondary criteria	Indicators
Economic	Life cycle costs	Capital costs	Average costs in £
		Operational costs	£/year
		Maintenance costs	£/year
		Decommissioning costs	Average costs in £
	Willingness to pay	For environmental benefits	Average £/unit of benefit
	Financial risk exposure	For capital investment	Risk of loss associated with investments – sensitivity analysis (qualitative)
		For other investments	As above
Social	Impact on human health	Risk of infection	Number of affected persons/100,000 population
	Acceptability to stakeholders	Perceived health and safety impact	% of 'users' perceiving a positive environmental impact, e.g. reduced flood risk, habitat creation, amenity benefit
	Public understanding and awareness	Stakeholder information	% knowledge in local community (will information be provided about the wider issues involved? Will awareness of 'water issues' be raised measurably?)
			Can the option be used to demonstrate a new method to other developers? (No. of site visits, enquiries)
	Social inclusion	Access to waterbodies/watercourse	Is access to a watercourse/waterbody increased? What are the perceived community/aesthetic benefits of clean rivers, or the creation of new water features?
Environmental	Resource utilisation	Land use	Land area used (km^2)
		Energy use	Energy use for construction (kWh)
			Energy use per unit of wastewater treated (kWh per m^3)
		Chemical use	Wastewater Treatment Plant (WTP) or on-site (herbicides) (Litres/year)
		Materials use (aggregates, plastics, metals etc.)	Total material requirement for construction
	Environmental impact	Impact on water	Watercourse quality (% of watercourse with 'good' ecological and chemical status)
		Impact on biological diversity	Number of key species at risk: habitat loss; sedimentation (soil/bank erosion in floods); low flows
			Number of key species introduced to area due to development: habitat creation
Technical	Performance	Quality of receiving water	Compliance with required standards in tests performed throughout the year (%)
			Number of water quality complaints per year (e.g. aesthetics)
		Floods	Number of floods/overloads per year from each drainage option.
			Number of customer's properties where risk of flooding is greater than the reference level.
	Reliability	System failure	Likelihood of system failing (quantitative)
	Durability	Durability of system: design life	Number of years system expected to operate successfully for
	Flexibility and adaptability	How flexible and adaptable is the system?	Level of accommodation in design: potential and ability to accommodate future changes (qualitative)
		Ability to add to or remove from the system	Cost (£) of adding/removing from system in response to change

IMPEDIMENTS TO THE INCLUSION OF SUSTAINABILITY CONCEPTS WITHIN WATER MANAGEMENT

The reluctance to accept BMPs amongst developers and others in the UK is in part a lack of experience and 'not wanting to be first' according to Everard and Street (2001). Acceptance depends heavily on the technical performance criteria, fulfilment of which requires the development of performance levels and indicators, long-term performance data and evaluation and monitoring of operational BMP systems. The importance of these factors to acceptance of BMP implementation by developers is an illustration of the 'technocratic culture' described by Brown *et al.,* (2005). A comparison of three independent research projects in Australia and the UK including the first mentioned above revealed 'significant and common socio-political impediments to improved practice'.

In Australia, a legislative requirement to prepare stormwater management 'plans' in collaboration with catchment stakeholders including the community did not effect widespread change and improvement, but largely the production of regulatory compliant 'plan documents' that did not drive the implementation of more sustainable stormwater management (Brown *et al.*, 2005). The study identified the requirements to effectively use capacity building strategies including political support, commitment to communities, transdisciplinarity and institutional capacity.

In Scotland, the 'Think Before You Flush' (TBYF) campaign was dropped by Scottish Water in 2003 in favour of the UK national 'bag-it-and-bin-it' national programme despite the considerable investment and research effort in the TBYF project. TBYF had shown the national programme to be fatally flawed due to the reluctance of many to have a bin located in the bathroom for sanitary waste (Ashley *et al.*, 2005), but this programme still being used in the UK today. Needless to say, habit change is not occurring.

In SWARD the TBYF research was expanded to investigate whether other options beyond the flushing of sanitary waste and solid waste stream disposal might be more sustainable (Ashley *et al.*, 2002). Here it was concluded that other sewer-related options could be effected which avoided the risk that those who had changed their behaviour might return to the flushing of sanitary waste in the future. The options considered were found to be more sustainable than current practices and the favoured option was the removal of stormwater as far as practicable from combined sewers, rather than the fitting of screens (which was found to be in the 'least sustainable' group of options). Despite these findings, the use of screens is widespread across the UK, which is due to prescription of their use as mandatory by environmental regulators, despite, ironically the regulator's duty to encourage sustainability.

This is illustrative of a number of key points about the difficulties of devising and implementing systems that are likely to be more sustainable:

- The regulatory, institutional and governance systems may not themselves understand the concepts, or if they do, may have other over-riding imperatives that mean they do not promote, or will not choose the more sustainable development option;
- The organisations charged to regulate, licence or otherwise control development activities may not have adequate systems in place to ensure that sustainability is properly included; they also may not have the in-house capacity (staff knowledge or time) to engage with the concepts and processes required;
- The boundaries and barriers between key stakeholders may prevent, prohibit or discourage the adoption of sustainable solutions.

As an example, in England and Wales, the Environment Agency (with a sustainability duty) gives over-riding priorities to 'environment' at the expense of economy and society and by its behaviour can encourage air pollution rather than water pollution, by transferability of burdens. The European Commission has so far largely avoided dealing with the air and land pollution consequences brought about by the implementation of water-related Directives.

Research into sustainable water techniques in UK house building has also demonstrated a lack of innovation due to standardization in practices (Guy and Shove, 2000; Myers, 2004), perceived customer requirements (Haworth, 2000; Myers, 2004) and lack of regulation in the sector (Adams and Watkins, 2002). It was concluded by Brown *et al.* (2005) that all professionals should understand and disseminate information from citizen to client, that boundaries of operation should be broadened to consider resource, waste and material flows in and out of the boundary and hence interrelationship with other sectors such as energy and transport should be strengthened, *i.e.* that radical changes in institutions and governance frameworks are required for the implementation of sustainable water systems.

There may be lessons to be learnt from business organisations many of whom are now making advances in delivering corporate sustainability through management of the business case for sustainability. The difficulty is how to operationally transform the vision and strategy into reality. Examples suggest that it is 'a culture that fosters innovation that is required above all else' (Schaltegger and Wagner, 2006). Economic success is also being recognised by successful businesses as being dependent on environmental and social issues that are 'non-market'. The UK regulated industries, such as the WSPs, have yet to determine how sustainability can best be incorporated into their business activities, a requirement under the UK's Water Act 2003, however, many of the barriers are institutional and regulatory rather than technological or methodological (Davis *et al.,* 2005) and legislation will be required.

SUSTAINABLE URBAN ENVIRONMENTS AND THE WAND CONSORTIUM

In 2001, the UK Engineering and Physical Science Research Council (EPSRC) established a coordinated national research programme entitled 'Sustainable Urban Environments' (SUE) to strengthen the capability of the UK research base in sustainability issues within the urban environment, to develop a pool of multidisciplinary academic excellence and to promote a strategic sustainability research agenda (Rodgers, 2005).

WaND (Water cycle management for New Developments) is part of the SUE consortium and deals with the provision of water, stormwater and wastewater services in new developments, such as those proposed in the South East of England (www.wand.uk.net). The project is led by Imperial College and comprises more than 30 research teams with some 25 stakeholders. The overall aim is to support the delivery of integrated, sustainable water management for new developments by provision of tools and guidelines for project design, implementation and management. It is due to complete in 2007.

So far detailed reviews of current and innovative water management practices and technologies have been carried out. These identify key design and performance criteria and will flag up the suitability of these techniques for new developments and the extent of their potential contribution to sustainability. A technology library has been developed and is being populated with the results of these reviews and will also incorporate expert judgement and priorities from end-users. Work is also underway to identify and evaluate the social, economic and health characteristics and potential implementation constraints on these technologies. An open architecture decision support platform or workbench has been developed with the potential to incorporate a series of support tools, described later in the paper, relevant to water systems (Morley *et al.*, 2004). Access for users to appropriate tools is facilitated by the development of an electronic portal. The tools are being applied to a common case study based on a set of agreed development scenarios.

The concept of WaND and its workbench/portal is that it allows a wide range of models, tools, maps, spreadsheets, guidelines and manuals to be associated together in a common location, with the portal giving structured access relevant to users' needs. The core water system model components, however, are threefold according to scale. These are denoted level I, II and III. Examples of levels I and II are given below.

Level I tools deals with larger-scale decisions to establish the 'best' location for new development based on encompassing the relevant technical and environmental issues related to urban water, as well as broader requirements related to the economic and social characteristics and constraints of the region. The Decision Support System (DSS) is based on a fuzzy inference methodology applied spatially using GIS, and is able to prioritise areas suitable for new urban areas and to screen out unsuitable sites using multi-criteria analysis. The effectiveness of the tool has been tested on a case study in the UK (Kokkalidou *et al.*, 2006).

Results suggest that the overall suitability evaluation is mostly influenced by the choice of site attributes taken into account and to a lesser degree by analytical processes. It is concluded that the tool is flexible and user-friendly and can significantly assist in the assessment and visualisation of the effect of spatial characteristics and user preferences on the siting of new urban developments. The proposed system architecture integrates three widely used software platforms (ArcView GIS, Matlab and Microsoft Excel) into a flexible and user-friendly decision support tool that can easily be adapted to different spatial decision environments (Liu *et al.*, 2005).

A similar technology is used for one of the level II tools. This is also based on soft computing to assist in the optimal siting of water systems within the context of new urban developments, through the creation of suitability maps. Again the system is based on multi-criteria decision analysis and fuzzy logic, but also allows the inclusion of uncertainty and of the decision-makers' attitude towards risk throughout the decision making process. This concept was originally applied to siting of stormwater systems (Makropoulos *et al.*, 2001) and demand management (Makropoulos and Butler, 2004), but most recently has been applied to determine the location of wastewater treatment plants – both on- and off-site.

Finally at level II, an optioneering tool has been developed to allow decisions to be made regarding the performance sustainability of a range of water systems and management approaches in new developments. Version 1 of this model (Sakellari *et al*, 2005) allows the user to assess sustainability using the SWARD criteria, based on a series of 'hard-wired' system scenarios. Version 2 develops the concept of systems. This has been coupled to a genetic algorithm routine to allow optimisation of any desired variable (e.g. cost, environmental impact), both at individual household level and at development scale.

One of the challenges of WaND is to identify who the stakeholders are and to ascertain their needs, perceived or otherwise (Markopoulos *et al.*, 2005).

The SWARD framework included fixed criteria and example indicators, directed only at one major stakeholder group: the WSPs, which were broadly acceptable in that context (Ashley *et al.*, 2004). In terms of WaND there are more main stakeholders: Government (main, regional and local), housing developers (and their consultants), WSPs, the Environment Agency, the public (householders) and the research and education community, for whom the DSS will be further advanced.

The boundaries for assessment within WaND are currently based on: the geographical area (the development), the resource and material flows into and out of that area and the present conditions. Review of these boundaries is being carried out during the course of the WaND project although there are also specific scenarios that have been defined to help with the temporal boundaries. Technical and other indicators have been defined individually for each WaND 'technical' work package. Primary criteria are those specified in SWARD, as agreed by stakeholder consultation, although these are open to review as the research progresses. Secondary criteria and indicators are drawn from a variety of sources but with special attention to those specified in the WFD and other 'industry standards'. Each constituent work package has set up its own specific secondary criteria and indicators that inform the same core primary indicators across the work packages. Otherwise different options cannot be compared. Table 4 shows examples from two of the WaND work packages (WP2 for BMP performance and WP7 on-site wastewater treatment) related to the SWARD core criteria. In the first stage of WaND assessment, for a screening and optioneering level of analysis, technical indicators are expressed qualitatively. Once options have been refined, more detailed quantitative and qualitative information is required for the technical process indicators to be used in multi-criteria decision making.

Table 4: How the constituent work packages (WP) in WaND define and use individual secondary criteria and indicators.

WP2 BMPs: secondary criteria and indicators (examples)	SWARD core criteria (Table 2) (These criteria used in the assessment of relative sustainability)	WP7 on-site wastewater treatment options: secondary criteria and indicators (examples)
Water quantity: extreme/peak events volume and flow rate	Performance of the system	Water Quality: Biological Oxygen Demand, Suspended Solids, turbidity, coliforms & *E. coli*, based on a range of (non-UK) guidelines for re-use
Health impacts: risk of infection Exposure to toxic compounds Safety (accidents/mortality)	Impact on risks to human health	Indicator organisms: indicator organisms in toilet bowl compared with flushing with potable water and potential for re-growth
Whole life cost: land take based on land area use	Life cycle costs	Whole life cost: capital, operational, maintenance costs, energy use, life span
Stormwater re-use: change in quantity of supply	Resource utilisation	Impact on downstream waste treatment: energy use

The current decision making process among feasible alternatives is driven mainly by socio-economic factors, which interact at all stages. The expression of preference by the decision maker through the use of preset criteria should be independent of technologies and options if these are to be effectively compared. Information for assessing whether the specific options meet the decision makers' aspirations within a given context will be included in the information attached to the DSS(s). This information, however, should, ideally, not include any in-built *a priori* value judgements about the option as this cannot be considered out of context. Information related to the DSS will also guide decision makers in their considerations of the complexity of the decision. The more complex the decision, the greater will be the number of stakeholders and conflicting viewpoints. Guidance emerging from current research as to means of stakeholder engagement in decision processes will be incorporated. A simplified illustration of the WaND decision making process is illustrated in Figure 3.

The approach adopted in WaND is to develop an intelligent explorer, accessible to a wide range of users, to act as a portal for (or gateway into) the WaND deliverables based on who the user is and what the relevant decision process is. Four stakeholders have been trialled: developers, planners, WSPs and house owner/buyers. This required collection and collation of ontologies for the domain to represent the decision web and then encoding these in computational form, leading to the establishment of a knowledge base (Mounce *et al.*, 2006). This can then be queried via a browser or portal, with routes defined for different users.

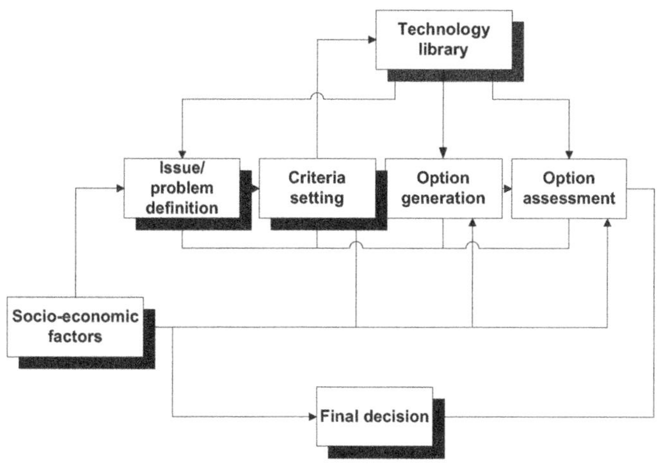

Figure 3: The decision making process within the WaND decision support system.

Figure 4 summarises how the decision making framework of criteria and indicators has been applied to increasingly complex and transdisciplinary projects, from the comparison of two options of the disposal of sanitary solids in the TBYF campaign through the development of SWARD for water industry professionals and its application in conjunction with different actors, to the WaND project with its numerous considerations of spatial, temporal and sectoral boundaries. The WaND project aims to bring together a broad spectrum of current knowledge in more sustainable practices for management of the water cycle and introduce it to practitioners in such a way that its application becomes a realistic proposition. It is often the case that practitioners are so focussed on the present and performance within their own professional boundaries that cross-sector and future needs are neglected. A project like WaND has a duty to provide ideas and tools to both show users what is now possible and to inspire them to see how this can be embedded in their work. Future cross boundary evaluations in conjunction with other service provision is a necessary aspiration if issues of sustainability to be properly addressed.

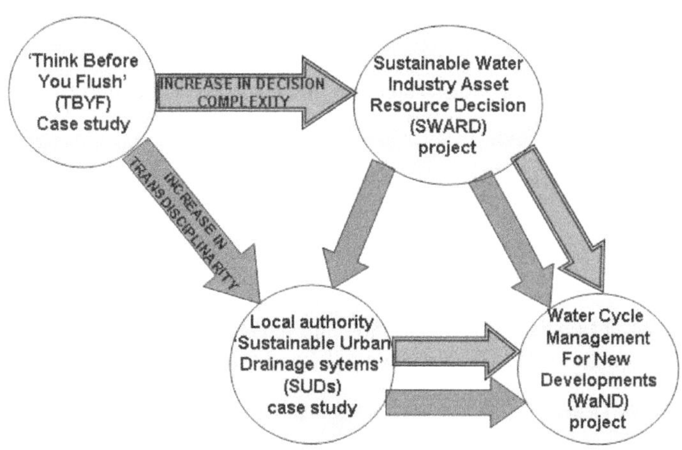

Figure 4: The development of the criteria and indicator framework through increasingly complex and transdisciplinary projects.

CONCLUSIONS

The journey to the delivery of systems that are likely to be sustainable is a long and complex one. Let us hope that we have the time to find our way. As yet the approaches being taken are perhaps too myopic in that they concentrate on single sectors: water, energy, transport etc., whereas the ultimate goal is to make human existence and activities sustainable by joining each of these together (Chapter 23). This is obvious but is only an aspiration, as the single sector areas are complex enough in themselves. This section has presented the results of a journey from a single issue: flushing of sanitary wastes (TBYF), through a tool for a single user, the water service provider (SWARD), to a more comprehensive perspective for new developments (WaND). Throughout, the studies have concentrated on water systems. The main lessons are summarised below.

- There appears to be no option at present to the use of criteria and indicators as a means to somehow 'measure' components of sustainability (Chapter 7). These measures have generally to be agreed by all the stakeholders, but may be weighted, vetoed or ignored by decision makers as decisions still tend to be made remotely from most of those affected by them (Chapters 4 and 9).
- There are essential behavioural elements that need to be influenced in the achievement of more sustainable systems, both in use of water services and in acceptability of new technologies and constraints. Knowledge about and engagement of stakeholders, is evolving, but is still not good enough and there is a general neglect in building the capacity for what is needed among all actors (Chapter 6).
- Knowledgeable champions are needed to ensure that the right messages are given, the right people and organisations are empowered and engaged and the right solutions delivered. Examples have been provided of how champions can be misguided (BMPs in the UK) and how the loss of champions can lead to adoption of poor solutions (the 'Bag-it-and-bin-it' campaign).
- Poor or too constrained institutions, regulatory or governance systems can lead to the inability, unwillingness and lack of desire to adopt solutions that are more sustainable. Of most concern is the lack of vision of major governance institutions such as the EU in this area and single-minded organisations such as the Environment Agency in the UK. Over-arching legislation and its consequential impacts on all areas of environment, society and economy seem often to be not predicted, untested or not appreciated when new and significant changes are imposed through new legislation or regulations.

REFERENCES

Adams, D. and Watkins, C. (2002) *Greenfields, Brownfields and Housing Development*, Blackwell, Oxford.

Ashley, R., Souter, N., Butler, D., Davies, J., Dunkerely, J. and Hendry,S. (1998) Assessment of the sustainability of alternatives for the disposal of domestic sanitary waste, *Water Science and Technology*, **39**(5), 251-258.

Ashley, R., Souter, N. and Hendry, S. (2001) *The importance of public attitudes and behaviour for the management of urban water systems*, UNESCO Conference on Frontiers in Urban Water, Marseilles, June.

Ashley, R.M., Smith, H., Jowitt, P., Butler, D., Blackwood, D., Davies. J. Gilmour, D. and Foxon, T. (2001) Making more sustainable decisions for asset investment in the water industry – *Sustainable Water Industry Asset Resource Decisions – the SWARD project, in: Proceedings of the 1st National Conference on Sustainable Drainage*, Coventry, 18-19 June 2000, ISBN 1 903818060.

Ashley, R. and Hopkinson, P. (2002) Sewer systems and performance indicators – into the 21st century, *Urban Water*, **4**, 123-135.

Ashley, R.M., Smith, H., Jowitt, P.H., Butler, D., Blackwood, D.J., Davies, J.W., Gilmour, D. and Foxon, T. (2002) Making More Sustainable Decisions for Asset Investment in the Water Industry – Sustainable Water Industry Asset Resource Decision – The SWARD Project, in: *Proceedings of the 9th International Conference on Urban Drainage*, September 8-13, Portland, Oregon. ASCE. ISBN 0 7844 0644 8, Strecker, E. and Huber, W. (eds.).

Ashley, R., Blackwood, D., Butler, D. and Jowitt, P. (eds.) (2004) *Sustainable Water Services. A procedural guide,* IWA Publishing, London.

Ashley, R., Blackwood, D., Souter, N., Hendry, S., Moir, J. Dunkerley, J., Davies, J., Butler, D., Cook, A., Conlin, J., Squibbs, M., Britton, A. and Goldie, P. (2005) Sustainable disposal of domestic sanitary waste, ASCE *Journal of Environmental Engineering,* **131**(2), 206-215.

Brown, R.R., Sharp, L. and Ashley, R.M. (2005) Implementation impediments to institutionalising the practice of sustainable urban water management, in; *Proceedings of the 10th International Conference on Urban Drainage*, Copenhagen/Denmark, 21-26 August.

Brunner, N. and Starkl, M. (2004) Decision aid systems for evaluating sustainability: a critical survey, *Environmental Impact Assessment Review*, **24**, 441-469.

Davis, R., Read, C. and Burston, P. (2005) *Regulating for a sustainable water sector*, Green Alliance. ISBN 0 9549757 4 X.

Ellis, J.B. (2000) Infiltration systems; a sustainable source-control option for urban stormwater quality management? *Journal CIWEM*, **14**(Feb), 27-34.

Everard, M. and Street, P. (2001) *Sustainable Urban Drainage Systems,* 2020 Vision Series No.3, The Natural Step.

Foundation for Water Research (FWR) (1996) *Assessing the benefits of surface water quality improvements*, FWR, London.

Foxon, T.J., McIlkenny, G., Gilmour, D., Oltean-Dumbrava, C., Souter, N., Ashley, R., Butler, D., Pearson, P., Jowitt, P. and Moir, J. (2002) Sustainability criteria for decision support in the UK water industry, *Journal of Environmental Planning and Management*, **45** (2), 285-301.

Gouda, H., Gilmour, D. and Ashley, R. (2001) Application of life cycle analysis approaches to develop the most sustainable way of dealing with solids in sewer systems, in: *Proceedings of COST 624 Working Group 3/5 Meeting*, Bologna, April.

Guy, S. and Shove, E. (2000) *A Sociology of Energy, Buildings and the Environment*, Routledge, London and New York.

Haworth, D. (2000) Water Conservation and Housing, in: Edwards, B. and Turrent, D. (eds.) *Sustainable Housing, Principles and Practice*, E and FN Spon, London and New York, pp. 0-65.

Kapelan, G., Savic, D C. and Walters, G A. (2005) *Decision-support tools for sustainable urban development*, Proc Institution of Civil Engineers: Engineering Sustainability, Vol 158, ES3, 135-142.

Kokkalidou, A., Makropoulos, C.M. and Butler, D. (2005) Supporting the siting of new urban developments for integrated urban water resource management, *NATO Advanced Research Workshop on Integrated Urban Water Resources Management*, Senec, Slovakia, October.

Lawrence, J.G. (1998) Getting the future you want: the role of sustainability indicators, in: *Community and Sustainable Development: Participation in the Future*, Warburton, D. (ed), Earthscan, London.

Liu, S., Butler, D., Makropoulos, C. and Memon, F. (2005) An Excel-Matlab based decision making framework and its application in urban water management, *CCWI Conference,* Exeter.

Makropoulos, C., Butler, D. and Maksimovic, C. (2001) GIS-supported stormwater source control implementation and urban flood risk mitigation. In: *Advances in Urban Stormwater and Agricultural Runoff Source Controls* (eds. J. Marsalek *et al.*), NATO Science Series, Kluwer Academic Publishers, IV 6, 95-105.

Makropoulos, C. and Butler, D. (2004) Planning site-specific water demand management strategies, *Journal of Chartered Institution of Water and Environmental Management,* **18**(1), 29-36.

Makropoulos, C K., Morley, M., Memon, F A., Butler, D., Savic, D. and Ashley, R M. (2005) A decision support framework for sustainable urban water planning and management in new urban areas, *Proc. 10th Int. Conference on Urban Drainage.* Copenhagen. August.

Malmqvist, P-A., Heinicke, G., Kärrman, E., Stenström, T A. and Svensson, G. (eds.) (2006) *Strategic Planning of Sustainable Urban Water Management*, IWA,. ISBN 13: 9781843391050.

Mounce, S., Hurley, L., Ashley, R M., Sharp, E., Makropoulos, C and Morley, M. (2006) The development of a flexible, knowledge based framework, for evaluation of the relative sustainability of water cycle management systems, *7th International Conference on Urban Drainage Modelling and 4th International Conference on Water Sensitive Urban Design.* Melbourne, Australia, April.

Morley, M.S., Makropoulos, C.K., Savic, D.A. and Butler, D. (2004) Decision-Support System Workbench for Sustainable Water Management Problems, *Proc. International Environmental Modelling and Software Society Conference*, IEMSs 2004, Osnabrück, Germany, June.

Myers, D. (2004) *Construction Economics: A New Approach*, Spon Press, London.

National SUDS Working Group (2003) Framework for Sustainable Drainage Systems (SUDS) in England and Wales. May.

Pahl-Wostl, C. (2002) Towards sustainability in the water sector- the importance of human actors and processes of social learning, *Aquatic Sciences,* **64**, 394-411.

Rodgers, C D F. (2005). Editorial. *Sustainable Urban Environments*: Part 1. Proc Institution of Civil Engineers: Engineering Sustainability. Vol. 158, Issue ES2. 55-56.

Sahota, P S. and Jeffrey, P. (2005) *Decision-support tools: moving beyond a technical orientation.* Proc Institution of Civil Engineers: Engineering Sustainability, Vol 158, ES3, 127-134.

Sakellari, I., Makropoulos, C.K., Butler, D and Memon, F.A. (2005) *Modelling sustainable urban water management options.* Proc Institution of Civil Engineers: Engineering Sustainability, Vol 158, ES3, 143-153.

Schaltegger, S., Wagner, M.(eds.) (2006) *Managing the Business Case for Sustainability*, Greenleaf pub. ISBN 10: 1-874719-95-0.

Water UK (2000) *Toward Environmental Sustainability: UK Water Industry Environmental Sustainability Indicators* (London, Water UK).

WFD (2000) Directive 2000/60/EC of the European Parliament and of the Council of 23 October 2000 establishing a framework for Community action in the field of water policy. Official Journal L 327, 22/12/2000, pp. 1 - 73

Whitelegg, K. and Weber, M. (2002) National Research Activities and Sustainable Development. A survey and assessment of national research activities in support of sustainable development, *IPTS Technical Report Services EUR 20389 EN.* European Joint Research Centre, Institute for Prospective Technological Studies, Seville, Spain.

Wrisberg, N., Udo de Haes, H.A., Triebswetter, U., Eder, P. and Clift R. (eds.). (2002) *Analytical tools for environmental design and management in a systems perspective*, Kluwer,. ISBN 0-4020-0626-8.

Part 6

Summary and prospective

Chapter 26

DayWater: summary and future development

Daniel R. Thévenot, Miriam Förster, Jean-Claude Deutsch, Govert Geldof, Gilbert Svensson, Peter Steen Mikkelsen, D. Michael Revitt, Emmanuel Aftias, Tomas Metelka, Heiko Sieker, Michel Legret & Maria Viklander

DAYWATER SUMMARY

Problem formulation

Why were we addressing decision support in urban stormwater management issues? Our problems are clearly related to hydraulics *i.e.* flooding or the need to remove stormwater from the wastewater sewer system thus reducing the number of combined sewer overflows (CSOs) and improving the efficiency of wastewater treatment plants (WWTPs) by providing a more stable water flow and quality. Additionally, the need to reduce diffuse pollution, as mentioned in the EU Water Framework Directive (2000), and the need to reduce impacts in the environment have to be addressed together with the public requirement for using stormwater for recreational purposes etc.

Innovative objectives (scientific & technical)

During the fifth Framework Programme for Research and Technology Development (RTD), the DayWater project (2002-2005) developed a prototype of an adaptive decision support system (ADSS) related to urban stormwater pollution source control. This decision support system (DSS) can be used for stormwater, treated locally in urban areas, using various source control techniques (Figure 1) such as swales, soakaways, constructed wetlands, ponds, porous paving etc. (Scholes *et al.*, 2003).

One of the innovative characters of the DayWater project (Chapter 2) has been the combination of management and technical models, just as for WATSIDE, a Decision Methodology for Water Supply Integrated Development (Thomas, 2001). In order to promote source control measures for integrated urban stormwater management, the major challenges of DayWater were to combine:

(1) scientific knowledge on source control techniques, their efficiency and cost, on risk and vulnerability, and on chemical hazard… with
(2) operational tools for source control technique dimensioning, for source and flux modelling and for risk and vulnerability screening, and with
(3) practical knowledge collected both by partners and end-users on different aspects of urban water, on negotiation between involved stakeholders and on decision-making procedures.

The overall objective was a newly developed user-friendly software architecture containing a front page known as HYDROPOLIS (Chapter 3).

Figure 1: Examples of stormwater source control techniques, either called Best Management Practices (BMPs) or Sustainable Urban Storm Drainage (SUDS): swale, porous pavement and detention ponds.

Access to the DayWater ADSS

The user has two different ways to use the ADSS (Figure 2). This can be by browsing the site and selecting data from the different components and saving them in a personal library. The alternative approach consists of being advised by the ADSS, while following a tree of questions (Chapter 4). The pre-defined answers are linked to Key Terms and allow a direct link to the specific components of interest for the user. This latter procedure is called 'dynamic thinking'. The user can also select key-terms directly out of a list in a procedure referred to as 'specific information'. Such Key Terms allow smart filtering of the knowledge base (Figure 2).

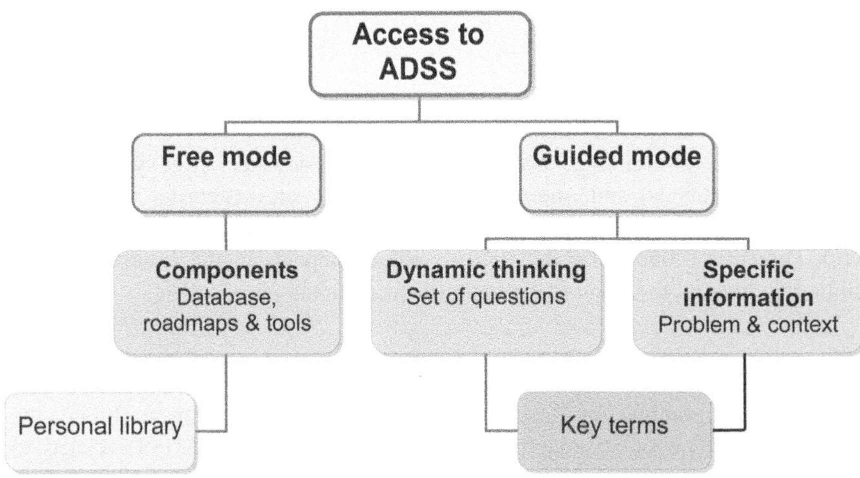

Figure 2: Modes of operating the DayWater Adaptive Decision Support through either 'free' or 'guided' modes with the latter allowing smart filtering of the knowledge base (Chapters 3 and 4).

Following a first step of identification and analysis of the stormwater management problem and context, most decision-making procedures (DMPs) include a project construction step, *i.e.* the selection of techniques, which appear possible at the given site and in the socio-economical context, before an actual comparison of alternative solutions (Figure 3). The DayWater ADSS intends to assist decision makers during each of these three DMP steps:

- a knowledge base facilitates the awareness building of the decision makers, using either the free or the guided mode (Figure 2) and presents available Best Management Practices (BMPs) (Figure 4),
- it helps the project construction and the identification of techniques compatible with the site and the stormwater management problem (Figure 4),

- it finally facilitates the discussion and negotiation between stakeholders through a comparison of selected alternatives, previously designed by various tools (Figure 5).

Figure 3: Successive steps in decision making for stormwater source control projects.

Stormwater problem identification and analysis

Several database structures have been developed in order to facilitate the collection of data relevant to stormwater source control, e.g. BMPs, costs, risks etc. (Figure 4). The case studies were provided by end-users involved in the DayWater project, and new information can be directly added to the login-protected web site. Similarly, partners or end-users can enter relevant links to web sites or up-load documents into the policy instrument database. The stakeholder database is pre-filled, but can be updated at any time.

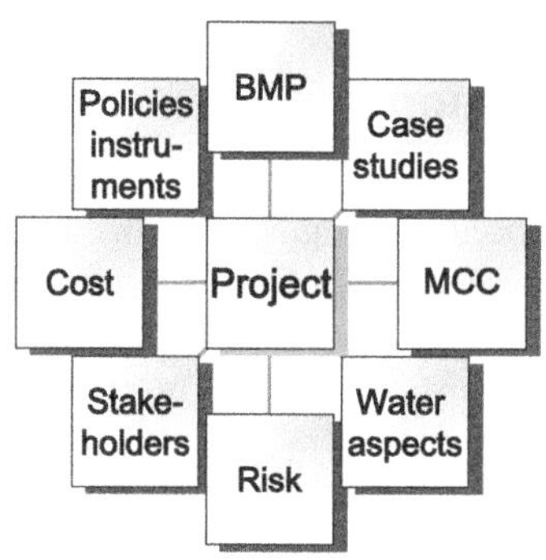

Figure 4: Databases, roadmaps and tools used for project construction (Chapters 6 to 15).

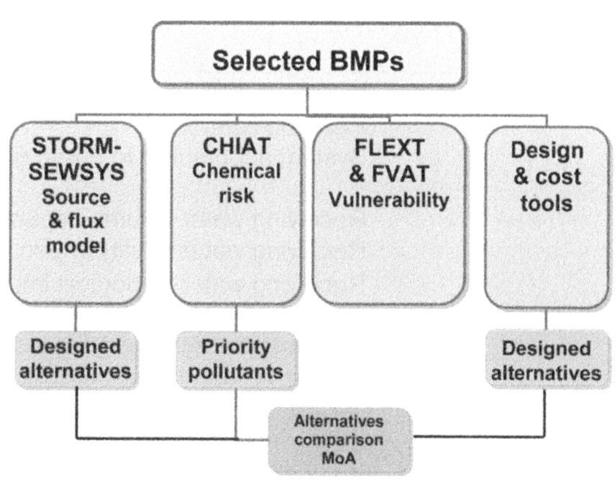

Figure 5: DayWater tools for designing BMPs, defining project alternatives and comparing them with the Matrix of Alternatives (MoA) (Chapter 4 to 15).

In order to select an optimal solution, the user should take the context into account (Lems and Valkman, 2004). The urban dynamic approach can help such analysis. The 'water aspects' ADSS component (Figure 6) explores the different facets of urban stormwater management (USWM) by broadening the perspective of the user on USWM and related problems. It provides a methodology to structure UWSM related social processes and finally, aspects of water can be used to categorize the indicators and criteria.

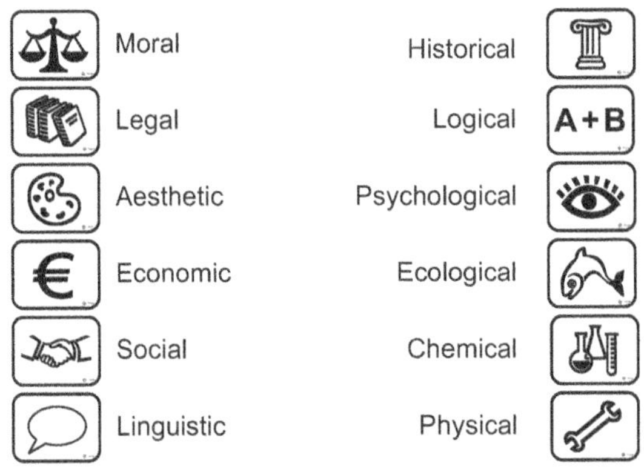

Figure 6: Water aspects, as presented in the DayWater ADSS web tool (Chapter 6).

Tools to support project construction

A survey of the applicability, pollution abatement and storage performance of different stormwater (best) management practices (BMPs) within Europe has been carried out (Figure 4). A list of criteria and indicators has been built in order to take into account all aspects of stormwater (Table 1). Collected information on design, operation and maintenance (O&M), costs, BMP performance, environmental advantages and sustainability have either been transformed into specific tools (pollution abatement, design or cost assessment) or are available via a web based catalogue within the ADSS (Chapter 8). Specific efforts have been directed at the integration of pollutant accumulation in snow and its transport during snow melting periods, as well as cold climate issues relevant when applying BMPs under winter conditions (Chapter 22). A multi-criteria comparator (MCC) approach, allowing the comparison of BMPs, has been developed, using default score values for indicators (Table 1) in order to support an objective comparison (Ellis *et al.*, 2004) (Chapter 9). The user also has the opportunity to adapt the scores and the weighting if preferred.

Table 1: Criteria and indicators within the BMP database (Chapter 8) and multi-criteria comparator (MCC) (Chapter 9).

Criteria	Indicators
Technical	Flood control
	Pollution control
	System adaptability to urban growth
Environmental	Receiving water volume impact
	Receiving water quality impact
	Receiving water ecological impact
Operation and Maintenance	Maintenance and servicing requirements
	System reliability and durability
Social and Urban Community Benefits	Public heath and safety risks
	Sustainable development
	Public/community information and awareness
	Amenity and aesthetics
Economic	Life cycle costs
	Long term affordability
Legal and Urban Planning	Adoption status
	Building development issues and stormwater regulations

The Chemical Hazard Identification and Assessment Tool (CHIAT) provides a framework for identification of chemical hazards related to micropollutants in stormwater runoff (Eriksson *et al.,* 2005) (Figure 7). The methodology has five steps: (i) source characterisation, (ii) recipient, receptor and criteria identification, (iii) hazard and problem identification based on inherent properties of the pollutants, (iv) hazard assessment where predicted environmental concentrations are compared with predicted no effect concentrations and (v) an expert judgement step where different stakeholders can influence the selection of priority pollutants. An important outcome of the methodology development is a list of Selected Stormwater Priority Pollutants (SSPP) identified during the DayWater project (Eriksson *et al.,* 2007) (Chapter 11).

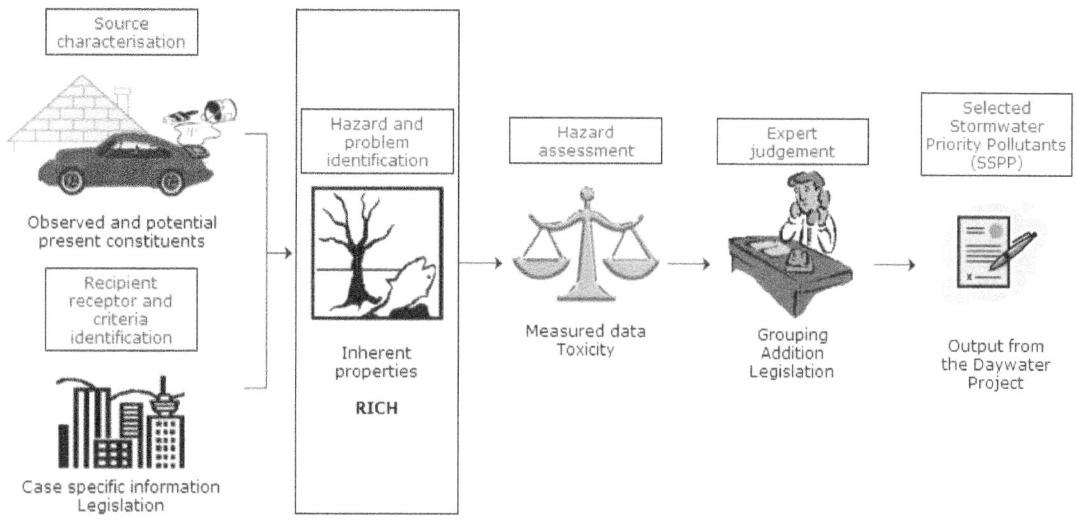

Figure 7: Chemical Hazard Identification and Assessment Tool (CHIAT) (Chapter 11).

The application of a battery of biotests (bacteria, algae and macroinvertebrates), in combination with chemical analysis, to urban stormwater sampled at Nantes (France), Wuppertal (Germany), Luleå and Stockholm (Sweden) has demonstrated the occurrence of chemical hazard and its large variability both within storm events and with regard to the biota used for such hazard assessment (Chapter 20).

BMP design and comparison of alternative solutions and their respective impacts

When possible BMPs have been identified and selected for the actual site and context of the stormwater project, they need to be designed either individually or as part of a treatment train. In addition to local or national regulations for building-up such alternative solutions, several tools have been developed within the DayWater project (Figure 5).

A prototype version of the stormwater sources and flux model (SFM), for quantification of stormwater-carried material flows in urban areas, provides quantitative and qualitative inputs to the ADSS and its components (Figure 8). Both parts of this SFM have been integrated (Chapter 14). The SEWSYS software assesses flow quality and quantity based on land use (GIS), using data collected within Europe, taking into account the selected stormwater priority pollutants (SSPP). Using the quality parameters derived from SEWSYS, the hydrological model STORM generates water flows and pollutant fluxes, depending both upon land use, selected BMPs and local hydrology.

A prototype of the Flood Vulnerability Analysis Tool (FVAT) has integrated various visions of flood vulnerability and shown that such risk analysis should not be disconnected from possible interactions with territorial identity (Chapter 13). It is an open tool, which can easily incorporate additional data describing the flood vulnerability of urban areas and available methods and techniques, which have been implemented to reduce it.

A transparent, knowledge-based and spatial DSS called Flext has been developed as an instrument to automate the evaluation of on-site stormwater management measures (Chapter 15); this instrument incorporates an expert system (ES) since the solutions for disconnecting stormwater from the conventional discharge systems require dynamic, experiential and professional knowledge.

Following either a free browsing of the knowledge base or a guided ADSS operation, the user is invited to identify possible alternatives for stormwater source control, using either one or a combination of BMPs. Such multi-purpose alternatives may be further assessed, using selected criteria, and then ranked using a Matrix of Alternatives (MoA), after each criterion has been scored and weighted (Figure 5). Such a procedure is the basis of negotiations between all stakeholders involved in the same USWM project (Chapter 4).

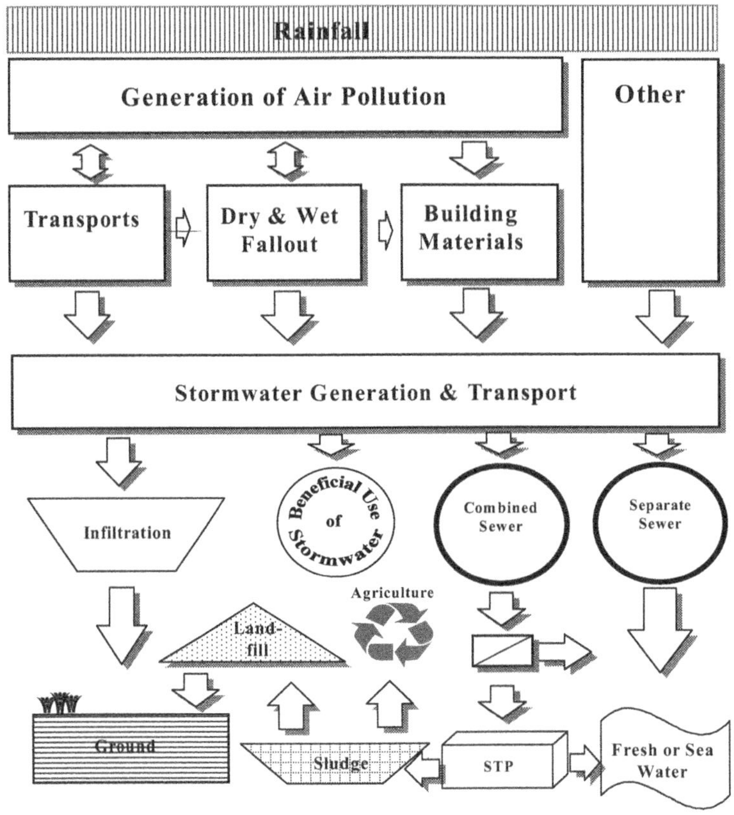

Figure 8: Schematic presentation of DayWater source and flux model (SFM) coupling SEWSYS and STORM models (Chapter 14).

Stakeholder experiences

The DayWater end-users (14 core and approximately 300 extended end-users) are important stakeholders in urban stormwater management and combine field knowledge with practical expertise. Through discussions with end-users involved in the testing of either individual ADSS components or the whole system, the following benefits resulting from the ADSS development have been identified:

- it is generally accepted that structural or non-structural BMPs can be efficient in controlling either urban stormwater pollution or flooding, or even both issues,
- it is also recognised that some devices fail to control pollution or flooding, mostly due to poor maintenance resulting from a lack of understanding of their operation,
- it is often recognised that decision makers are reluctant to replace standard down-pipe treatment systems by source control; stakeholders are numerous, needs are often ill defined, local or national regulations are complex and needed competencies are numerous for designing an efficient system and convincing decision makers,
- thus, major efforts of the DayWater partners have been spent on building a knowledge base aimed at increasing the awareness by citizens or technicians (Figure 6), in order to show the benefits of 'day lighting' stormwater and addressing, from the beginning of an USWM project, the importance of sustainability issues, including not only cost and maintenance, but also decision making based on a consensus reached by all involved stakeholders.

As an example of such strong interactions between DayWater core end-users and scientific partners, Chapter 16 presents the evolution of the feelings of 'Syndicat Mixte Marne Vive' (Living Marne River syndicate) during the ADSS development. These include very high expectations at the start of the project, followed by difficulties in presenting the end-user needs or 'profile' through the initial set of questions to the ADSS user engaged in the 'Guided Tour'. The difficulties in understanding the complexity of the ADSS were also highlighted due to possible conflicts between the language used by scientists and practitioners.

Three chapters present end-user experience when testing some of the ADSS components:

(1) Focussing on the importance of awareness building in numerous 'Water Aspects' within the 'Urban Dynamics' component of the ADSS (Figure 6 and Chapter 6) and of public participation in USWM projects, the Nijmegen municipality (The Netherlands) presents the factors of success and failure in the implementation of urban stormwater source control measures (Chapter 17).

(2) The city of Göteborg (Sweden) illustrates the use of the urban runoff pollution flow model SEWSYS (Figure 8) for identifying the major pollution sources within a catchment and the treatment measures which are expected to be most efficient (Chapter 18).

(3) The operation of the CHIAT methodology (Figure 7) for selecting the priority pollutants for the city of Stockholm (Sweden) is described in Chapter 19. It shows that this selection can, to a large extent, be performed based on scientific knowledge but that the final selection will be based on more or less subjective opinions.

The increasing focus on environmental concerns has led to the introduction of new techniques in urban planning and the concept of sustainable development is becomimg increasingly important in urban stormwater management (Chapter 25). This evolution makes it necessary to develop a more integrated hydrological modelling approach, by considering the whole urban water cycle including the different stormwater flow paths (concentrated in both rain and wastewater sewers) and by taking into account the spatial variability of geometrical and physical characteristics of the heterogeneous urban environment. Some recent hydrological research efforts have been dedicated to this aim (Jia *et al.*, 2001; Rodriguez *et al.*, 2003 and 2005), by developing a distributed approach to hydrological modelling, which is able to reproduce in detail hydrological processes at a small spatial scale. This can be applied not only to runoff, but to other relevant hydrological variables such as the soil hydric status (Chapter 21). Such hydrological approaches, with the use of urban GIS, make it possible to model the introduction of BMPs at both the local and catchment scales, and to analyse the influence of such techniques on both the quantitative and qualitative aspects of the catchment hydrological behaviour.

CONCLUSIONS AND PERSPECTIVES

The challenge of the DayWater project has been to develop and combine scientific knowledge, operational tools and policies into user-friendly software architecture. An important innovation evolves from the integration of all these research areas into a learning process tailored to enable convergence of ideas and lines of thought. It is obvious that the integration of the end-user knowledge was a further challenge of the ADSS development. The DayWater project involves both types of integration. It results in an ADSS prototype, which should facilitate the awareness building and participation of numerous stakeholders in the decision and realisation processes. Furthermore, the ADSS prototype promotes transparent negotiation and decision procedures on urban water related aspects. Examples of case studies and BMPs illustrate successful projects involving a large spectrum of stakeholders including land or equipment owners, local government officials, environmental officers and/or urban planners. These features support a more sustainable approach to urban water management and city governance.

Although dissemination of the ADSS will require assistance by scientific partners and translation tasks, the present prototype status has been found valuable by a large number of USWM technicians or decision makers both within and outside the DayWater consortium. Furthermore the scientific partner in charge of developing the ADSS, *i.e.* DHI Hydroinform a.s., has accepted to maintain the web based tool for some time and assist partners and users when necessary. Indeed, the DayWater ADSS is freely accessible, as 'guest' from www.daywater.cz pages and any end-user willing to actually operate this prototype may ask for user's rights either to the DayWater coordinator (daywater@cereve.enpc.fr) or to any scientific partner of this RTD programme. Urban stormwater managers are invited to browse and use the various DayWater ADSS components and tools and test how these databases and tools effectively facilitate valuable source control options for complying with the EU Water Framework Directive (2000).

Both during the DayWater project and after its completion, fruitful exchanges have occurred between DayWater and the Swedish Mistra Urban Water Management programme on chemical and microbial risk assessment. CHIAT will be further developed together with Stockholm Water and the Health and Environment Administration within the 'New toxins – New tools' project funded by Stockholm City under the so-called "environmental billion".

The work developed within the DayWater project, involving the prediction of the relative performance efficiencies of fifteen different BMPs (Scholes *et al.*, 2007), has been continued and extended in the Source Control Options for Reducing Emissions of Priority Pollutants (Score-PP) project, which is an EC funded STREP within FP6. The DayWater approach has been applied to the physico-chemical and biological characteristics of over 40 priority pollutants and associated compounds to interpret their behaviours within BMPs during stormwater treatment. The stormwater research developed within DayWater is also currently being utilised within the SWITCH project (an EC Integrated Project within FP6) on a global basis and within the COST Action 636, which investigates xenobiotics in the urban water cycle at a European scale.

The use of structural BMPs, such as retention and detention basins or porous pavement, as well as non-structural BMPs such as street cleaning, generate substantial amounts of solid waste, sediments and sludge, containing part of the pollutants drained by runoff waters. Studies are currently being carried out in order to characterise these contaminated by-products (Durand *et al.,* 2005) and to evaluate strategies for their treatment and re-use with respect to sustainable development approaches (Pétavy *et al.,* 2007).

The cluster of 6 EU RTD projects, including DayWater, within the CityNet project (citynet.unife.it/) EVK1-2002-00570 addressed issues of water management in urban areas (water supply, sewerage, drainage) including urban/rural interfaces (raw water sources, receiving waters, groundwater). The partnership, consisting of 60 research institutions and 56 end-users, thus contributed significantly to the European R&D capacity and implementation potential in urban water systems. From 2001 to 2006, the consortium demonstrated the benefit of successful scientific exchange between projects, including common publications, conferences and workshops such as those held during the 10th International Conference on Urban Drainage (10ICUD) on August 2005 in Copenhagen.

REFERENCES

Durand, C., Ruban, V. and Amblès, A. (2005) Characterisation of complex organic matter present in contaminated sediments from water retention ponds, *J. Anal. Appl. Pyrolysis*, **75**, 17-28.

Ellis, J.B., Deutsch, J.C., Mouchel, J.M., Scholes, L. and Revitt, D.M. (2004) Multicriteria decision approaches to support sustainable drainage options for the treatment of highway and urban runoff, *Science of the Total Environment*, **334-335**, 251-260.

Eriksson, E., Baun, A., Mikkelsen, P.S. and Ledin, A. (2005) Chemical hazard identification and assessment tool for evaluation of stormwater priority pollutants. *Water Science and Technology*, **51**(2), 47-55.

Eriksson, E., Baun, A., Scholes, L., Ledin, A., Ahlman, S., Revitt, M., Noutsopoulos, K. and Mikkelsen, P.S. (2007) Selected stormwater priority pollutants - a European perspective. *Science of the Total environment*, **383**, 41-51.

EU Water Framework Directive (2000) Directive 2000/60/EC of the European Parliament and of the Council of 23 October 2000 establishing a framework for Community action in the field of water policy.

Jia, Y., Guangheng, N., Kawahara, Y. and Suetsugi, T. (2001) Development of WEP model and its application to an urban watershed, *Hydrol. Process.*, **15**, 2175-2194.

Lems, P. and Valkman, R. (2004) Contextual water management, from necessity to opportunity, Proceedings Novatech'04, 5th International conference on sustainable techniques and strategies in urban water management, 6-10 June 2004.

Pétavy, F., Ruban, V., Viau, J.Y. and Conil, P. (2007) Conception of a pilot plant for the treatment of runoff water sediments,. Proceedings Novatech'07, Lyon, France.

Rodriguez, F., Andrieu, H. and Creutin, J.D. (2003) Surface runoff in urban catchments : morphological identification of unit hydrographs from urban databanks, *Journal of Hydrology*, **283**, 146-168.

Rodriguez, F., Morena, F. and Andrieu, H. (2005) Development of a distributed hydrological model based on urban databanks - Production Processes of URBS. *Water Science and Technology*, **52**, 241–248.

Scholes, L.N.L., Revitt, D.M. and Ellis, J.B. (2003) A European project (DayWater) investigating the integration of stormwater source control into sustainable urban water management strategies, *Journal of Health, Social and Environmental Issues*, **4**(2), 37-40.

Scholes, L, Revitt, D.M. and Ellis, J.B. (2007) A systematic approach for the comparative assessment of stormwater pollutant removal potentials. *Journal of Environmental Management*, doi:10.1016/j. jenvman.2007.03.003.

Thomas, J.-S. (2001) WATSIDE: a decision methodology for Water Supply integrated development, PhD Thesis, RMIT University, Australia.